ORBITAL WORKSHOP
E

C AIRLOCK MODULE

D INSTRUMENT UNIT

E. ORBITAL WORKSHOP

1 OWS HATCH
2 NONPROPULSIVE VENT LINE
3 VCS MINING CHAMBER AND FILTER
4 STOWAGE RING CONTAINERS (24 PLACES)
5 LIGHT ASSEMBLY
6 WATER STORAGE TANKS (10 PLACES)
7 TO13 FORCE MEASURING UNIT
8 VCS FAN CLUSTER (3 PLACES)
9 VCS DUCT (3 PLACES)
10 SCIENTIFIC AIRLOCK (2 PLACES)
11 WMC VENTIATION UNIT
12 EMERGENCY EGRESS OPENING (2 PLACES)
13 M509 NITROGEN BOTTLE STOWAGE
14 SO19 OPTICS STOWAGE CONTAINER
15 S149 PARTICLE COLLECTION CONTAINER
16 SO19 OPTICS STOWAGE CONTAINER
17 SLEEP COMPARTMENT PRIVACY CURTAINS (3 PLACES)
18 M131 STOWAGE CONTAINER
19 VCS DUCT HEATER (2 PLACES)
20 M131 ROTATING CHAIR CONTROL CONSOLE
21 POWER AND DISPLAY CONSOLE
22 M131 ROTATING CHAIR
23 WMC DRYING AREA
24 TRASH DISPOSAL AIRLOCK
25 OWS C&D CONSOLE
26 FOOD FREEZERS (2 PLACES)
27 FOOD PREPARATION TABLE
28 M171 ERGOMETER
29 MO92 LOWER-BODY NEGATIVE PRESSURE
30 STOWAGE LOCKERS
31 EXPERIMENT SUPPORT SYSTEM PANEL
32 BIOMEDICAL STOWAGE CABINET
33 M171 GAS ANALYZER
34 BIOMEDICAL STOWAGE CABINET
35 METEOROID SHIELD
36 NONPROPULSIVE VENT (2 PLACES)
37 TACS MODULE (2 PLACES)
38 WASTE TANK SEPARATION SCREENS
39 TACS SPHERES (22), PNEUMATIC SPHERE
40 REFRIGERATION SYSTEM RADIATOR
41 ACQUISITION LIGHT (2 PLACES)
42 SOLAR ARRAY WING (2 PLACES)

APOLLO TELESCOPE MOUNT

1 COMMAND ANTENNA
2 TELEMETRY ANTENNA
3 SOLAR ARRAY WING 1
4 SOLAR ARRAY WING 2
5 SOLAR ARRAY WING 3
6 SOLAR ARRAY WING 4
7 COMMAND ANTENNA
8 TELEMETRY ANTENNA
9 SUN-END WORK STATION FOOT RESTRAINT
10 TEMPORARY CAMERA STORAGE
11 QUARTZ CRYSTAL MICROBALANCE (2 PLACES)
12 ACQUISITION SUN SENSOR ASSEMBLY
13 ATM SOLAR SHIELD
14 CLOTHESLINE ATTACH BOOM
15 EVA LIGHTS (8 PLACES)
16 SUN-END FILM TREE STOWAG
17 HANDRAIL
18 SO82-B EXPERIMENT APERTURE DOOR
19 HA-2 EXPERIMENT APERTURE DOOR
20 SO82-A FILM RETRIEVAL DOOR
21 SO82-A EXPERIMENT APERTURE DOOR
22 SO54 EXPERIMENT APERTURE DOOR
23 FINE SUN SENSOR APERTURE DOOR
24 SO56 EXPERIMENT APERTURE DOOR
25 SO52 EXPERIMENT APERTURE DOOR
26 HA-1 EXPERIMENT APERTURE DOOR
27 SO55A EXPERIMENT APERTURE DOOR
ERTURE
OOR
33 RACK
34 CHARGER-BATTERY-REGULATOR MODULES (18 PLACES)
35 HANDRAIL
36 CMG INVERTER ASSEMBLY (3 PLACES)
37 CONTROL MOMENT GYRO (3 PLACES)
38 SOLAR WING SUPPORT STRUCTURE (3 PLACES)
39 ATM OUTRIGGERS (3 PLACES)

SPACE STATIONS

JPMIC3-05

SPACE STATIONS

THE ART, SCIENCE, AND REALITY OF WORKING IN SPACE

DR. GARY KITMACHER

RON MILLER

ROBERT PEARLMAN

FOREWORD

NICOLE STOTT

NASA ASTRONAUT AND ARTIST

Smithsonian Books

Washington, DC

This 2018 edition published by Smithsonian Books by arrangement with Elephant Book Company Limited, Southbank House, Black Prince Road, London, SE1 7SJ.

This book may be purchased for educational, business, or sales promotional use. For information, please write: Special Markets Department, Smithsonian Books, P. O. Box 37012, MRC 513, Washington, DC 20013

Published by Smithsonian Books
Director: Carolyn Gleason
Senior Editor: Christina Wiginton
Project Editor: Jaime Schwender

Editorial Director: Will Steeds
Project Editor: Chris McNab
Designer: Nigel Partridge
Proofreader: Gregory McNamee
Picture Researcher: Susannah Jayes
Color Reproduction: Pixel Studios Ltd.

Elephant Book Company Limited wishes to thank the following for their help in preparing this book: Carolyn Gleason and Jaime Schwender from the Smithsonian; Nicole Stott for graciously providing the foreword; Susannah Jayes for her tireless work as picture researcher.

ISBN: 978-1-58834-632-2

Library of Congress Cataloging-in-Publication Data

Names: Kitmacher, Gary, author. | Miller, Ron, 1947- author. | Pearlman, Robert (Robert Zane), author.
Title: Space Stations: the art, science, and reality of working in space / Gary Kitmacher, Ron Miller, and Robert Pearlman.
Description: Washington, DC : Smithsonian Books, [2018] | Copyrighted by Elephant Book Company Limited. | Includes bibliographical references and index.
Identifiers: LCCN 2018004838 | ISBN 9781588346322 (hardcover)
Subjects: LCSH: Space stations--History. | Space stations--Popular works. |
Outer space--Exploration--Popular works.
Classification: LCC TL797 .K5745 2018 | DDC 629.44/2--dc23 LC record available at https://lccn.loc.gov/2018004838

Manufactured in China, not at government expense
22 21 20 19 18 5 4 3 2 1

Front cover: Image courtesy of Lockheed Martin Corporation.

THE AUTHORS

GARY KITMACHER: Gary Kitmacher has served at the Johnson Space Center in Houston Texas since 1981, working for NASA since 1985. He currently works in the ISS Program Office, which oversees international collaboration and communication of the program's story to the world. He served as the Space Station Man-Systems Architectural Agent, defining the requirements and designing the station's modules, nodes and cupola, and later leading design efforts for manned Moon and Mars habitats. From 1993 to 2000, Dr. Kitmacher served in the Space Shuttle-Mir international program as the US manager for the NASA–Mir Priroda Project, leading US efforts to develop new systems and integrate payloads on the Mir Orbital Station; subsequently he served as the US manager for Mir Operations and Integration. In the Space Shuttle Program, Dr. Kitmacher managed the commercially developed Spacehab module, served as the mission manager for the early Spacehab missions and was the lead manager for the Spacehab-2, STS-60 mission. He served as a subsystem manager for Crew Equipment and Stowage on the shuttle and for Crew Health Care Systems on the International Space Station.

RON MILLER: Ron Miller is an award-winning illustrator, and the bestselling author of more than 50 books, including the Hugo-nominated *The Grand Tour*, *Cycles of Fire*, *In the Stream of Stars*, *The Art of Space*, *The History of Earth*, *Spaceships*, and *Aliens*. His "Worlds Beyond" series received the prestigious American Institute of Physics Award of Excellence and *The Art of Chesley Bonestell* (Paper Tiger, 2001) received a Hugo Award. Before becoming a freelance illustrator in 1977, he was the art director for the Smithsonian National Air and Space Museum's Albert Einstein Planetarium. His primary work today entails the writing and illustration of books specializing in astronomical, astronautical, and science-fiction subjects. Miller is considered to be an authority on Jules Verne. Miller has also been a production illustrator for motion pictures, notably *Dune* and *Total Recall*. He is a contributing editor for *Air & Space/Smithsonian* magazine, a member of the International Academy of Astronautics, and a Fellow and past Trustee of the International Association of Astronomical Artists.

ROBERT PEARLMAN: Robert Pearlman is a journalist, space historian, and the founder and editor of collectSPACE.com, a website devoted to space history with a particular focus on how and where space exploration intersects with pop culture. In the 1990s, his Ask An Astronaut website preceded NASA's online efforts to connect the public with the men and women who have flown in space and was honored as Kid Site of Year in 1996. Also in 1996, Pearlman was hired by Space Adventures, Ltd., the first company to launch privately-financed tourists to the ISS as their first Marketing and Public Relations Director, a position held until 2003. As Online Program Director, Pearlman led the expansion of the National Space Society's online presence, and in 1997, he was recruited by Apollo 11 moonwalker Buzz Aldrin to develop the website accompanying Aldrin's first novel. In 1999, Pearlman co-founded the astronaut-endorsed Starport.com, later acquired by Space.com (Pearlman was then hired by Space.com as a producer and tasked with managing their community projects). Pearlman serves on the History Committee of the American Astronautical Society and on the nominating induction committee for the Astronaut Hall of Fame in Florida. He is a director emeritus of the National Space Society and a former national chair of the Students for the Exploration and Development and Space (SEDS). In 2001, his work on collectSPACE earned Pearlman the Collector of the Year Award from the Universal Autograph Collectors Club (UACC). In 2009, Pearlman was inducted into the U.S. Space Camp Hall of Fame.

FOREWORD

NICOLE STOTT: A veteran NASA astronaut, Nicole Stott's experience includes two space flights and 104 days living and working in space on both the International Space Station (ISS) and space shuttle. She performed one spacewalk, was the first person to fly the robotic arm to capture the free-flying HTV cargo vehicle, was the last crew member to fly to and from their ISS mission on a space shuttle, and was a member of the crew of the final flight of the space shuttle *Discovery*, STS-133. A personal highlight of Nicole's spaceflight was painting the first watercolor in space. Nicole is also a NASA aquanaut, who in preparation for space flight and along with her NEEMO9 crew, lived and worked during an 18-day and longest saturation mission to date on the Aquarius undersea habitat. As an artist, and now retired from NASA, Nicole combines her artwork and spaceflight experience to inspire creative thinking about solutions to our planetary challenges, to raise awareness of the surprising interplay between science and art, and to promote the amazing work being done every day in space to improve life right here on Earth.

CAPTIONS

Endpapers: An annotated diagram of the Skylab space station.

Half-title: An early space wheel design concept, shown under construction.

Title page: A view out across Earth and into space from a window on the International Space Station.

CONTENTS

FOREWORD: NICOLE STOTT

NASA ASTRONAUT, AQUANAUT, AND ARTIST
STS-128, ISS EXPEDITION 20 & 21, STS-129, STS-133

The year 2018 marked 18 years of sustained human presence in space—18 years of crews, composed of astronauts from 16 partner countries, cooperating, living, and working together in space. And not just in space . . . thousands of people from the different space programs around the world have labored purposefully together right here on Earth in support of the common International Space Station (ISS) mission and the crews orbiting above. The ISS is by far the most complex international partnership that has ever been undertaken by humans in pursuit of a common goal. All of that work together in space has been for the benefit of all of us on Earth. In space we have managed to build a mechanical environment that mimics what our own "Spaceship Earth" does for us naturally. More important, we have demonstrated how planetary partnerships and long-term thinking can result in mutually beneficial success for all involved.

The year 2018 also saw the 50th anniversary both of the Apollo 8 mission's iconic "Earthrise" image and of Arthur C. Clark and Stanley Kubrick's science fiction classic *2001: A Space Odyssey*. Through Clark's visionary writing and Kubrick's progressive filmmaking came a creative vision of a future in space, one that remains inspiring and thought-provoking today. We still aspire to the technical beauty and functionality of the gigantic twin-wheeled space station and to the ability to travel, explore, and inhabit places distant from our own planet. I am still very hopeful for reality continuing to copy the fiction. I'm also thankful that what was thought at best unlikely or at worst impossible in 1968—a crew of Russians and Americans working peacefully and productively together—we have seen in subsequent space station programs like Mir and the ISS. The "Earthrise" image, meanwhile, shared with all of us a beautiful image of our home and our powerful connection to it. "Earthrise" profoundly communicated our interconnectivity, our interdependence, our reality—we live on a planet and we are all Earthlings.

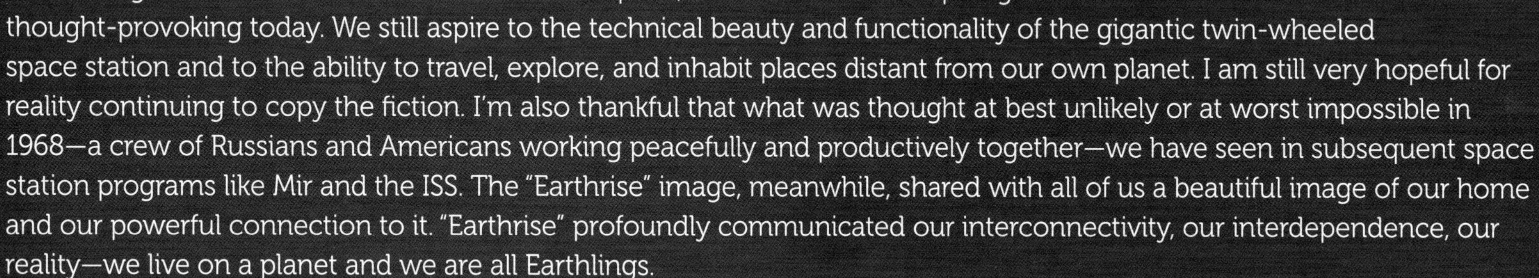

As a fan of the amazing things that come to us through the intersection of art and science, I am so happy that the authors of this wonderful book chose to represent the human story of space stations—not just the engineering associated with our space stations, but their relationship to humanity. This book beautifully presents the inspirational science fiction–based vision of our future like we find in *2001*, our reality as evoked by the "Earthrise" image, and the international partnership that has been achieved through space station programs such as the ISS. The way we have lived and worked together on our space stations is a fine example of how we can live in union here on Earth, and therefore anywhere else we choose to explore and settle. It is in the meeting of space exploration with art, science, and engineering that we can find a common humanity.

Chapter One

SPACE STATIONS: A PREHISTORY

As soon as human beings looked up into the night sky, they imagined what it might be like to travel to other visible worlds, such as the Moon. With the advent of astronomical telescopes, the mental ambitions grew to include other planets, such as Mars. Would it one day be possible, scientists reflected, that we might live on such planets? There was also the possibility, equally intriguing and equally challenging, of living in the vast expanses of space itself.

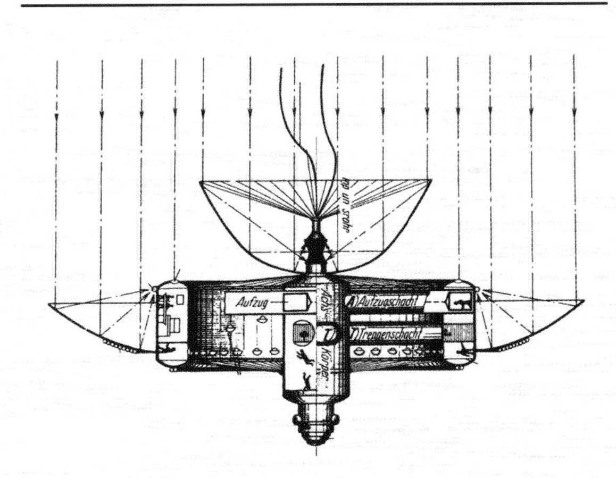

(Left) This illustration accompanied Hermann Noordung's *The Problem of Space Travel* (1920), the first serious study for an orbiting space station. Here we see, at the lower left, the station itself, along with its auxiliary solar generator ("machine room") at the top and an observatory at the right.

(Above) A cross-section of Noordung's 1929 "rotary house" shows the relationship of the habitable section with the large solar mirrors.

Space Stations: A Prehistory

THE WORLD CHANGED FOREVER ON A FALL NIGHT IN THE YEAR 1609. UNTIL THEN, EVERYONE ASSUMED THAT EARTH WAS UNIQUE IN THE UNIVERSE. THERE WERE NO OTHER WORLDS THAN THIS ONE AND THE WORD "PLANET" REFERRED ONLY TO A SPECIAL CLASS OF WANDERING STAR.

About ten months earlier, an Italian scientist named Galileo Galilei had heard about a new tubular and optical device invented in the Netherlands. When you looked through the tube it had the remarkable property of enlarging whatever you pointed it at. The invention was called a "telescope." The properties claimed for this invention were so remarkable that Galileo doubted what he heard. It was not until a French colleague, Giacomo Badovere, confirmed the rumors that Galileo decided to try building one of the instruments himself. After some experimentation, he found himself with a tube made of sheet lead, into either end of which he fixed a glass lens.

GALILEO'S DISCOVERY

The telescope had been invented originally as an aide to mariners and soldiers, but on the evening of November 30, 1609, Galileo did what had not occurred to anyone to do. He pointed the device at the night sky. And at that moment humanity's perception of its place in the universe—indeed, the very nature of the Earth itself—changed forever.

Galileo's discovery that there existed worlds in space other than the Earth were sensational, and raised many questions. What were these worlds like? What curious manner of creatures might live on them? And, perhaps most important of all, could humans ever visit these worlds?

The latter question led more or less directly to the invention of the spaceship, and in turn the space station. Scientists reflected upon every conceivable way and means of achieving space travel, from high-altitude balloons and improbably powerful catapults to attaching a brave astronaut to a gaggle of Moon-bound geese. But while inventive minds devoted themselves to methods of traveling to the Moon and planets, few gave any thought to an idea that seemed as impossible as it was pointless: living between worlds in space itself.

In *The Stone from the Moon*, Otto Willi Gail described the space station Astropol and a mirror reflecting heat and light onto Earth's polar regions.

"DEVELOPMENT OF THE SPACE STATION IS AS INEVITABLE AS THE RISING OF THE SUN."

WERNHER VON BRAUN

FLOATING IN SPACE

The idea of living in space would be like expecting great explorers such as Magellan or Cook to be satisfied with floating between continents, never setting foot on an alien shore. But the challenge was there, and science soon answered the question *how*? That occurred when Isaac Newton defined the third law of motion—that for every action there is an equal but opposite reaction—and in turn revealed the secret of rocket propulsion.

In the last decades of the nineteenth century, a few free-thinking pioneers began to discuss the possibilities of a station in space. Among these was Edward Everett Hale, who described an artificial Earth satellite that performed most of the functions of a modern space station, such as weather observation, navigation, and communication. Later writers agreed on those possibilities, while others thought such stations might be stepping stones into deeper space. For instance, space stations could be places where spaceships refueled or where passengers and crew could transfer to planet-bound spacecraft.

One of the first close looks by a scientist at the design of a large-scale space station was by Konstantin Tsiolkovsky. As early as 1894, he described what he called a "space cottage" in orbit around Earth. In all respects a space station, it would have been used for astronomical observations. He equipped it with a greenhouse for growing plants and trees as sources of oxygen and food. A year later, he wrote about an artificial satellite in which scientists could conduct scientific experiments.

In 1895, in *Dreams of Earth and Sky*, Tsiolkovsky wrote of a station orbiting the Earth

In the early 1950s, science fiction illustrator Frank R. Paul predicted space stations telecasting images of the Sun to observers on the Earth.

A nineteenth-century French cartoon depicted, in a typically imaginative fashion, interplanetary way stations for future travelers across the heavens.

at a distance of about 2,000–3,200 km (1,300–2,000 miles). "Little by little appear colonies with supplements, materials, machines, and structures brought from Earth." In his 1929 book, *The Aims of Astronautics*, Tsiolkovsky wrote: "So far, we cannot even dream of landing on large heavenly bodies. . . . Even a landing on a smaller body like our Moon is something that belongs to the very remote future. What we can realistically discuss is going to some of the minor bodies and moons, for instance..." And if travel to a body as small as an asteroid is possible, he said, why not to an artificial asteroid? "How," he asked, "is such a dwelling to be constructed? It is cylindrical, closed at each end with half-spherical surfaces. . . . To make the thickness of its walls a practical proposition, the dwelling is built for several thousands or hundreds of persons. . . . A third of the surface turned toward the sun consists of latticed window panes. . . ." This tube would be 2–3 m (6–9 ft) in diameter and perhaps up to 3 km (1.8 mi) long or even longer. It would be divided into 300 compartments, each sufficient for a family and a garden capable of supplying their needs.

Tsiolkovsky's tubular space colony would of necessity have been weightless; there was no provision for the creation of artificial gravity. If weightlessness were to prove to be a problem, he suggested an alternative design that would take the form of an enormous cone. The base of the cone—consisting of a transparent, spherical surface—would be turned toward the Sun, providing light and warmth for the interior. The cone itself would rotate on its longitudinal axis to provide a sense of gravity on its inside surface, covered with soil. No one before had ever thought about space stations on such a lavish scale or suggested that they might have a permanent population.

In 1923, the seminal astronautics pioneer Hermann Oberth published *Die Rakete zu den Planetenräumen* (The Rocket into Planetary Space), which contained the first serious proposal for a crewed space station to appear in scientific literature rather than fiction. His station would be permanently occupied. Orbiting the Earth at a distance of 1,000 km (621 mi), it would have been regularly supplied by rockets from Earth and rotated to produce an artificial gravity for the crew, who would engage in scientific observations and the refueling of interplanetary spacecraft. It was, in fact, in this book that Oberth coined the word "space station" (*Weltraumstation*). A few years later, Oberth published *Wege zur Raumschiffahrt* (The Way to Spaceflight), a landmark publication in which he went into specific details for several orbital space stations, including the construction of enormous mirrors, reflecting sunlight to the Earth's surface.

THE ENGINEERS

At around the same time Oberth was speculating, the Austrian Baron Guido von Pirquet laid the mathematical foundations for the space station as a way station to space. He said that it was far more practical to launch spacecraft to the Moon and planets from a space station than to launch them directly from the surface of the Earth. He proposed that three different types of space station could be developed. One would be dedicated to Earth observation and would orbit at an altitude of 750 km (466 mi). Another would orbit at 5,000 km (3,107 mi) and act as a launching platform for spacecraft bound for the planets. The third would occupy a highly elliptical orbit that would intersect with the orbits of the other two stations.

ROTATING HABITAT

The first detailed engineering study for the design of a space station, and certainly the most influential, was published in 1928 by "Hermann Noordung." This was a pseudonym of one Captain Herman Potočnik, who described an elaborate orbiting space station concept called the "rotating habitat" in a slim volume called *Das Problem der Befahrung des Weltraums* (The Problems of Space Flying).

The rotating habitat was to be a doughnut-shaped structure not too dissimilar from that proposed later by Wernher von Braun and others. Noordung's rotary house was the progenitor of a long line of space station designs that extended well into the 1960s and even beyond. (A young Wernher von Braun may have even taken inspiration from Noordung's book to write a short story, "Lunetta," in 1929, which described a trip to a space station.)

Twenty years after Noordung's book appeared, H. E. Ross, in collaboration with engineer-artist R. A. Smith, presented a space station design to the British Interplanetary Society that was in reality an updated version of the "rotary house." Their design corrected many of Noordung's errors in addition to adding a great many improvements. Thinking about space stations was gradually inching toward theoretical and practical reality.

COPERNICUS, BRAHE, KEPLER, NEWTON

A revolution in scientific thinking began in 1507 with Copernicus. He wanted to simplify Ptolemy's explanation of the universe known to the Greeks. Copernicus said it was simpler if the Sun were at the center and the planets orbited the Sun. Earth, Copernicus said, was the third planet from the Sun. Others set out to prove him right. Tycho Brahe sought to better predict the movement of the planets. He did not believe in Copernicus' Sun-centered universe.

Johannes Kepler, Brahe's assistant, sought to explain the *harmony of the spheres* using data that Brahe had collected over decades. Kepler showed that elliptical orbits, with the Sun at a focus, and the speed of the planet's movement related to the Sun's distance, explained Brahe's data perfectly. Galileo, meanwhile, put his telescope, a new invention, to good use, discovering new stars, moons orbiting Jupiter, the phases of Venus, and seas and mountains on the Moon. He observed that the planet Saturn was not round and was not a sphere, but that it had lobes to either side. The universe, Galileo showed, was *imperfect*, a thought regarded as heresy. But his work inspired others to look and dream further.

Although Danish astronomer Tycho Brahe never accepted the heliocentric theory, his careful observations nonetheless helped to lay its foundations.

Polish clergyman-astronomer Nicolas Copernicus realized that the simplest explanation for the movement of the planets was to assume that they, along with the Earth, orbited the Sun.

The universe of Ptolemy had the Earth at the center with the Sun and planets circling it and the stars in a distant shell surrounding everything. This model of the universe was to prevail for many centuries, until it was unsettled and finally overturned by science.

Kepler heard of Galileo's discoveries, and in 1630 he wrote the story of an imaginary trip to visit the Moon. He described "the Earth as an insignificantly small world amongst the stars" and said that "one day ships would be built that could travel through space and once they are built, men will step forward to sail them."

Isaac Newton later read Kepler's laws of planetary motion and learned of Galileo's demonstration, from the leaning tower of Pisa, that when objects were dropped their time of descent was independent of their mass. In 1665, while Newton sat in his garden, he observed an apple fall. Newton thought about Galileo's falling, accelerating stones and about the force drawing apples and stones toward Earth—gravity. He was able to show that the same force equally attracted the Moon toward Earth and the planets toward the Sun. Kepler had defined the laws that described the planets' movements. Newton now defined the laws that explain the motions of falling rocks, planets, and projectiles in his *Principia* in 1687. He postulated that the force of gravity would describe the trajectories of future trips beyond the Earth.

German mathematician Johannes Kepler described the physical laws that governed orbiting bodies, including satellites and space stations.

An image of Isaac Newton, the fallen apple prompting him to ponder gravitational theory and develop the laws of motion that underpin much physical science.

PROJECTILES, GRAVITY, AND MOTION

Newton was the first to explain the nature of gravity; it was a force depending on two masses attracting one another. He described the reactions of the two masses as changes in motion. Their change in motion, an acceleration, could be described precisely enough to predict the motions. It worked for a falling apple and it worked for planets in orbit.

Newton also defined three laws of motion: (1) a body at rest will remain at rest unless a force acts upon it; (2) the force equals the mass multiplied by the acceleration; the acceleration is proportional to the force applied and the mass; (3) for every action there is an equal and opposite reaction. From these laws, Newton was able to derive the law of universal gravitation, that gravitation is an attraction proportional to the two attracting masses and inversely proportional to the distance between the two masses. Newton was able to explain not only why Kepler's laws of planetary motion worked the way Kepler had observed, but almost every motion in the universe.

Newton's laws explained the motion of projectiles. Throw a projectile into the air; its rise is slowed by the force of gravity. When it reaches its high point, it begins to fall back. The speed increases as it falls, and the rate at which the speed increases is constant. Hurl

Two objects will travel together at the same speed in a vacuum, as Verne's astronauts in *From the Earth to the Moon* (1865) found when they tried to toss a dead dog overboard

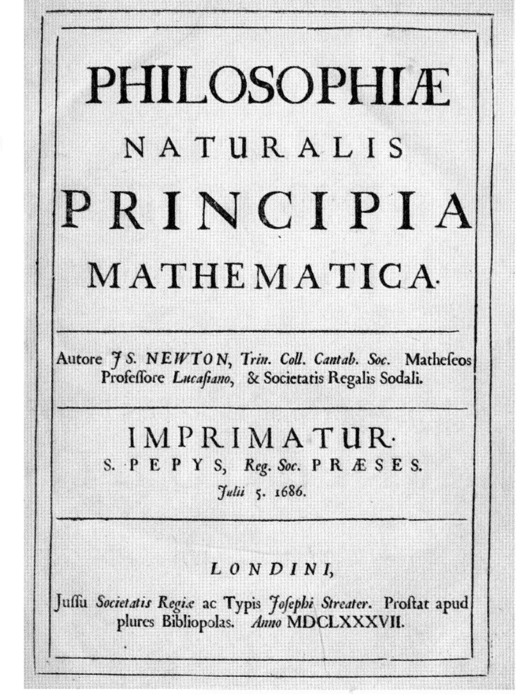
PHILOSOPHIÆ
NATURALIS
PRINCIPIA
MATHEMATICA.

Autore *J S.* NEWTON, *Trin. Coll. Cantab. Soc.* Matheſeos Profeſſore *Lucaſiano*, & Societatis Regalis Sodali.

IMPRIMATUR.
S. PEPYS, *Reg. Soc.* PRÆSES.
Julii 5. 1686.

LONDINI,

Juſſu *Societatis Regiæ* ac Typis *Joſephi Streater*. Proſtat apud plures Bibliopolas. *Anno* MDCLXXXVII.

The title page of Newton's *Principia* (1687), in which he described the Second and Third Laws of Motion, laying the groundwork for modern rocketry and astronautics.

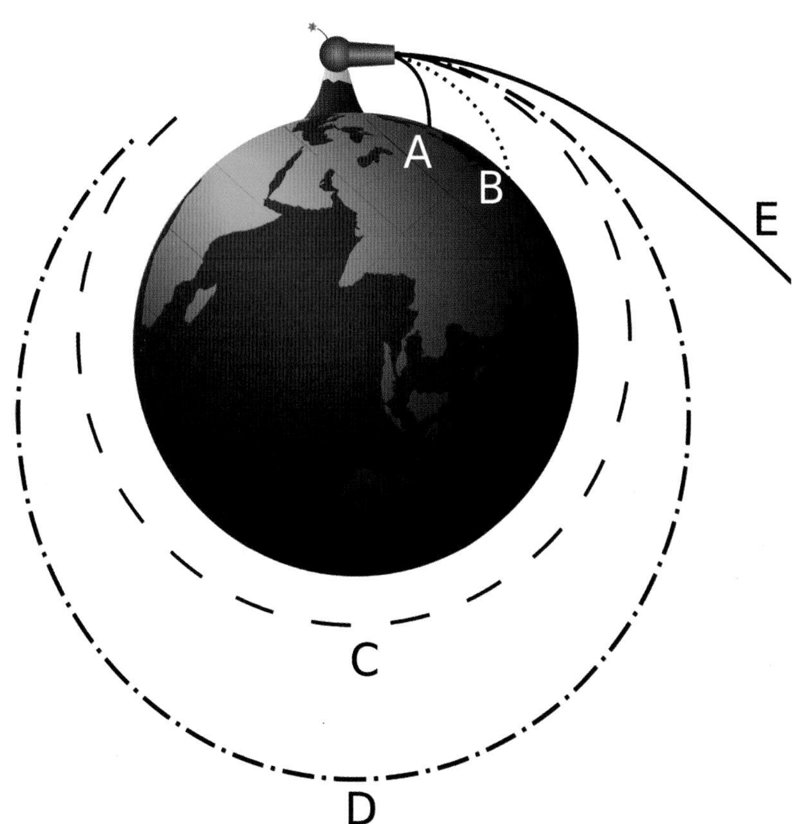

Newton used the example of a cannonball fired from a high mountain to explain the principle of orbiting. If the speed of the projectile is not great enough, it will fall back to the Earth (A and B). But the more speed the projectile has, the further it will travel before falling to the ground. Eventually, given enough speed, the curve of its fall will match the curve of the Earth (C and D) and it will orbit the planet, always falling toward it but never reaching it. Too much speed, and the projectile will escape the orbit (E).

a projectile horizontally, it also falls as a result of the force of gravity. The rate at which it falls increases at the same constant rate. The faster a projectile is hurled horizontally, the further it will travel before falling back to the ground. Because the Earth is a sphere, if the projectile is hurled fast enough, it can make it all the way around the Earth before it hits the ground. When the projectile can make it all the way around the Earth it is said to be in orbit.

Isaac Newton explained the motions of the planets based on their gravitational effects on one another. Many people think of space as a place where there is no weight; where astronauts float weightless. This is a common misunderstanding. There is gravity in space and everywhere in the solar system. The perception of weightlessness is not due to the lack of gravity. It is the perception of an object in freefall as it orbits in airless space.

A German print from the mid-nineteenth century shows how the understanding of planetary orbits had firmly embedded itself in astronomy by the 1800s.

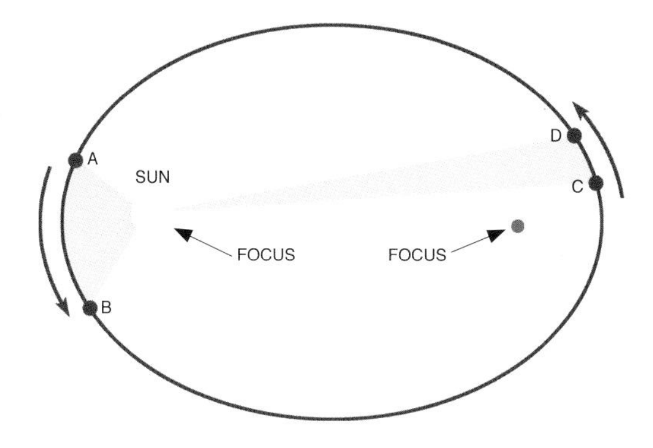

Johannes Kepler showed that the planets orbited in elliptical patterns, traveling fastest when they were nearest the Sun and slowest when they were furthest away.

THE GENESIS OF AN IDEA

While it might be possible to place some sort of object, perhaps even one inhabited by people, into orbit around the Earth, what would be the point? Surely it would make more sense to travel to another world than to forever remain suspended above this one. An early thinker who answered that question was an American Unitarian minister named Edward Everett Hale in 1869.

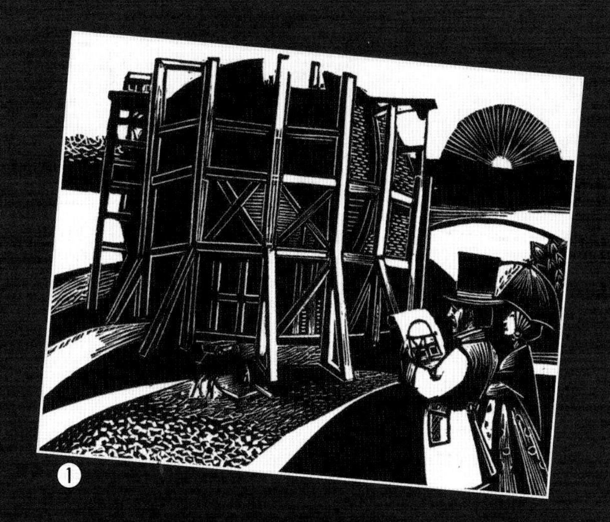

1

2

In his short story "The Brick Moon" (1869), Hale foresaw many of the modern purposes for artificial Earth satellites. He suggested that they could be used as navigational aids, for mapping and surveying, for military reconnaissance and observation, for communications, meteorology and—most important so far as we are concerned—as a potential habitat for humans. While Hale was not the first to describe an artificial Earth satellite (that honor belongs to Isaac Newton), he was the first to realize the practical purposes of a satellite and the first to describe one inhabited by human beings.

In Hale's story, a group of American entrepreneurs decide to launch a satellite into Earth orbit. Its purpose is to act as a navigational aid for mariners, who would be able to calculate longitude accurately from the precisely known position of the satellite. If the initial experiment were successful, three more moons would be launched, which would allow one of them always to be visible from anywhere on the Earth. The artificial moon was to be a hollow sphere made of brick, 60 m (200 ft) in diameter. Inside are 13 smaller hollow spheres of brick, resulting in a structure of great strength and low weight. A pair of gigantic flywheels, powered by a force of a waterfall, would catapult the Brick Moon into orbit.

An unfortunate accident, however, results in the premature launch of the moon, along with the workman and their families who are still on board. After some months, astronomers eventually locate the moon in their telescopes and are astonished to find that it is no longer red, but green. They see small black specks moving upon it. The specks prove to be the workmen, who miraculously survived the launch. The astronomers establish communications by means of a kind of semaphore. They learn that not only has life survived on the Brick Moon, it is flourishing. There is rainfall and plants are evolving with startling rapidity ("Write to Darwin that he is all right. We began with lichens and have come as far as palms and hemlocks").

The "Bricks," as the inhabitants of the moon become known as, transmit their observations of the Earth to scientists. They report on storms below and inform geographers about the nature of the Arctic, Antarctic, and the interior of Africa.

In the novel, Earth-dwellers devise a method to send supplies to the inhabitants of the Brick Moon (Hale even accounts for protection against the heat created by atmospheric friction as the packages are launched into space). These attempts are not always successful. One container bursts along the way and the disappointed Bricks have to report that "Nothing has come through but two croquet balls and a china horse." Hale's story, while meant to be humorous and satirical, was well received by scientists. Hale was sent a note of congratulations from Asaph Hall, the discoverer of Phobos and Deimos, the twin moons of Mars.

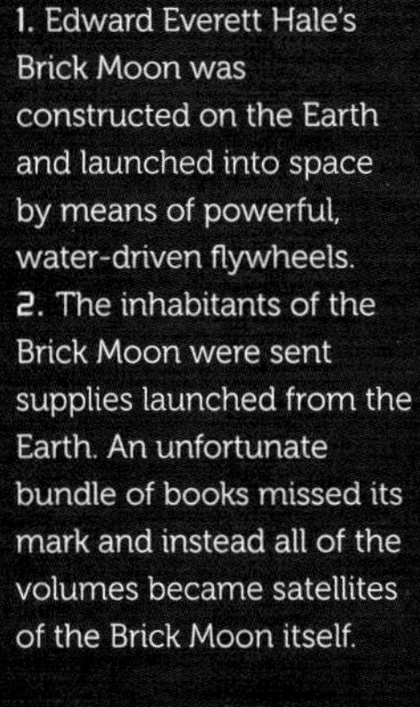

1. Edward Everett Hale's Brick Moon was constructed on the Earth and launched into space by means of powerful, water-driven flywheels.
2. The inhabitants of the Brick Moon were sent supplies launched from the Earth. An unfortunate bundle of books missed its mark and instead all of the volumes became satellites of the Brick Moon itself.

3

3. The Brick Moon soon developed its own weather system and climate, and plant life eventually evolved and even flourished on the little moon.
4. The cover of a modern edition of *The Brick Moon* illustrates how its inhabitants communicated with the Earth by spelling out words in fabric letters 6 m (20 ft) high.

4

KURD LASSWITZ

In 1897, the German author Kurd Lasswitz published the science fiction novel *Auf zwei Planeten* (On Two Planets). In this visionary novel, a crew of particularly adventurous balloonists attempting a transpolar flight are caught in a magnetic field that carries them precipitously upward to a space station hovering 6,115 km (3,800 miles) above the North Pole. Martians operate the station as a base from which their spaceships can safely approach and land upon the Earth, since the rapid rotation of the Earth—1,609 kmh (1,000 mph) at the equator—makes landing anywhere else but at the relatively motionless poles impossibly dangerous.

Lasswitz described the space station as a wheel 120 m (360 ft) wide, hovering 6,437 km (4,000 miles) above the pole. The station derives all of the energy it requires from the Sun. Surrounding the ring are broad, flat disks extending the radius another 200 m (656 ft). These act as flywheels helping the station to maintain its attitude. The main ring contains the Martians' living quarters as well as facilities for docking their spaceships.

Although an English translation of the novel was not published until 1971, *Auf zwei Planeten* was translated and published throughout continental Europe, where it became an important influence on the many engineers and scientists who were just then beginning to take the concept of space flight seriously. One of these was a young German named Wernher von Braun. "I shall never forget," he wrote, "how I devoured this novel with curiosity and excitement as a young man. . . . From this book the reader can obtain an inkling of that richness of ideas at the twilight of the nineteenth century upon which the technological and scientific progress of the twentieth is based."

The cover of a German edition of *Auf Zwei Planeten* (On Two Planets) by Kurd Lasswitz, published in 1948.

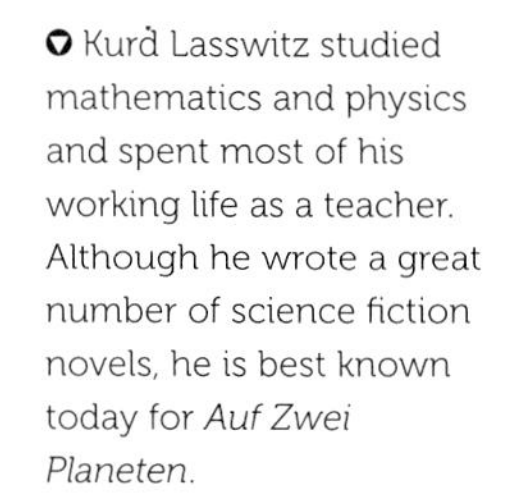

Kurd Lasswitz studied mathematics and physics and spent most of his working life as a teacher. Although he wrote a great number of science fiction novels, he is best known today for *Auf Zwei Planeten*.

KONSTANTIN TSIOLKOVSKY

Lasswitz was followed by Konstantin E. Tsiolkovsky, a Russian mathematics teacher. In 1895, he published a science fiction novel entitled *Reflections on Earth and Heaven and the Effects of Universal Gravitation*. In it he described the use of asteroids and artificial satellites as bases for rocket launches. He proposed that artificial gravity could be produced by rotating a space station. In 1903, he suggested the possibility of an unmanned "satellite rocket" and in 1911 described a crew-carrying version. Between 1911 and 1926, Tsiolkovsky put many of these ideas on a strict mathematical basis, the first time anyone had done so.

Tsiolkovsky touched on the issue of how to create a permanently crewed habitats in Earth orbit. The first such mention was in "Free Space" written in 1883. Much of his early work concerned life-support systems and orbital construction. In his later years, in the 1920s, he wrote about different designs of orbital habitats, including cylinders, torroids, and cones. His "space habitats" would rotate to provide artificial gravity. The park-like interior had gardens and trees so that it would be self-sustaining. Much of his work was not published until his later years or after his death in 1935.

By the time this photo of Konstantin E. Tsiolkovksy was taken in 1897, he had already published a science fiction novel about the future of space flight.

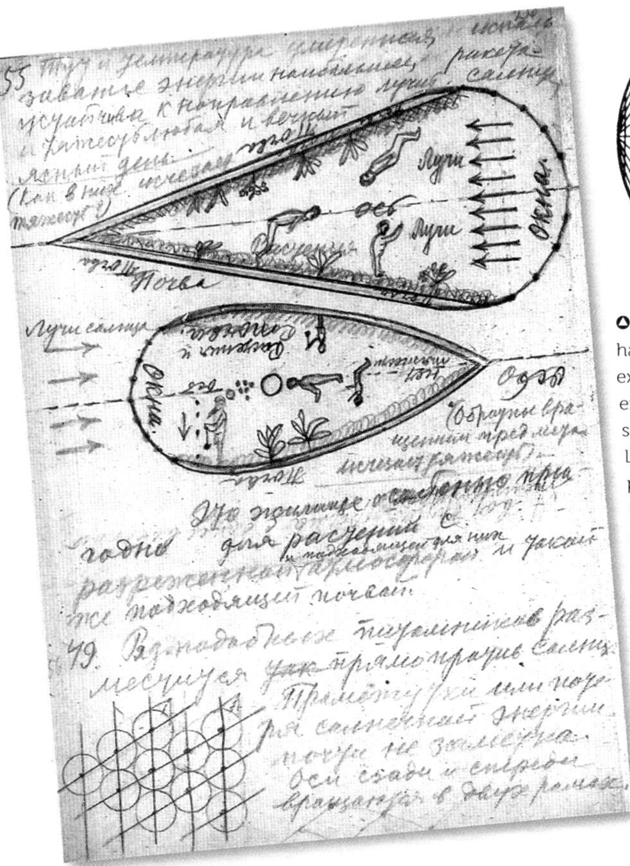

Tsiolkovsky drew this sketch of a space habitat in the 1880s. It evolved into the more detailed concept (far right) which he began developing in 1903.

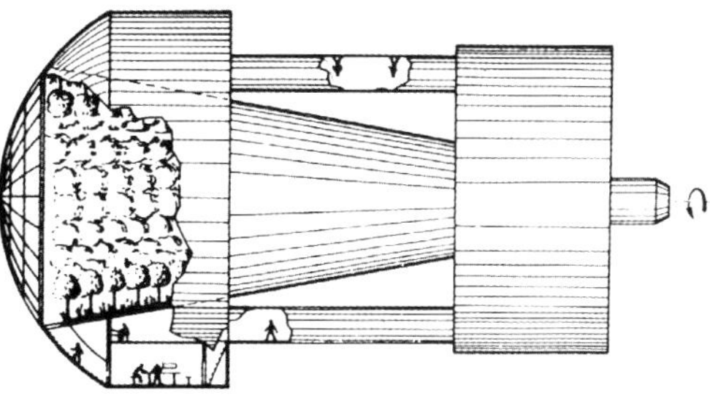

Tsiolkovsky's space habitat ideas were explored to the fullest extent by a popular science writer, B. V. Liapunov. Rotating to provide artificial gravity for its inhabitants, it was self-sustaining with gardens and trees that provided both oxygen and food.

CAPTAIN NOORDUNG AND HIS ROTARY HOUSE

Hermann Noordung was the pseudonym of Herman Potočnik, He studied engineering and received a commission as a lieutenant in the Austro-Hungarian Army, where he served during World War I, After contracting tuberculosis during the war, he retired with the rank of captain. Noordung was later awarded a doctorate in engineering and spent the rest of his life studying the problems of rocket travel and living in space.

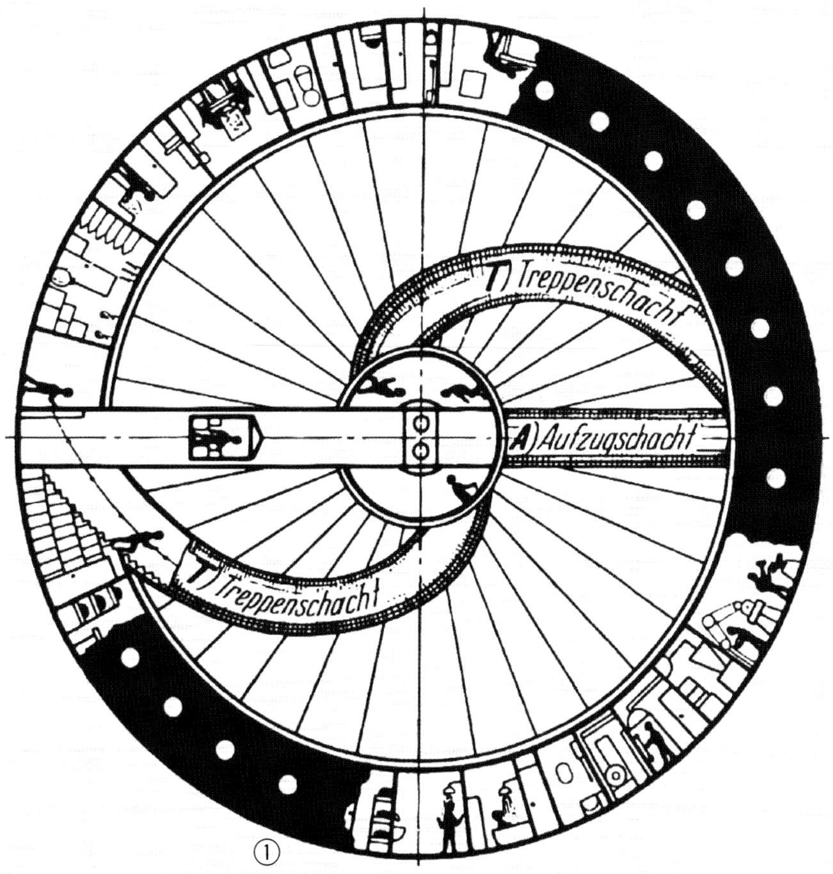

①

②

1. A plan view of Noordung's 30 m (100 ft) diameter space station, showing its different compartments, as well as its elevators and the spiral staircases that allowed astronauts climbing them to always keep their heads and feet in line with the artificial gravity created by the wheel's rotation.
2. Hermann Noordung—aka Captain Herman Potočnik—studied engineering before his military service.
3. The cover of the original edition of Noordung's classic book, *The Problem of Space Flight* (1929).
4. A detail of the station's solar mirrors. These focused sunlight, creating steam that in turn powered electric generators.
5. Frank R. Paul recreated an illustration from Noordung's book for the cover of *Science Wonder Stories* (August 1929), the first appearance of a space station in an English-language publication.

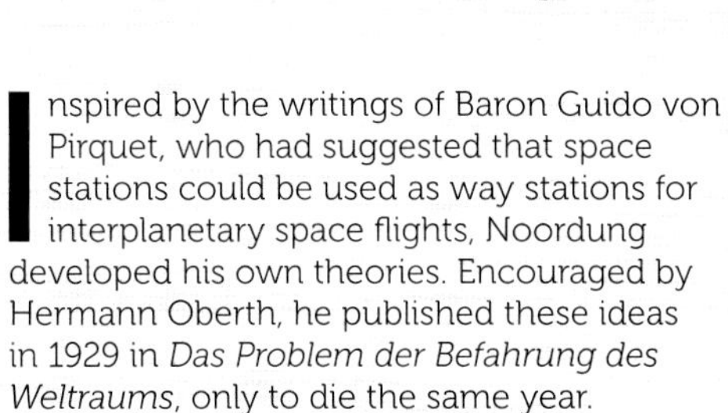

Inspired by the writings of Baron Guido von Pirquet, who had suggested that space stations could be used as way stations for interplanetary space flights, Noordung developed his own theories. Encouraged by Hermann Oberth, he published these ideas in 1929 in *Das Problem der Befahrung des Weltraums*, only to die the same year.

What makes this book scientifically important is its extensive and detailed treatment of the engineering aspects of the construction of a space station. While other authors had discussed the importance of space stations as a concept, few, if any, had addressed the problem of just what form such a station would take, or how it would be built.

Although Noordung made numerous errors, he got many details right. For instance, his space station included an airlock, and its electricity was created using a mirror that reflected sunlight, spinning a turbine that generated energy. His space station also rotated to provide artificial gravity. Noordung considered every detail of the space station's construction, and much of what he wrote anticipated Skylab and the International Space Station (ISS) decades later. "The entire structure," he wrote, "including its equipment would have to be assembled first on Earth and tested for reliability. Furthermore, it would have to be constructed in such a manner that it could easily be disassembled into its components and if at all possible into individual, completely furnished 'cells' that could be transported into outer space by means of space ships and reassembled there without difficulty."

He described such minute details as how the interior of the station would be lit, its water recycled, and how its air would be subject to "a ventilation system where it is cleaned, regenerated, and heated." The crew would communicate with the Earth by means of radio and the station itself would be controlled by means of gyroscopic flywheels and small thrusters. The rotating habitat would contain cabins for the crew, "work and study areas, a mess hall, laboratory, workshop, dark room, etc., as well as the usual utility areas, such as a kitchen, bath room, laundry room and similar areas."

③

④

⑤

"FOR ME, A ROCKET IS ONLY A MEANS—ONLY A METHOD OF REACHING THE DEPTHS OF SPACE—AND NOT AN END IN ITSELF. THERE'S NO DOUBT THAT IT'S VERY IMPORTANT TO HAVE ROCKET SHIPS SINCE THEY WILL HELP MANKIND TO SETTLE ELSEWHERE IN THE UNIVERSE."

KONSTANTIN TSIOLKOVSKY

THE SMITH-ROSS SPACE STATION

In a paper titled "Orbital Bases," published in 1949, Harry Ross and Ralph Smith of the British Interplanetary Society (BIS) designed a space station based on the one originally described by Hermann Noordung 20 years earlier.

①

②

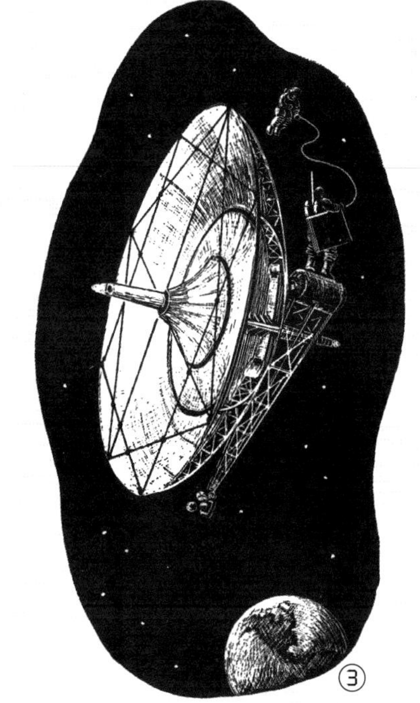

③

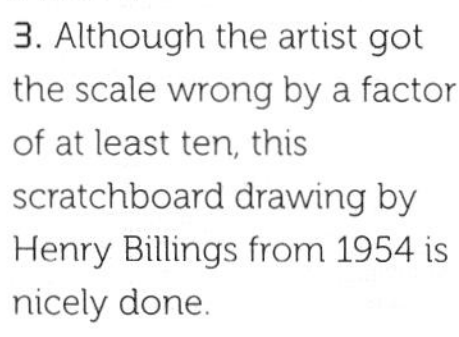

1. This painting of the space station under construction was done by one of its designers, R.A. Smith, who was not only an engineer but an accomplished artist as well.
2. Although never quite capturing the public imagination like Wernher von Braun's Wheel, the Smith-Ross station nevertheless managed to inspire a number of illustrators. This interpretation was done by the American artist, Fred Wolff, in 1954.
3. Although the artist got the scale wrong by a factor of at least ten, this scratchboard drawing by Henry Billings from 1954 is nicely done.

In the late 1940s, engineer H. E. Ross and engineer-artist Ralph A. Smith of the British Interplanetary Society began to speculate on the need for a crewed artificial satellite or "orbital base." Only above the Earth's atmosphere, they argued, could humans carry out research on solar radiation and cosmic rays, something they felt would be critical to the development of human space flight. In addition, such stations could observe weather conditions on the Earth or be way-stations for spacecraft bound for the Moon or other planets. Another valuable function would be in communications, since a series of space stations could provide global television and radio coverage. In describing the Ross-Smith station, Arthur C. Clarke, in *The Exploration of the Moon* (1954), suggested that the "economic consequences of this will be enormous, and the orbital radio stations may by themselves pay for the initial development of astronautics."

Astronauts would assemble the Ross-Smith station in orbit from prefabricated parts lifted from the Earth by a series of supply rockets. The station itself would consist of three main units. The first was the ring-shaped 30 m (98ft 6in) diameter living quarters. This rotated to provide artificial gravity. The second and perhaps most prominent feature was a 55 m (180 ft) diameter parabolic mirror. This was part of the solar-power generating system that provided electricity for the station. Sunlight focused on a system fluid-filled pipes created steam that in turn powered eight turbines that generate electric power. The habitat consisted of two concentric galleries, which were subdivided into crew quarters, laboratories, workshops, a galley, radio room, storage, etc. The station would have a permanent crew of 24 people, who would consume 32–54 tonnes (35–60 tons) of water and oxygen and 10.78 tonnes (11.9 tons) of food every year. In the central hub would be the water reclamation plant, air tanks, radio equipment, and the flywheels that would be used to adjust the station's attitude. The authors compared the living and working space as being equivalent to a single-story building 137 m (450 ft) long, 4.9 m (16 ft) wide, and 3 m (10 ft) high.

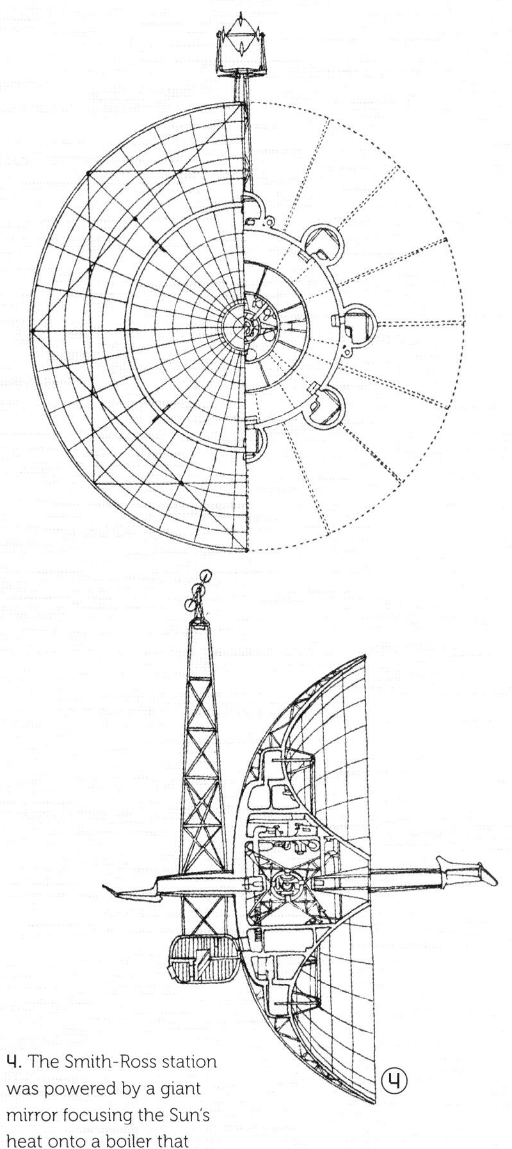

4. The Smith-Ross station was powered by a giant mirror focusing the Sun's heat onto a boiler that generated steam for turbines. The mast carried communications antennae as well as the airlock, and could be spun in the opposite direction of the station itself.

5. The space station would be supplied by cargo rockets from Earth, carrying food, medicine, scientific equipment, construction materials, and personnel to the base.

The third feature of the space station was a large boom extending from the point on the hub opposite that of the mirror. One end of this held the station's radio antennae, while at the other end was the station's airlock. Normally, the boom remained stationary while the station rotated. To use the airlock, an astronaut transferring from a nearby spaceship would first enter it. This would be safe and easy since the airlock and spaceship were motionless relative to one another. The boom would then be slowly accelerated until its rotation matched that of the station. The astronaut could then move from the airlock into the station. The rotation of the station was necessary in order provide a sense of gravity for the crew. It did this by rotating once every seven seconds. This would create Earth-normal gravity—or 1 g—throughout most of the station.

"MAN MUST AT ALL COSTS OVERCOME THE EARTH'S GRAVITY AND HAVE, IN RESERVE, THE SPACE AT LEAST OF THE SOLAR SYSTEM."

KONSTANTIN TSIOLKOVSKY

The authors left it to their readers to imagine what life might be like on board such an alien structure, "with its upcurved floors that give the impression that one's colleagues are about to fall down, and its concave billiards table that looks impossible but on which the balls roll quite normally" (H. E. Ross, "Orbital Bases," 1949).

ASTROPOL

In Otto Willi Gail's 1926 novel, *Der Stein vom Mond* (The Stone from the Moon), the space station made its first significant appearance in science fiction since Lasswitz's *Auf zwei Planeten*. Gail describes the space station Astropol as being in a polar orbit 95,000 km (60,000 mi) above the Earth. A fleet of eight spaceships handles the traffic between Earth and Astropol, which also acts as a way-station for spacecraft bound for Venus. Astropol, which is constructed of metallic sodium, was built around a spaceship as its core (in much the same way the giant space station proposed by Darrell Romick in 1956 was to have evolved). It is shaped like a squat cylinder nearly 120 m (394 ft) in diameter with a counter-rotating observatory at the end of one axis and an airlock for spaceships at the other. The crew quarters are in a rotating "gravity cell" located at the end of a long tether attached to the station.

Otto Willi Gail (1896–1956) was a famous German science writer and science fiction author who was influenced by the many theorists and experimenters working in his country during the turbulent era of the 1920s and 1930s.

The Stone from the Moon contained a detailed description of the space station Astropol, including numerous details about its construction and its operation. These illustrations by Frank R. Paul are from an English translation of the story published in *Science Wonder Quarterly* in 1930.

JONATHAN SWIFT'S ISLAND IN THE SKY

While E. E. Hale was the first to describe a true space station in fiction, a nod must be given to Jonathan Swift's satirical novel, *Gulliver's Travels*, which was published in 1726. The book's hero goes through numerous adventures in various strange countries before finding himself on the flying island of Laputa. Swift describes Laputa as being an artificial structure about 7 km (4.5 mi) in diameter—shaped like a dome with a flat bottom—and levitated above the surface of the Earth by means of a giant magnet. Laputa does not orbit our planet—in fact, it is never more than a few miles above the island of Balnibarbi, over which it perpetually hovers, since it is the mutual repulsion of the minerals in that island and Laputa that keeps the latter airborne. Yet it is still one of the first descriptions of a self-sustaining artificial world suspended above the Earth.

An illustration from a French translation (1875) of *Gulliver's Travels*. Here Gulliver waves his handkerchief to get the attention of people looking down from Laputa.

Hayao Miyazaki's animated film *Castle in the Sky* (1986) featured this floating island, inspired by the story of Laputa.

This map, included in the original edition of *Gulliver's Travels*, shows the position of Laputa and its movements as it passes above Balnibarbi.

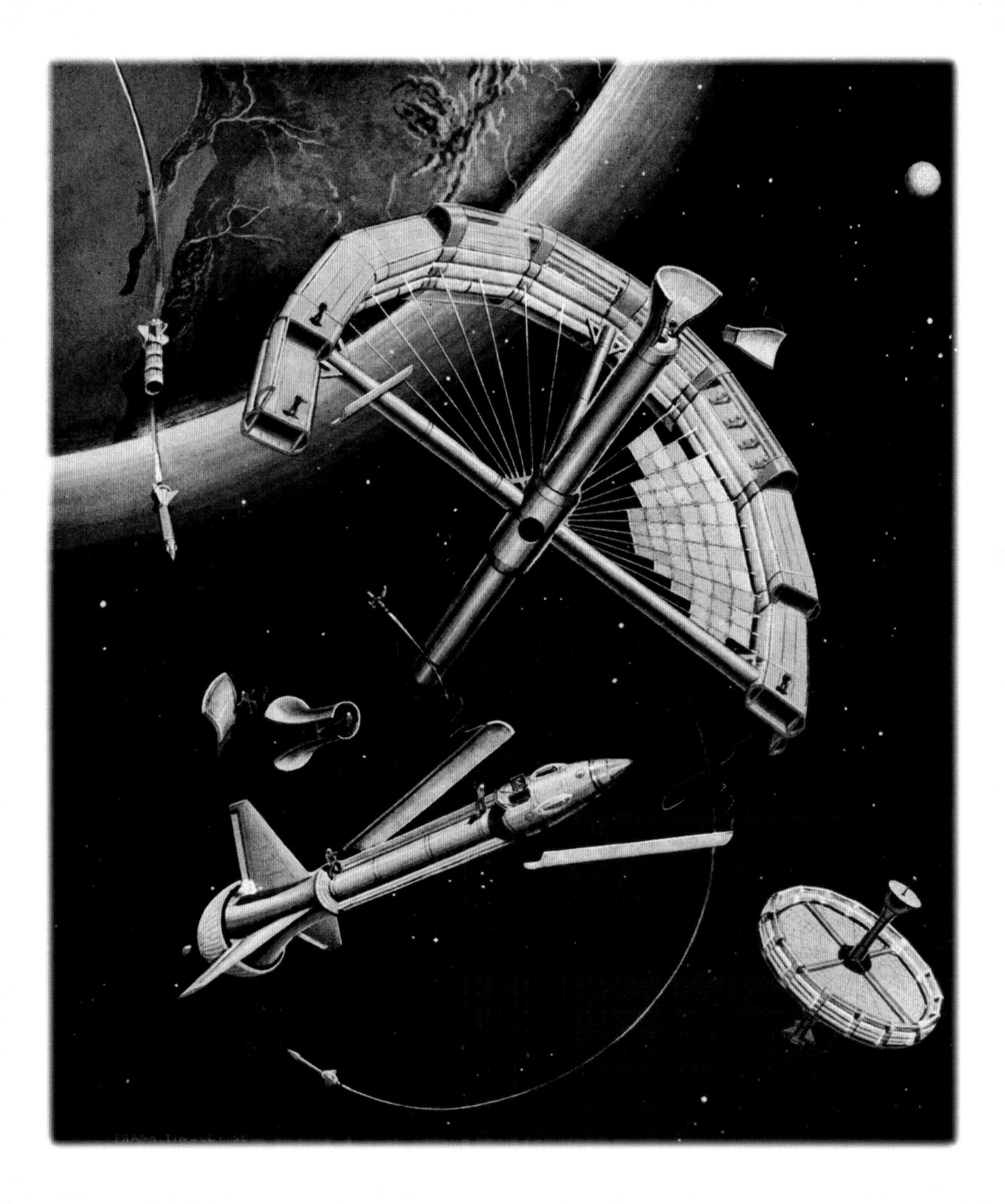

Chapter Two

PLANNING TO LIVE IN SPACE

Scientists and engineers developed significant rocket technology during World War II, primarily as weapons of war. Yet the evolution of ever-more powerful rockets meant that the notion of a permanently crewed orbiting craft out in space no longer seemed fanciful. Indeed, in the aftermath of the war many considered that artificial satellites were not only possible but also absolutely necessary, and a stepping stone to the realization of a space station.

(Left) Renowned author and illustrator Frank Tinsley created this visionary concept for a solar-powered space station in 1958.

(Above) Artist Jack Coggins was among the few to suppose that space station crews might not require artificial gravity. His cylindrical space station is shown here under construction.

Planning to Live in Space

IN 1945, *LIFE* MAGAZINE REVEALED "THE ASTONISHING FACT THAT GERMAN SCIENTISTS HAD SERIOUSLY PLANNED TO BUILD A *SUN GUN*," WITH WHICH THEY HAD PLANNED TO INCINERATE ENTIRE SWATHS OF LAND. THIS WOULD HAVE BEEN ACCOMPLISHED BY MEANS OF A GIGANTIC ORBITAL MIRROR THAT WOULD "FOCUS THE SUN'S RAYS TO A SCORCHING POINT ON THE EARTH'S SURFACE . . . TO BURN AN ENEMY CITY OR TO BOIL PART OF AN OCEAN."

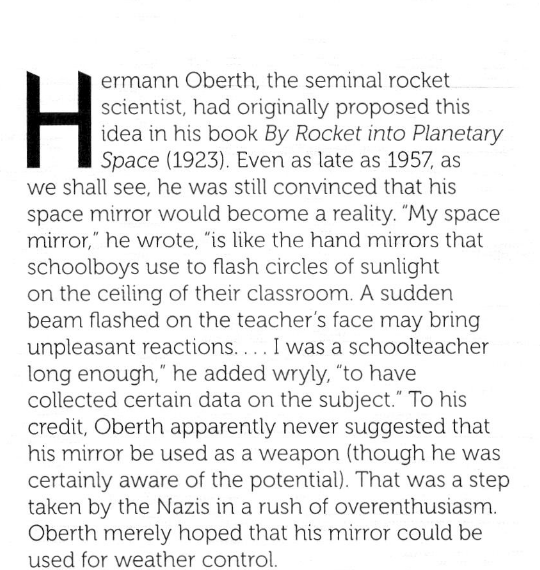

Hermann Oberth, the seminal rocket scientist, had originally proposed this idea in his book *By Rocket into Planetary Space* (1923). Even as late as 1957, as we shall see, he was still convinced that his space mirror would become a reality. "My space mirror," he wrote, "is like the hand mirrors that schoolboys use to flash circles of sunlight on the ceiling of their classroom. A sudden beam flashed on the teacher's face may bring unpleasant reactions. . . . I was a schoolteacher long enough," he added wryly, "to have collected certain data on the subject." To his credit, Oberth apparently never suggested that his mirror be used as a weapon (though he was certainly aware of the potential). That was a step taken by the Nazis in a rush of overenthusiasm. Oberth merely hoped that his mirror could be used for weather control.

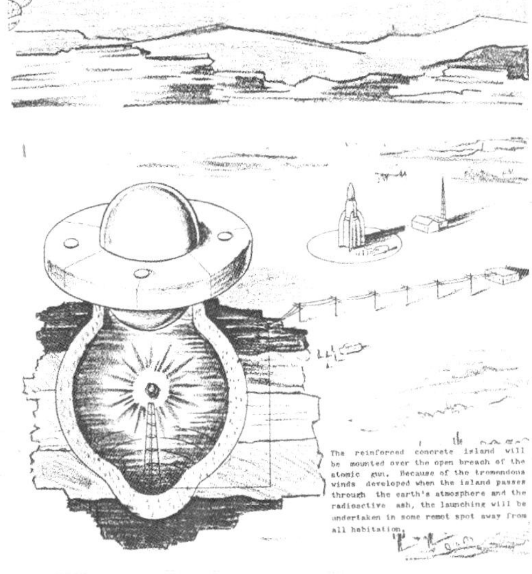

In the late 1950s, engineer Constantin Lent proposed the idea of launching a space station into orbit by means of a nuclear explosion.

SPACE MIRROR

Engineers would have assembled Oberth's giant mirror from prefabricated sections. When the structure was finished, the result would have been a disk one mile in diameter. The surface would be a concave, parabolic curve like that found in a shaving or makeup mirror. The space station would be placed in a polar orbit at a distance of 8,207 km (5,100 mi). *Life* magazine assumed that the mirror would be a crewed station, equipped with 9 m (30 ft) diameter docking ports for supply rockets, hydroponic gardens to provide oxygen and food for the crew, and solar-powered generators for electric power.

The construction of the mirror would begin with the launch of a single unmanned rocket. Once in orbit, it would unreel six long cables, each only 12.7–38 mm (½ to 1½ in) thick. The rocket would then be set spinning on its axis, which would cause the cables to extend radially, like the arms of a rapidly spinning ice skater.

These cables would span 145 km (90 mi). Astronauts—working like spiders creating a web—would use this network as the foundation for a series of hexagonal cells, each of which would be several miles wide. Each of these cells would contain a movable circular mirror made of thin sodium-metal foil. The rotation of the entire structure would keep it taut. The area of the finished mirror would be vast in its extent: 70,000 km² (27,000 square mi). The pressure of sunlight on the vast surface would be used to maneuver the craft in orbit, with steering accomplished by adjusting the angles of the individual mirrors.

Oberth thought it might take 10–15 years to assemble a complete mirror at a cost of $3 billion. *Life's* experts believed that there was a fundamental flaw in the scheme. That was the fact that an image of the Sun cannot be projected to a point. They observed that since the Sun appears in the sky as a disk and not as a point, the best that any systems of mirrors or lenses can produce is an image of that disk. All that is required is an ordinary magnifying lens and a sunny day for you to see this for yourself. If you hold the lens so that you focus sunlight onto a sheet of paper, you will be able to produce an intensely bright spot of light. This is in fact an image of the Sun projected onto the paper. If you hold the lens steady long enough, you may even set the paper on fire. This might seem to argue in favor of Oberth's proposal. The problem, though, lay in how far the mirror would be from the surface of the Earth. The farther away the mirror (or a lens) the larger the projected image of the Sun will be. At a distance of many thousands of miles, the image of the Sun would be so large that it couldn't possibly do any damage.

Oberth disagreed with this argument. The mirror, he said, would not *need* to focus sunlight to a point to create catastrophic damage. Indeed, the focused "spot" on the Earth's surface would be about 5,180 km² (2,000 square mi) in area. The heat and light in this region, he said in *Man in Space* (1957), would be "no stronger than the normal at the equator." But, he continued, if "the mirror were double the size mentioned . . . the irradiation would be four times as strong . . . [and] the temperature on the surface . . . would be 200°C (392°F)." Perhaps not enough to burn cities and melt battleships, but more than

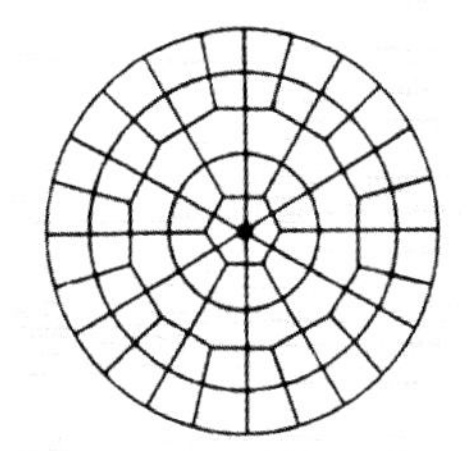

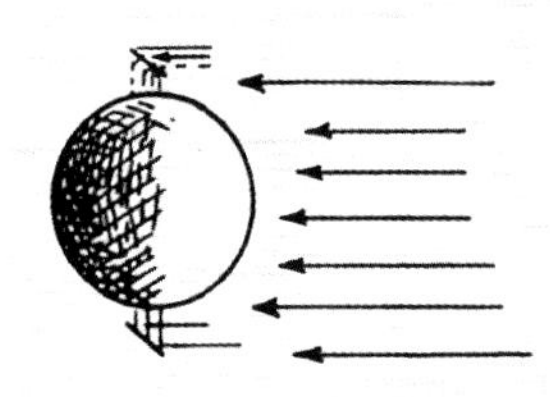

Oberth created these drawings to illustrate his giant space mirrors and how they could be used to reflect the Sun's heat and light onto the Earth's cold arctic regions.

In 1951 Austrian-born American physicist Fred Singer proposed the Minimal Orbital Unmanned Satellite, Earth (MOUSE), a satellite that would contain Geiger counters for measuring cosmic rays and other instruments for measuring the environment of space.

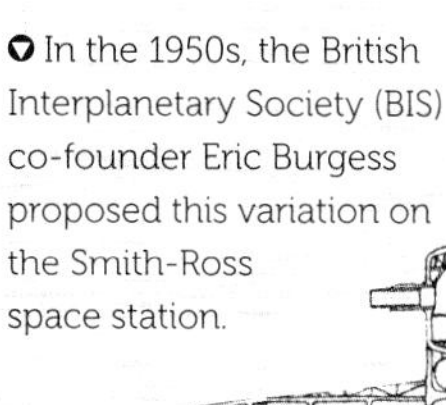

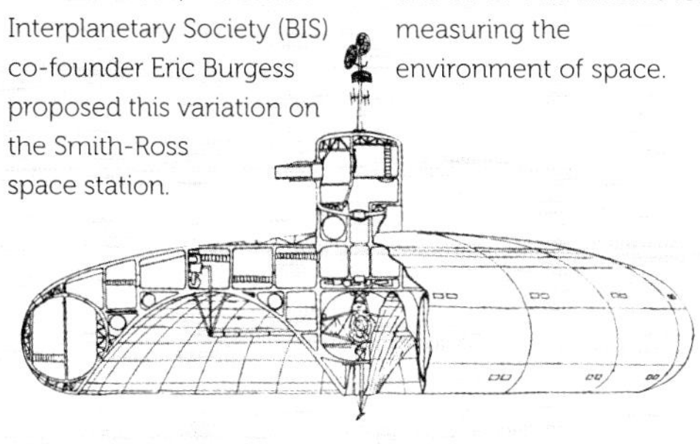

In the 1950s, the British Interplanetary Society (BIS) co-founder Eric Burgess proposed this variation on the Smith-Ross space station.

sufficient to render a region lifeless. Oberth also revisited his space mirror concept in *Man Into Space*. "I am certain," he wrote, "that my space mirror will one day be a reality."

THE ICON

In 1952, a space station design was revealed to the public that was to dominate and even define the very concept of an inhabited orbiting Earth satellite for more than a generation.

Collier's was one of four top-circulation general magazines that flourished during the 1940s and 1950s. The magazine was rivaled only by *Life*, *Look*, and *The Saturday Evening Post*. Postwar concerns about the potential military use of space had led its editors to investigate the feasibility of space travel in the near future. To this end, the magazine sponsored a gathering of the world's greatest space experts who, in a series of illustrated articles published from March 22, 1952, through April 30, 1954, outlined one of the first comprehensive scenarios ever conceived for the exploration of space.

The symposium was headed by former German ballistic missile designer Wernher von Braun, then technical director of the US Army Ordnance Guided Missile Development Group. He was joined by Fred L. Whipple, chairman of astronomy at Harvard University; Joseph Kaplan, professor of physics at UCLA; Heinz Haber, of the US Air Force Department of Space Medicine; and Willy Ley, an authority on space travel and rocketry who had worked on some of the earliest German rocket experiments and who had fled Germany when the Nazis came to power.

The symposium told astonished *Collier's* readers that the United States could send a 50-man expedition to the Moon by 1964 and a manned mission to Mars soon afterward. Both of these events, however, were entirely dependent on the construction of a giant, orbiting space station.

The *Collier's* team was entirely serious. There was no practical barrier to accomplishing their plans. The necessary technology already existed. "Speculations regarding . . . future technical developments have been carefully avoided," von Braun explained. "Only by stubborn adherence to the engineering solutions based exclusively on scientific knowledge available today, and by strict avoidance of any speculations concerning future discoveries, can we bring proof that this fabulous venture is fundamentally feasible." What von Braun intended to prove was that space travel was more a matter of will than of material. His point was not to show that space travel was a possibility of some distant future, but that it was possible in 1952. The first step would be to launch a "baby satellite." This would be the 3 m (10 ft) nose cone of a large rocket that would carry three unfortunate rhesus monkeys into space. It would orbit the Earth at an altitude of 322 km (200 mi) for 60 days, after which it would re-enter the atmosphere and burn up (after the monkeys were given a merciful dose of lethal gas).

Following the success of these experiments, the crewed space program would begin. Its workhorse would be a monster, three-stage rocket, 80 m (265 ft) tall and weighing 6,350 tonnes (7,000 tons). By comparison, the Apollo program's Saturn V was 110 m (363 ft) tall and weighed 2,913 tonnes (3,211.5 tons). The third, crew-carrying stage was in reality a winged aircraft carrying ten astronauts (some of whom, the authors pointed out, would be women). This would be used to ferry into orbit the material needed for the construction of the 76 m (250 ft) diameter space station that was a necessary part of von Braun's scenario. He planned for at least ten or twelve ferries to be in operation during the construction of the space station. At the height of the work there would be as many as one launch every four hours! Once the station was completed and in service, only one supply or personnel launch every three days would be necessary.

Previous space station proposals, like that of the BIS, largely approached the subject as an isolated problem. The von Braun scenario, however, was the first to consider an orbital station as an integral part of a complete space program, one that progressed step by step from Earth satellites to crewed space stations to flights to the Moon and Mars.

> "EVERYTHING IN SPACE OBEYS THE LAWS OF PHYSICS. IF YOU KNOW THESE LAWS, AND OBEY THEM, SPACE WILL TREAT YOU KINDLY. . . . MAN BELONGS WHEREVER HE WANTS TO GO."
>
> WERNHER VON BRAUN

VON BRAUN'S SPACE WHEEL

The completed wheel-shaped station created by Wernher von Braun would accommodate several hundred crew members and was scheduled for 1963 (later revised to 1967 when the magazine articles were published in book form). The tire-like rim of the station would be 10 m (30 ft) in diameter and would have three floors. The centrifugal force generated by the rotation of the station would create an artificial gravity, so that "up" would be toward the hub and "down" would be away from it. Von Braun divided the interior into numerous sections. There would be a communications center and a meteorological observatory that would send regular weather reports back to the Earth. An electronic computer would take up a large part of the upper deck: the *Collier's* experts were writing long before the advent of the transistor and solid-state electronics. Another two decks would be devoted to astronomical observations and research. An outer layer of thin metal covering the station would create a "meteor bumper" that would protect its

Wernher von Braun (left) shares models of his space station and ferry rocket with the editor of *Collier's* magazine, Cornelius Ryan. It was Ryan who popularized von Braun's ideas, bringing them from the scientific drawing board to public awareness.

This iconic painting by Chesley Bonestell includes all of the fundamental elements of von Braun's space station model: the station itself, a supply rocket for ferrying supplies and personnel, a "space taxi," and an orbital space telescope.

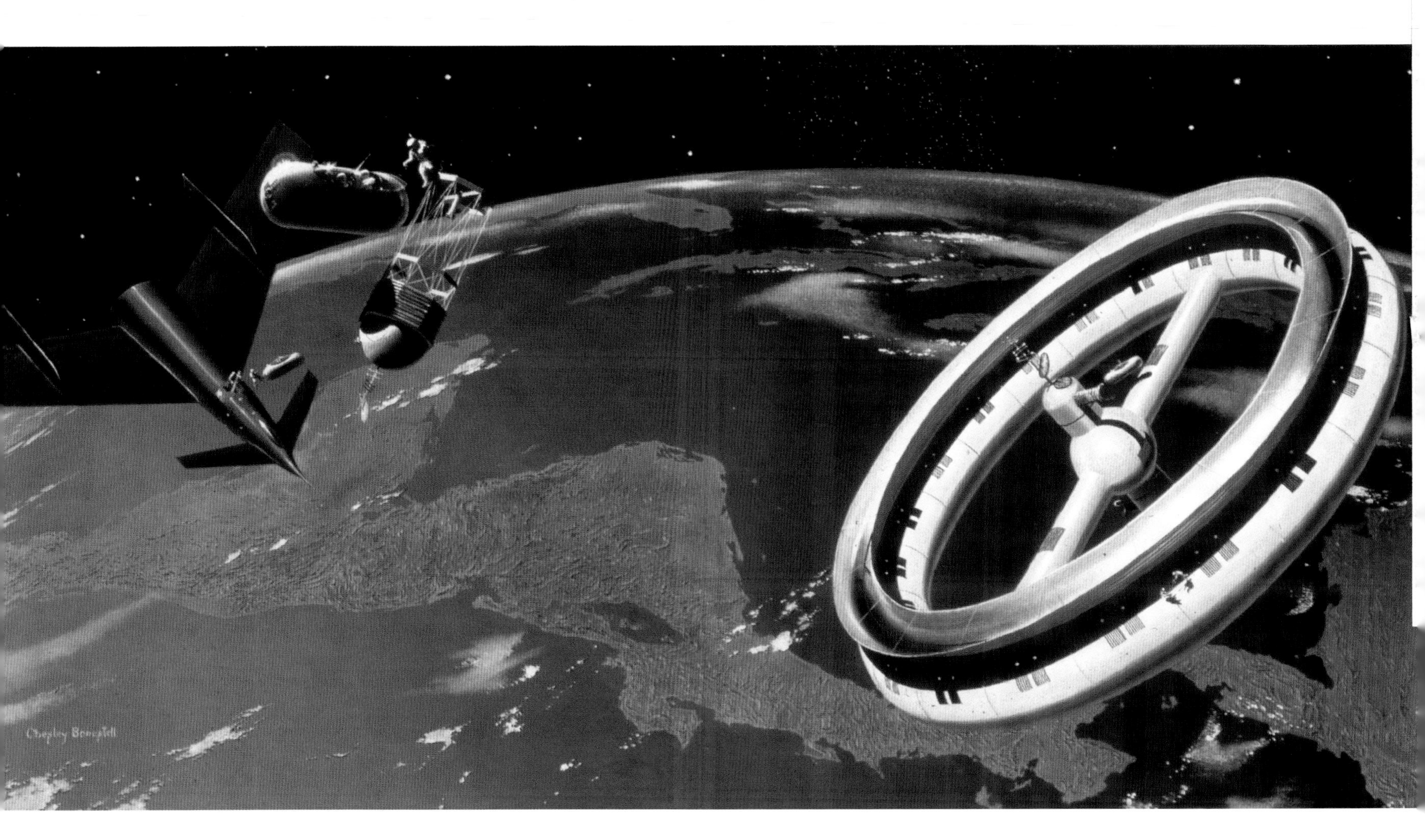

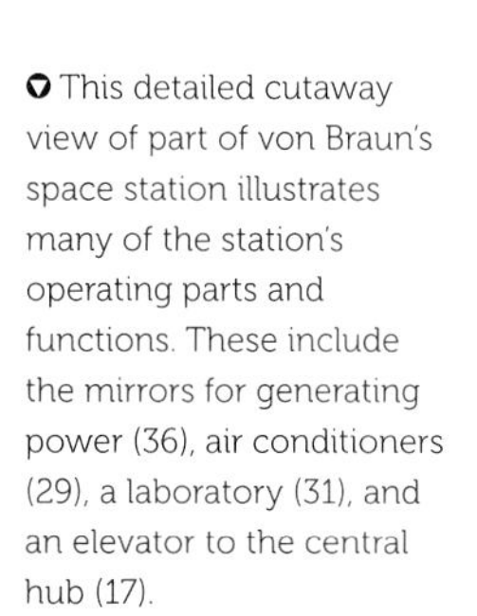

Wernher von Braun's design quickly became accepted as the gold standard for space station design, influencing artists around the world well into the twenty-first century. This interpretation was created in 1954 by illustrator Fred Wolff.

This detailed cutaway view of part of von Braun's space station illustrates many of the station's operating parts and functions. These include the mirrors for generating power (36), air conditioners (29), a laboratory (31), and an elevator to the central hub (17).

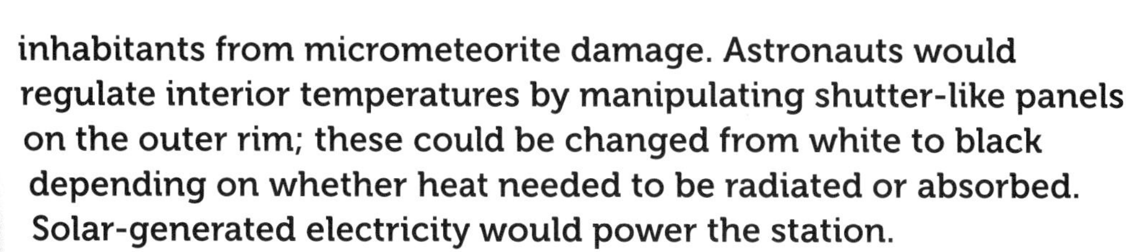

inhabitants from micrometeorite damage. Astronauts would regulate interior temperatures by manipulating shutter-like panels on the outer rim; these could be changed from white to black depending on whether heat needed to be radiated or absorbed. Solar-generated electricity would power the station.

Publishers reproduced Wernher von Braun's inspiring visions of future spaceflight—and the astonishing illustrations that went with them—in books and magazines all over the world, captivating the public. The spacecraft he designed in conjunction with Chesley Bonestell became the gold standard for spacecraft for many decades. They influenced everything from movies to TV (Disney's "Man in Space" series was the first), from pulp magazines to toys. The image of the wheel-shaped space station, as simple and elegant as the rings of Saturn, became ingrained in the public consciousness as being somehow "right."

THE SPACE CITY

At about the same time that Wernher von Braun was making headlines with his popular *Collier's* series and the books it spawned, a Goodyear Aircraft engineer named Darrell C. Romick presented several visionary papers in which he outlined a future world of ion propulsion, reusable launch vehicles, crewed lunar missions, and permanently occupied orbital colonies.

①

②

1. Romick proposed an evolutionary scheme for the exploration of space that began with unmanned satellites and ended with enormous populated space colonies.
2. Romick's space station would have been inhabited by 20,000 residents, living in the rotating wheel at one end.
3. A cross-section of the habitat, showing its divisions into discrete sections depending on their function.
4. The Romick space habitat as it would have looked after completion.

> "AT FIRST SPACE NAVIGATORS, AND THEN SCIENTISTS WHOSE OBSERVATIONS WOULD BE BEST CONDUCTED OUTSIDE THE EARTH, AND THEN FINALLY THOSE WHO FOR ANY REASON WERE DISSATISFIED WITH EARTHLY CONDITIONS, WOULD COME TO INHABIT THESE BASES AND FOUND PERMANENT SPATIAL COLONIES."
>
> J. D. BERNAL

In some ways, Darrell Romick's scenario was similar to that of Wernher von Braun's, especially in that he envisioned a fundamentally developmental approach to the conquest of space. Like von Braun, Romick's scheme depended on a three-stage space shuttle. Unlike von Braun's vision, however, Romick's piloted orbiter was more than a mere shuttle for transporting personnel and supplies. It was in actuality the nucleus of the grandest part of his plan.

ORBITERS

First, a series of orbiters would be launched into orbit. As each spacecraft arrived, waiting astronauts would attach them alternately nose-to-nose and tail-to-tail, eventually creating a long cylinder from the fuselages. Around this core would evolve an enormous space station, one that dwarfed the 76 m (250 ft) wheel von Braun wanted to create. Instead, Romick's plan would consist of a huge, zero-gravity cylindrical "dry dock" over 457 m (1,500 ft) long and 305 m (1,000 ft) in diameter, with a total volume of about 278 million m² (3 billion square ft). At one end was a 457 m (1,500 ft) rotating disc that would be inhabited by 20,000 people. A rotation rate of 2 rpm would provide 1 g of artificial gravity. "Movement at [this rate]," wrote Romick in a true masterpiece of understatement, "would give a somewhat majestic motion for this large wheel."

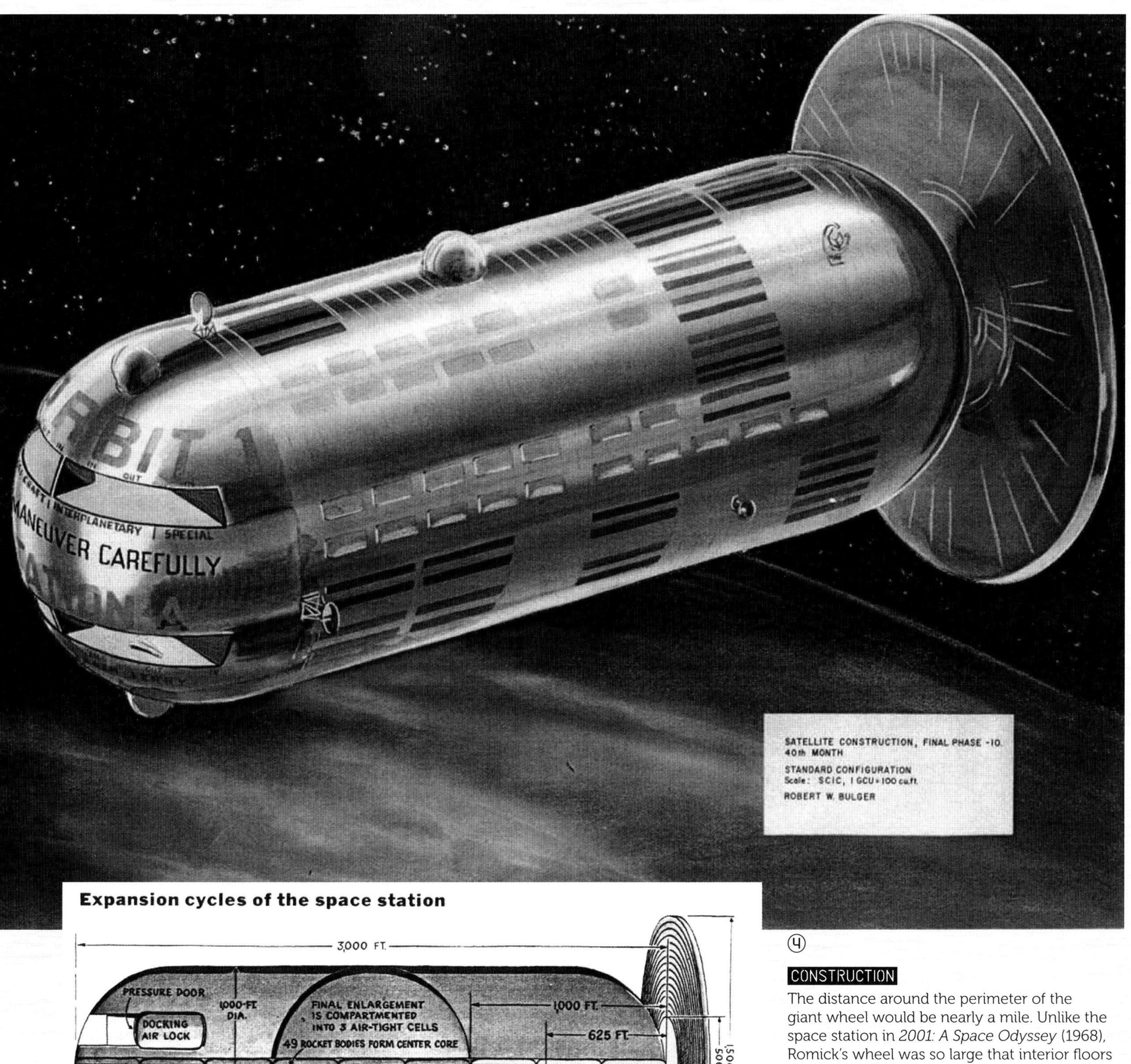

③

④

CONSTRUCTION

The distance around the perimeter of the giant wheel would be nearly a mile. Unlike the space station in *2001: A Space Odyssey* (1968), Romick's wheel was so large that interior floors would not appear to be curved. Indeed, as Romick pointed out, "a room would have to be thirteen feet [4 m] long to give a 1 degree total change in floor level." The entire inhabitable volume of the wheel would be over 4,000,000 m³ (141,000,000 cubic ft), compared to the 63,000 m³ (2.22 million cubic ft) of the von Braun torus.

Eighty-two floors would be between the central hub and the outer rim. These would contain apartments, stores, churches, schools, gymnasiums, and theaters as well as TV stations

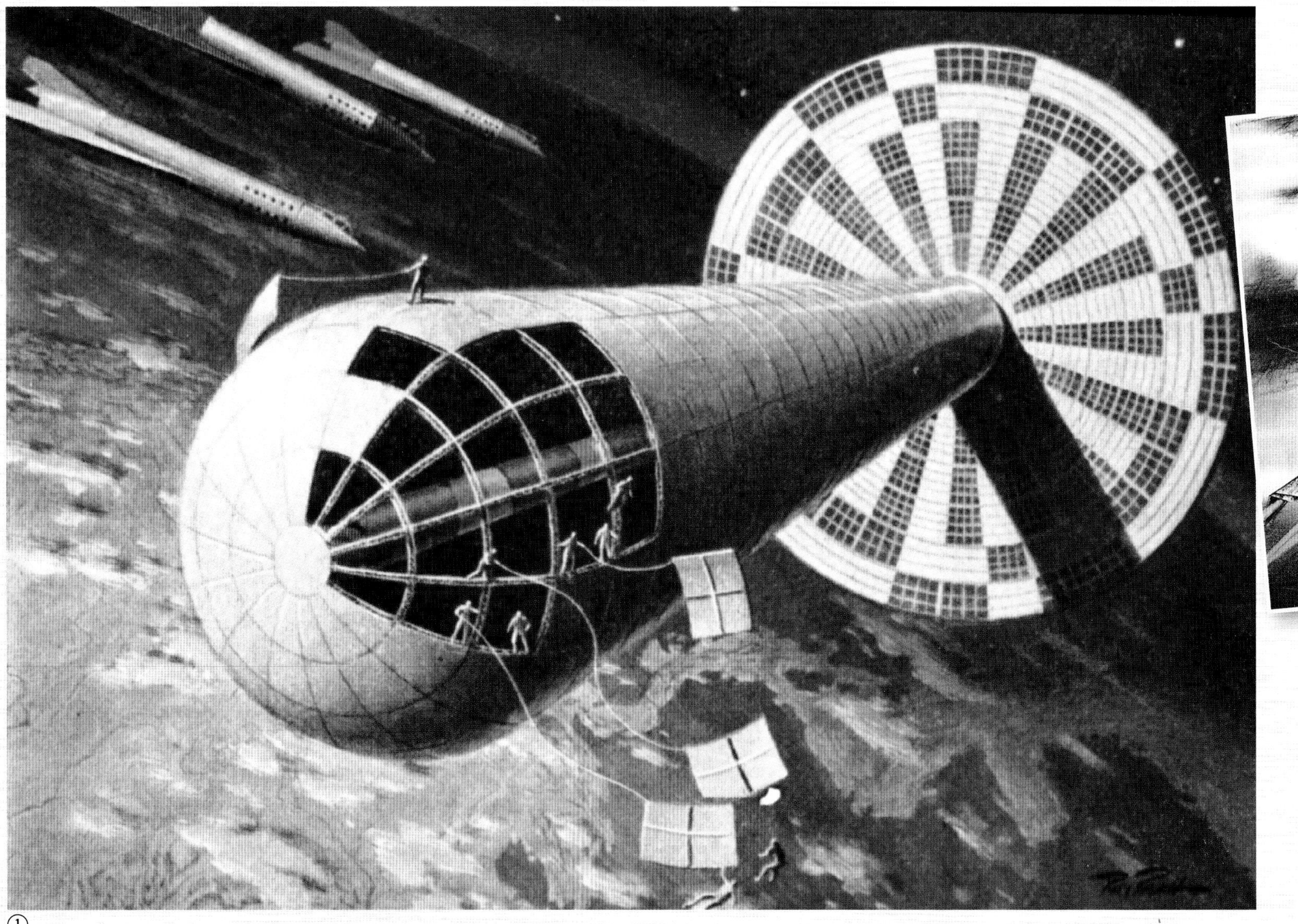

①

1. Romick's giant space station habitat is seen here nearing completion, with some of the last remaining panels being fitted. In the upper left corner of the picture, a fleet of supply rockets wait nearby.
2. To accomplish the construction of his station, Romick designed a multi-stage ferry rocket. Each stage was piloted and could be returned to Earth for reuse.
3. The core of the station is built up from a series of interconnected rocket fuselages, assembled end to end to form the space station's core section.
4. Once the core of the space station is completed, the station is built up in concentric layers, with the habitat wheel added to one end.
5. The core of the station is in an advanced stage of construction, with astronauts conducting space walks.. The linked rockets are now being sheathed and will form the spine of the cylindrical airtight dry dock.

and newspapers. Romick pointed out that his design provided for quarter-gravity and zero-gravity gymnasiums, "which would be most interesting and stimulating." A total of 5 hectares (12 acres) of solar panels would provide the electrical power for the space station. These would cover entirely the main cylinder of the station, there converting sunlight into electrical energy with great efficiency in the atmospheric clarity of space.

COMPLETION RATE

At a rate of two shuttle launches a day—a particularly ambitious schedule—Romick estimated that it would take about three years to bring the wheel to its final state of completion, though it would be functional during that entire time. Romick also looked further ahead conceptually than von Braun in assuming that people would travel into space not just as scientists or engineers, but also as immigrants. "The station," Romick claimed, would be "capable of providing safe, comfortable living accommodations similar to those on the ground for thousands of inhabitants, as well as affording them recreational facilities and cultural experiences not available on Earth. The view from the living-room window would be sensational. . . ."

"A large, inhabited satellite terminal," Romick concluded, sounding rather like a futuristic real-estate agent on the hard sell, "can certainly be constructed—a glistening 'city in the sky'. . . . And it can be a visible symbol, as it grows of the great promise of new service, new knowledge, inspiration and accomplishment for all mankind. Its potential seems unlimited—a stepping-stone to the Moon, to the planets, and beyond."

Scientific visionaries would conceive of nothing on such a grand scale of imagination until nearly two decades later, when Gerard O'Neill published *The High Frontier: Human Colonies in Space* (1976) and introduced an enthralled public to the important concept of artificial, self-supporting, orbiting habitats that would be the permanent home for tens and even hundreds of thousands of people out in the expanses of space.

"ONE OF TODAY'S OUTSTANDING OBJECTIVES FOR ENGINEERS AND SCIENTISTS . . . IS THE ESTABLISHMENT OF A PERMANENT AND INHABITED EARTH-SATELLITE TERMINAL."

DARRELL C. ROMICK

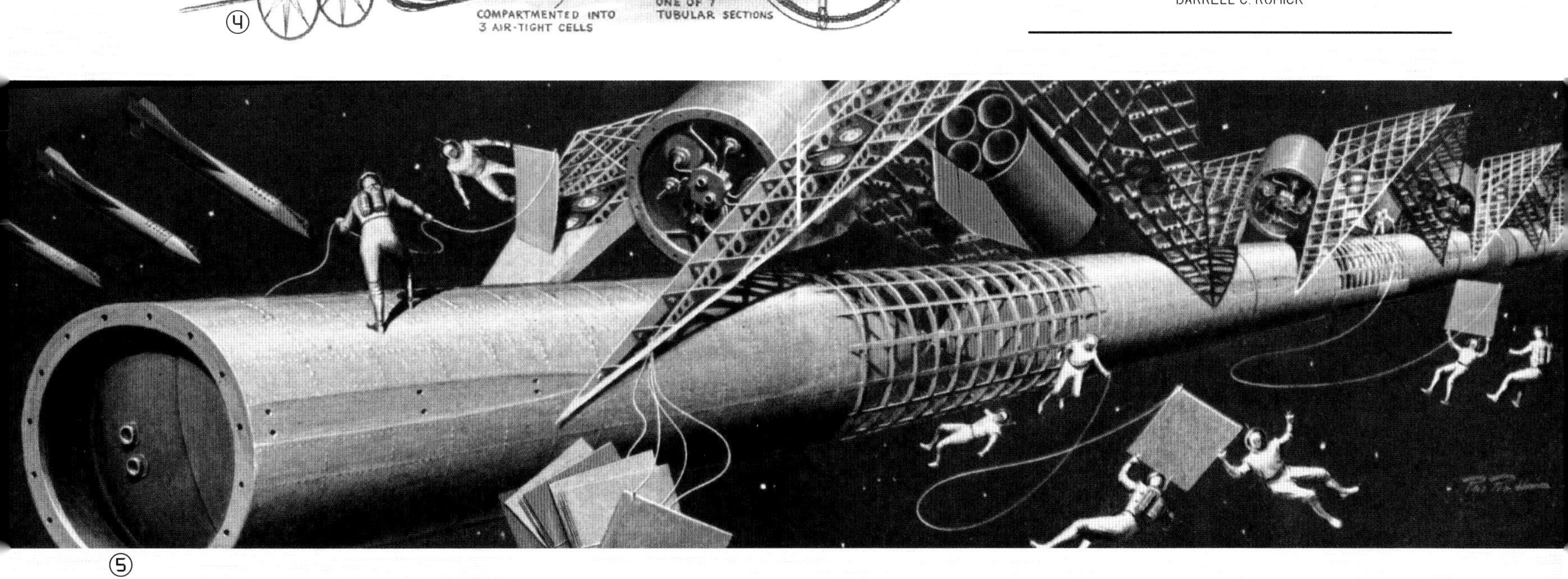

OTHER SPACE WHEELS

Noordung had invented the rotating toroidal space station, improved by Smith and Ross, but it was the elegant design created by Wernher von Braun and illustrated by Chesley Bonestell that became iconic. First published in 1952 in *Collier's*, one of the most popular American magazines of the time, it began appearing everywhere in one form or another, legitimately and otherwise, from the covers of science-fiction magazines and comic books to motion pictures and toys.

❶

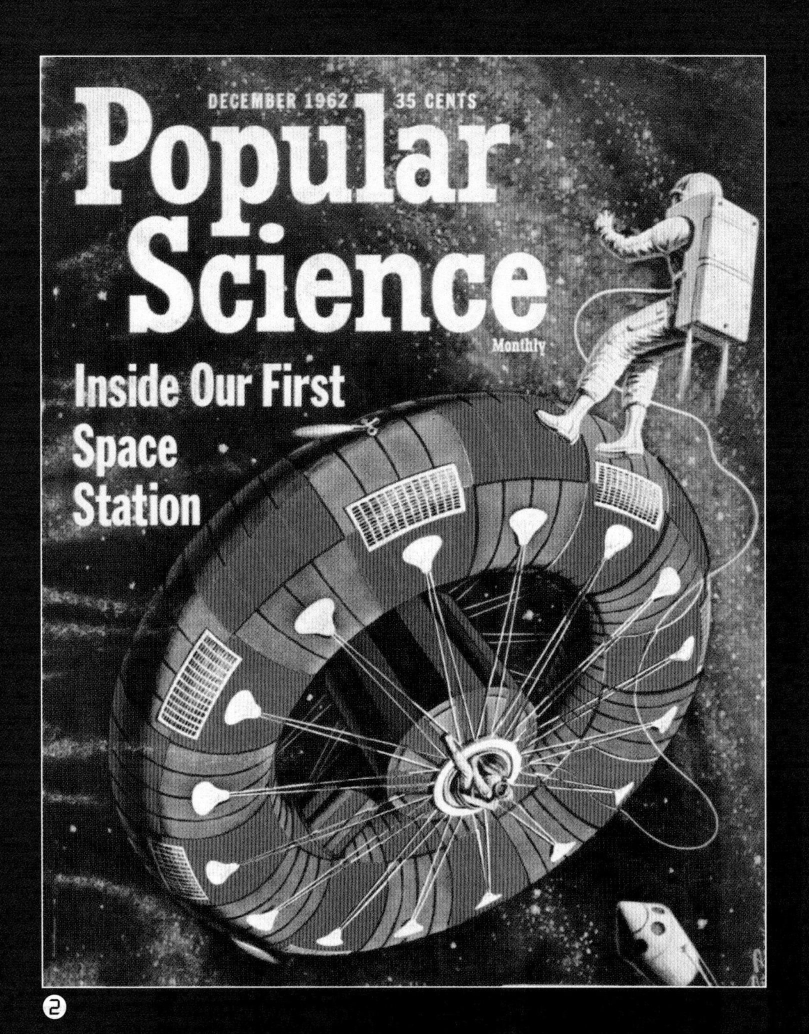

❷

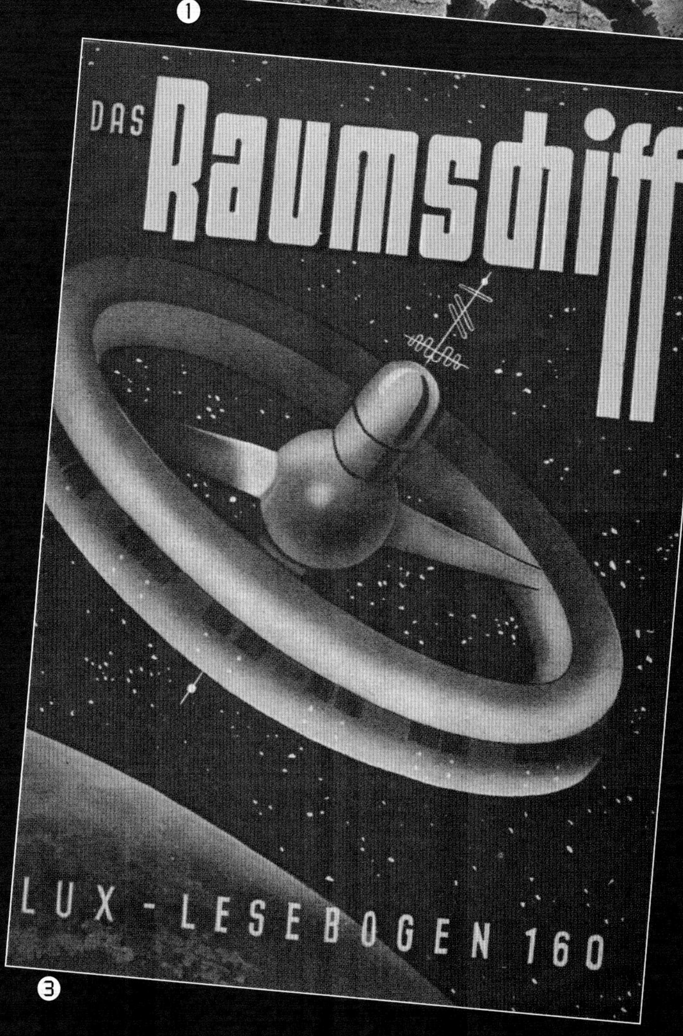

❸

For the "Man in Space" episodes of *Walt Disney's Wonderful World of Color* television series (1955), von Braun himself designed a variation of his now-classic space wheel. Even serious speculation about future space station by engineers and scientists usually assumed that they would be rotating rings of some kind. This was primarily driven by the assumption that its human occupants would require artificial gravity, and that rotational force was the way to supply this.

Artists specializing in technical subjects made a career of creating and publishing ideas for futuristic spacecraft. Among them was the prolific Frank Tinsley, whose work appeared often in popular magazines such as *Mechanix Illustrated*. These artists catered to the burgeoning public interest in space travel that was spreading with feverish intensity during the 1950s—an interest largely created almost single-handedly by von Braun and his *Collier's* team. Among Tinsley's original space-station creations were—needless to say—two designs inspired by the ubiquitous von Braun station. By the end of the 1950s and the beginning of the 1960s, engineers and aerospace companies began taking a serious look at the rotating, wheel-shaped concept and began developing designs of their own—perhaps the most unusual of which was an inflatable space station created—and even tested (on Earth) by the Goodyear Tire and Rubber Company. It was probably no coincidence that it resembled nothing more than an enormous inner tube.

1. This concept by artist Frank Tinsley was to have been an orbiting farm, growing algae beneath rows of greenhouses.
2. In 1962, the Goodyear Rubber Company proposed an inflatable space station. It would be launched deflated; once in orbit, it could be inflated to its full size.
3. The cover of *Das Raumschiff* (The Spaceship) presents a classic model of the wheel spaceship.
4. In the 1970s, the idea of enormous space colonies housing tens of thousands of inhabitants became popular. This illustration by David A. Hardy depicts one of these designs, constructed largely from asteroidal or lunar material.
5. Many artists were inspired by von Braun's design. Here one has depicted a cutaway of the famous station.
6. Artist Fred Wolff depicted a von Braun space station under construction in 1954, the supply spacecraft inspired by BIS designs.
7. Frank Tinsley's 1957 design for a rotating "space laboratory" included solar panels similar to those that power the ISS today.

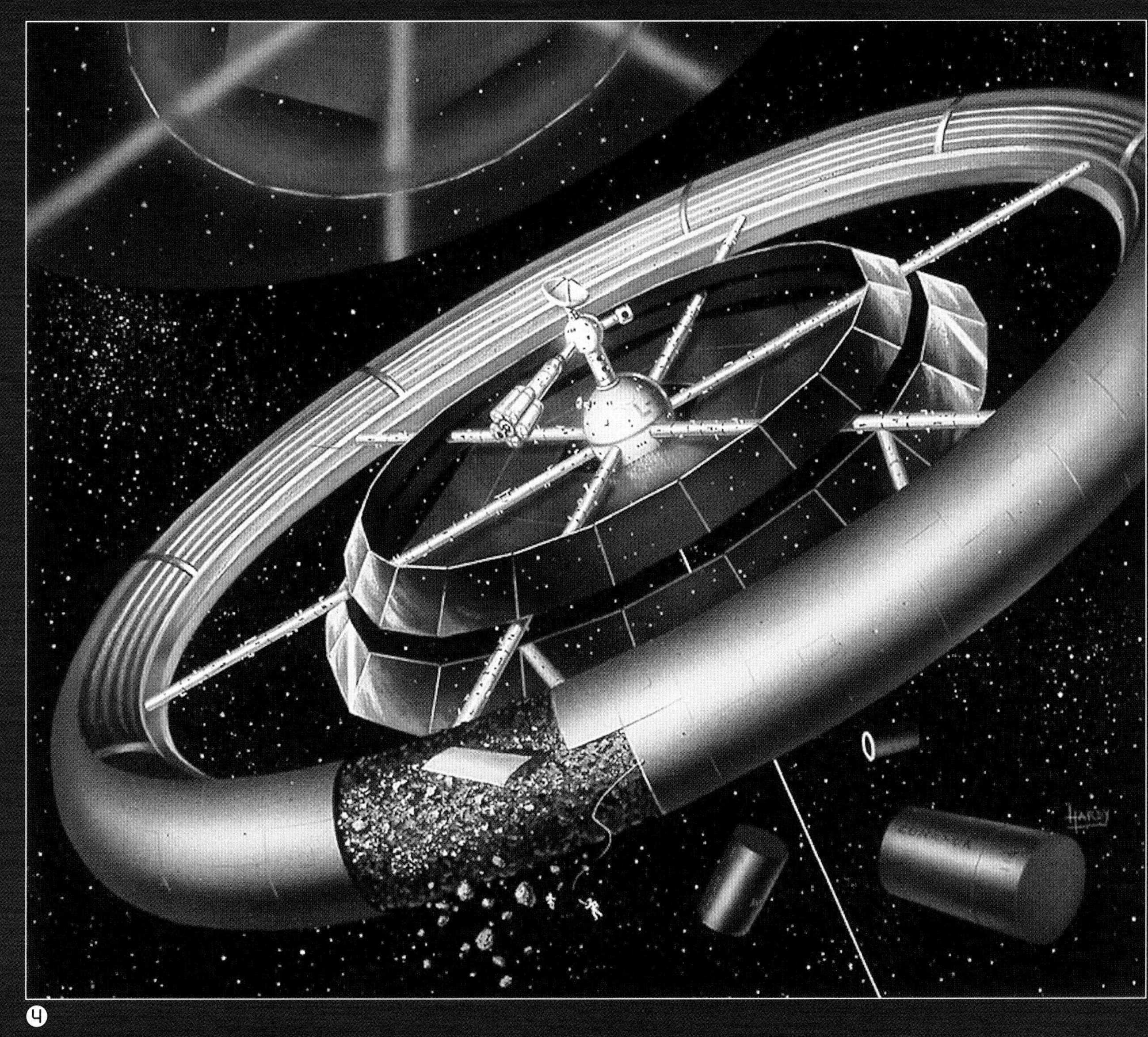

4

5

6

7

NEW CONCEPTS

While Wernher von Braun's space wheel had become the ideal model for space stations—at least in the minds of the public and popular culture at any rate—scientists, engineers, and even science-fiction authors and illustrators were looking at alternative designs. The driving idea behind this was the question: Was it really necessary to provide artificial gravity?

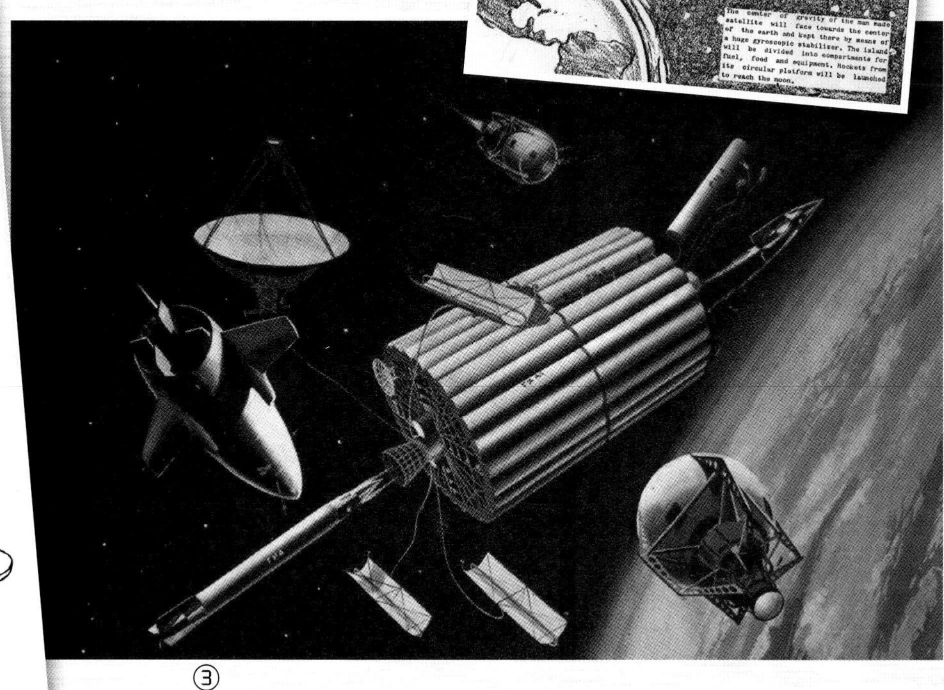

1. One of many unique concepts proposed by inventive engineer Constantin Paul Lent, this 1959 design would have been stabilized by a gigantic gyroscope.

2. This fine illustration from Patrick Moore's *Earth Satellites* (1956 depicts six stages in the development of the space station concept from 1929 up to von Braun's "late models."

3. This 1955 Italian design for a space station would have taken power from solar generators attached to the station by cables.

4 and 5. Artist Jack Coggins and author

The entire rationale for a rotating space station was the assumption that the health and physical well-being of the astronauts on board would suffer if they were to spend more than just a brief time in a weightless environment. But some came to question that basic assumption, especially since at that time there was little or no experimental evidence as to the effects of long-term weightlessness on the human body.

ROTATING CRAFT

What if the issue of providing gravity were to prove, in fact, not a problem at all? A non-rotating space station would eliminate a great many complex construction problems. For instance, how do you dock a spacecraft to a rotating space station? A non-rotating station could also be built in any shape required, allowing for a much more efficient use of space. The shape most often proposed was that of a sphere, which would offer the most interior volume with the least surface area. But other shapes were possible. In 1952, Constantin Paul Lent—a skilled draftsman and past vice president of the American Rocket Society—proposed a space station consisting of a row of repurposed rocket fuselages attached side-by-side, like a raft made of logs. In the same year, artist Jack Coggins and author Fletcher Pratt created a space station in the shape of a hexagonal prism that resembled an enlarged version of the later Skylab space station. Its crew would work in an entirely weightless environment, with equipment and instruments mounted in whatever position proved most convenient. "Inside," wrote Pratt, "this room will not look like anything on earth, because all the walls will be in use." From the outside, with observatories, space docks, solar collectors, and other equipment attached wherever convenient, the Coggins–Pratt station "will have a strange, lumpy look" very different from the elegant von Braun concept.

LIFE WITHOUT GRAVITY

As inelegantly serviceable—even clumsy—as these designs seemed at the time, they proved

"THE EARTH IS JUST TOO SMALL AND FRAGILE A BASKET FOR THE HUMAN RACE TO KEEP ALL ITS EGGS IN."

ROBERT HEINLEIN

④

⑤

⑥

Fletcher Pratt were among the first to suggest that weightlessness might not be a problem for the crew of a space station. Their 1952 design for a cylindrical, non-rotating space station anticipated Skylab by two decades. **6.** In 1952, technical illustrator Frank Tinsley proposed this spherical space station as a base for military operations.

to be prescient. It did indeed turn out to be unnecessary to provide artificial gravity, at least for the relatively short periods of time crew members spent aboard space stations. Astronauts staying aboard for months or even a year did have to contend with many health issues, some of them serious, which needed to be addressed. But space station designers found themselves free from the problems and constraints imposed by having a rotating spacecraft, with the result that modules and instruments could be added wherever they might be required and fragile solar panels could extend as far as needed (even over an acre—2,500 m^2 / 27,000 square ft—as they do on the ISS).

①

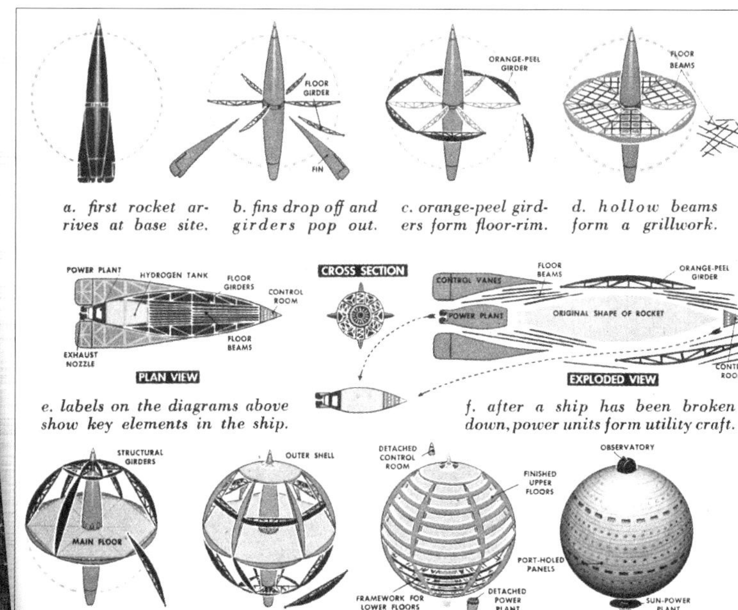

②

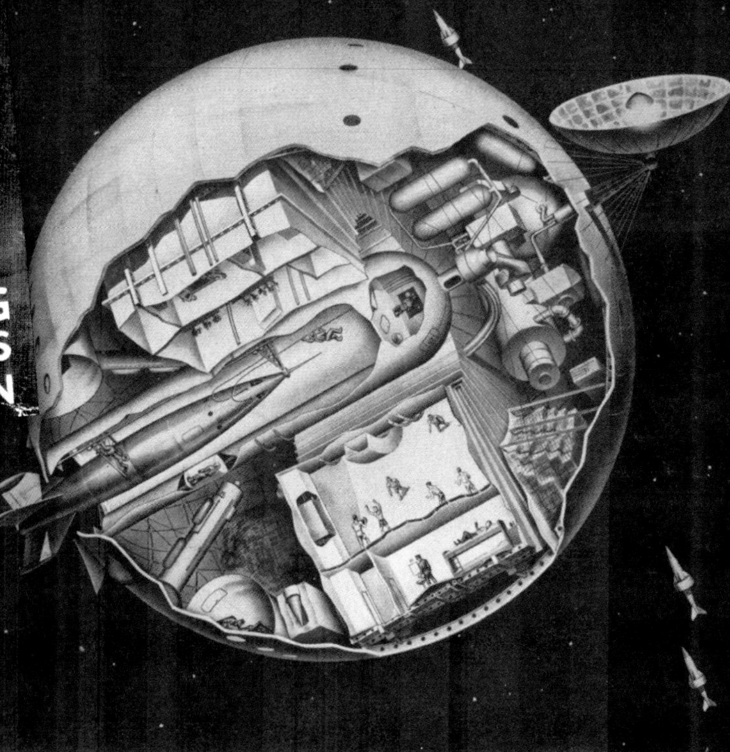

Drawings by JAMES CUTTER

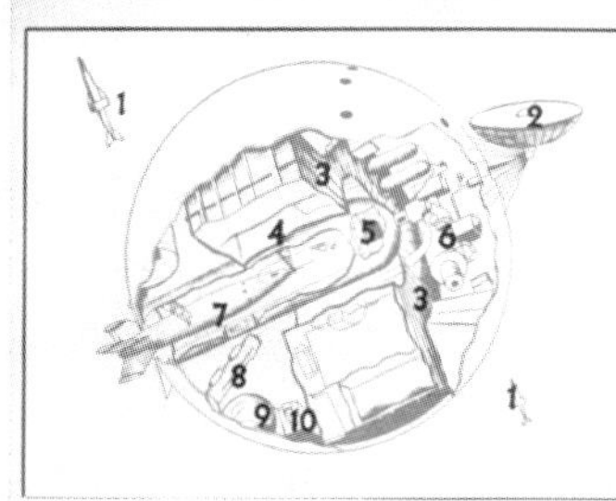

PLAN OF SATELLITE

1. "Wolf pack" of rockets.
2. Parabolic sun mirror to receive radiant energy.
3. Rotating "living zone."
4. Hydroponic gardens (inner stratum of "living zone").
5. Engineers' control room.
6. Engine to convert solar energy to electricity.
7. Supply rocket in hangar.
8. Missile launcher.
9. Observation and firing-control station.
10. Warhead storage.

③

1. Two spacecraft typical of those found throughout the 1950s—including a space station—are featured on the cover of this vividly illustrated 1952 German book. Its title translates into English as "Rocket Flight into Space: The Conquest of the Universe by the Humans."

2. In 1952, artist Frank Tinsley published this diagram illustrating the incremental steps required for the construction of his spherical space station, beginning with the body of a cargo rocket at its core.

3. This imagining of a spherical space station, published in 1950, was intended to rotate to provide artificial gravity. Like Tinsley's similar design, it was meant to operate as a military base armed with nuclear warheads.

4. This imaginative if unlikely space station design was featured on the cover of a popular book on spaceflight by Frank Ross published in 1954.

5. In 1951, artist Jack Coggins created this evocative illustration of a space station under construction, its crew carefully tethered to the structure for safety.

6. In 1952, Constantin Lent proposed this design for a solar-powered space station made from the shells of spacecraft launched from Earth.

7. This striking concept, published in 1960, was a little more imaginative than realistic. That it was to be part of a "United States Space Force" was typical of the Cold War origins of many of these designs.

SPACE SHIPS & SPACE TRAVEL
FRANK ROSS Jr
The scientifically accurate story of man's attempts and plans to travel into interplanetary space.
JACK COGGINS
THE ATOM GUN
TO LAUNCH A SATELLITE
©C.HUNT 1952

④ ⑤ ⑥ ⑦

THE RETURN OF DR. OBERTH

Just before the launch of the first Earth satellites, Hermann Oberth weighed in on the question of space stations. "Whenever work on a space station is started," he wrote ruefully, "it will undoubtedly be for military reasons." He did not believe that any one space station could fulfill all the required functions. A space station that was to act as a way station between Earth and the planets must orbit the Earth at a relatively low altitude, which would make it easier and less expensive to ship supplies to it from Earth. It would also need to travel from west to east in an equatorial orbit. But these conditions would make the station unsuitable for Earth observations. These would be better served by a geosynchronous satellite, positioned above the same point on the Earth below. Three of these orbiting 120 degrees apart would be able to see the entire Earth. The space stations Oberth visualized could hardly be more different than von Braun's simple wheels. Resembling a Christmas tree ornament more than anything else, Oberth's stations were delicate-seeming, nearly spherical structures. "I am not," he wrote

Fascinated with space from his early years, Hermann Oberth went on to become one of the founding fathers of modern rocketry and spaceflight.

Oberth's "springboard station," which included a pair of enormous, gimbal-mounted telescopes, was to be a way station for spacecraft headed for the Moon and the planets.

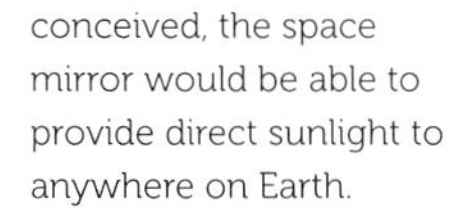

Hermann Oberth's orbiting space mirror as it would appear when completed. As originally conceived, the space mirror would be able to provide direct sunlight to anywhere on Earth.

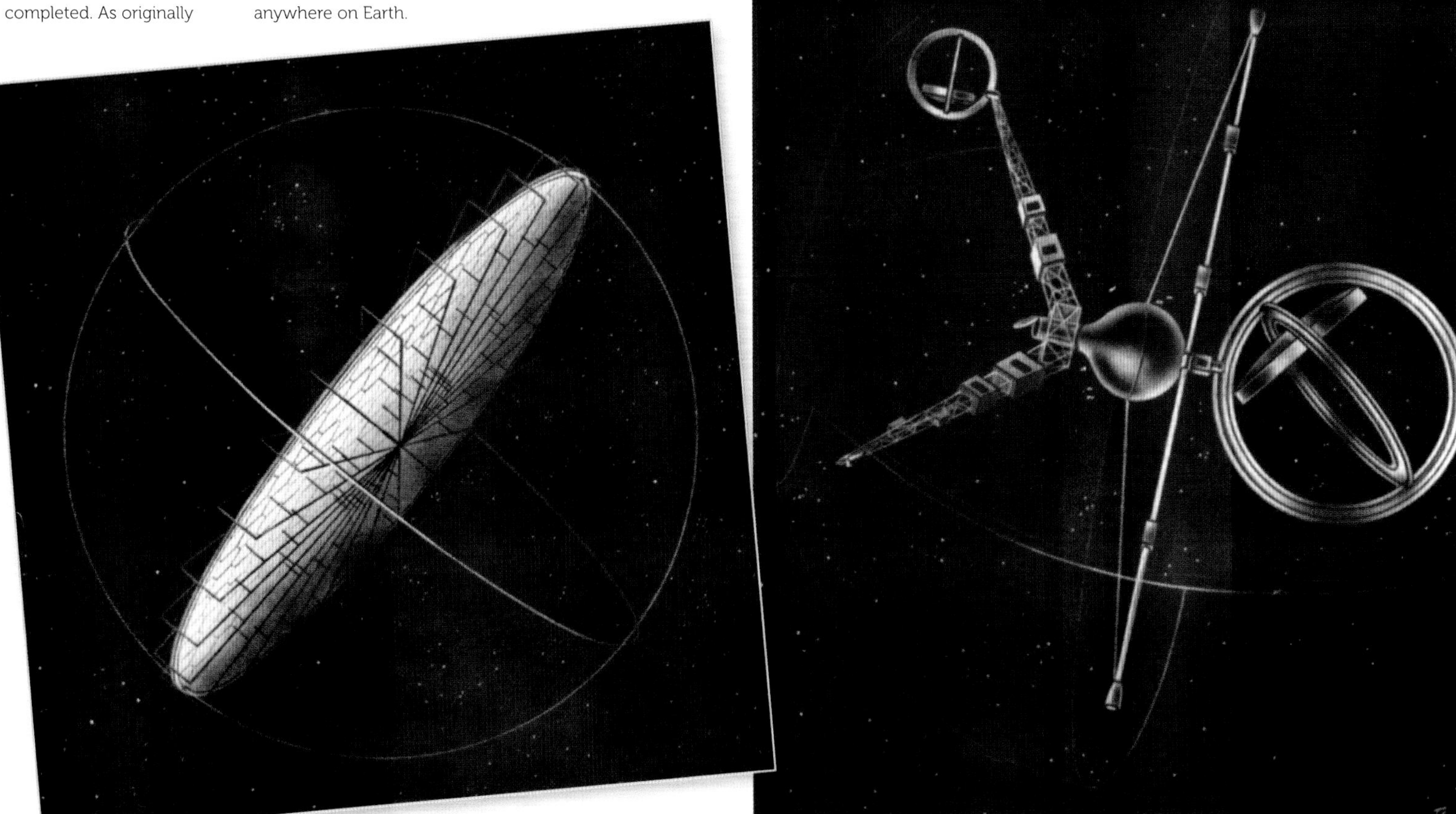

Hermann Oberth, foreground, posed in 1956 with some of the leading space pioneers of the day. From left to right: Dr. Ernst Stublinger (seated), Major General H. N. Toftoy, Wernher von Braun, and Dr. Eberhard Reed.

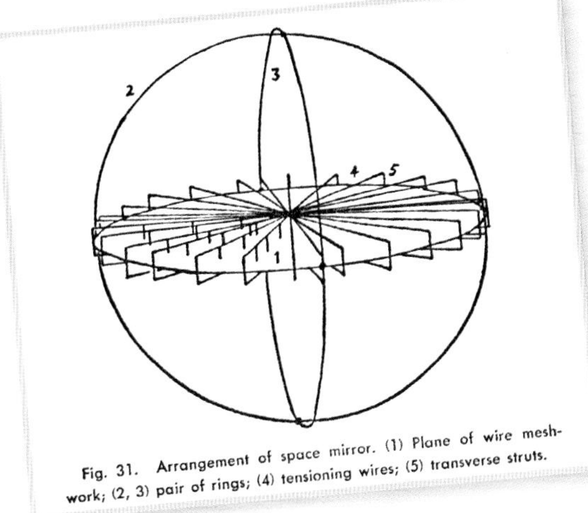

Fig. 31. Arrangement of space mirror. (1) Plane of wire meshwork; (2, 3) pair of rings; (4) tensioning wires; (5) transverse struts.

Demonstrating engineering vision, Oberth proposed that his space mirror be kept rigid by means of a system of support rings and tensioning wires.

in *Man in Space*, "in favor of compact designs which give an impression of solidity and recall heavy earthly buildings. Other laws prevail in space and there is no reason why the old architectural rules should be followed."

Oberth's final station design was nothing if not on a grand scale. Living quarters—each 12–20 m (40–65 ft) in diameter—were to be at either end of a rotating arm 4 km (2.5 mi) long. The arm holding a space telescope would be 16 km (10 mi) long. Surrounding the entire station would be a spherical cage of wires, to which "watchdog" bombs were attached. These were meant to protect the station against any enemies, natural or man-made.

Chapter Three

TOWARD THE SPACE STATION

ARTIFICIAL SATELLITES AND HUMANS IN SPACE

What is in the heavens above is a question that has haunted human civilization since the beginning of recorded time. Our understanding of the universe has been built over the millennia. Isaac Newton wrote: "if I have seen further than other men, it is by standing on the shoulders of giants."

(Left) The first human in space, Yuri Gagarin, rode in a spherical vehicle, Vostok 1, making a single orbit around Earth on April 12, 1961. He was ejected from the spacecraft to parachute to a successful landing near the village of Smelovka. (Above) From the 1920s, rocket-powered aircraft were developed by several countries and used operationally in World War II for interception. The Bell X-1 shown here made the first supersonic flight on October 14, 1947, with Chuck Yeager at the controls.

Toward the Space Station

THE ANCIENT GREEKS BELIEVED IN AN ORDERLY UNIVERSE WITH EARTH AT THE CENTER. THE SKY REVOLVED AROUND THE EARTH. STARS WERE LUMINOUS POINTS ON A SOLID CRYSTALLINE SPHERE AND PLANETS CIRCLED THE EARTH ON CRYSTALLINE SPHERES INSIDE THE SPHERE OF THE STARS. OVER TIME, THE SCIENTIFIC MODEL OF THE UNIVERSE WOULD SHATTER THIS FRAMEWORK.

This revolution began with observation and logic, and with the critical realization that Earth was round, not flat. In 585 BCE, Thales of Miletus described an eclipse as being caused by the motion of the Sun, Moon, and Earth. Pythagoras of nearby Samos, another Greek island, later believed the sphere to be the perfect form, and circular movement the perfect movement. Earth, he said, therefore, must be a sphere.

THE GEOMETRIC UNIVERSES

For Plato, founder of the Academy of Athens, the Earth was flat. Plato lived his entire life in Athens, and even though he traveled around the region, the flat horizon kept Plato from grasping the Earth as a sphere. But Plato's student Aristotle felt differently. While Plato dealt in the abstract, Aristotle believed in what he could see and measure. Aristotle argued that the North Star stayed at the same point in the sky, but its elevation changed as he moved north or south, which meant he must be moving on the surface of a sphere. In 340 BCE, Aristotle wrote that the circular shadow of Earth falling on the Moon also meant Earth must be round.

Eratosthenes saw that at noon on the summer solstice, the Sun was directly overhead in the Egyptian city of Syene; sunlight reached the bottom of a deep vertical well. But at this time in Alexandria, the Sun was 7.2° from vertical, so he therefore showed the distance between the two cities was $1/50$ the circumference of the Earth. He was 99 percent accurate.

The advances began to gather pace. In the third century BCE, Aristarchus accurately estimated the size of the Moon and the relative distances of the Moon and Sun. The last ancient Greek astronomer, Ptolemy, presented geometric models of orbital motion with Earth at the center. His model was in fact wrong, but it nevertheless predicted the motions of the planets. Ptolemy calculated the size of the universe as 250 million km (155 million mi) across with Earth at its center. He was also wrong—the universe is millions of times larger—but it was the biggest expanse anyone had ever imagined; it was beyond comprehension. Ptolemy wrote of looking back at Earth from the edge of the universe. His universe was completely wrong scientifically, but it remained in continuous use for the next 1,400 years.

The first large rocket used operationally was the German V-2 ballistic missile. Development began in the late 1930s and the first successful flights occurred in 1942. It entered military service in 1944.

> "THIS IS THE FIRST DAY OF THE ERA OF SPACE TRAVEL."
>
> MAJOR GENERAL WALTER DORNBERGER, AFTER THE FIRST SUCCESSFUL V-2 LAUNCH, OCTOBER 3, 1942

CHALLENGING THE GEOCENTRIC UNIVERSE

Between 1500 and 1700, there was a scientific revolution. Copernicus, Tycho, Galileo, Kepler, and Newton progressively and controversially moved Earth from its central position, showed that there are no crystalline spheres, and that the planets move not in circles but in egg-shaped ellipses. They also showed that the speed of a planet varies with its distance from the Sun and that a planet goes faster when it is closer to the Sun. They showed that gravity is the force that moves the planets.

Astronomers and scientists used telescopes to make systematic observations of the Sun, Moon, planets, and stars, discovering new stars and new satellites in orbit around the planets, determining that Venus has phases and that the Moon has seas and mountains. Simplistic ideas about the universe were relegated to history. In 1630 in *Somnium* (The Dream), Kepler wrote of "the Earth being an insignificantly small world traveling through the stars," and that "one day ships would be built that could travel through the void between the stars and once they are built, there will step forth men to sail them." What connects these men and the future world of space stations was that they were steadily unveiling the laws of physics and motion that would send humans and machines into space.

○ The Redstone family rockets. Left to right: (1) the Redstone ballistic missile (IRBM); (2) using more potent fuel the Juno carried the first US satellite. (3) A lengthened version launched the first two US astronauts.

○○ The Redstone was a short range missile based on the German V-2. Von Braun developed it for the US Army at Redstone Arsenal in Alabama, and it first flew successfully in 1953.

○ Sputnik 2 carried the first living thing into orbit—the rather unfortunate dog Laika, a stray from the streets of Moscow— on November 3, 1957. A cylindrical pressurized compartment transported the animal.

○○ Edward H. White II, pilot for the Gemini-Titan 4 (GT-4) spaceflight, floats in the zero-gravity of space during the third revolution of the GT-4 spacecraft, June 3, 1965. His face is protected from the Sun by a gold-plated visor.

SATELLITES IN SPACE

As we have already seen, the operation and use of rockets and spaceships were the work of four men over 50 years: Konstantin Tsiolkovsky, starting in the 1880s, Robert Goddard, Hermann Oberth, and Hermann Potočnik (Noordung) in the 1920s. Their work inspired a young German engineer, Wernher von Braun, a Nazi rocket scientist during World War II, to develop the first machine to reach space. On October 3, 1942, his V-2 ballistic missile reached an altitude of 100 km (60 mi).

Von Braun was a brilliant manager and engineer, but he was also a dreamer, a visionary who looked beyond Earth at the possibilities in space. He wanted to go into space—not just his rockets. So while he was working professionally for the US Army after the war, developing new missiles based on the V-2, he began a campaign to publicize the idea of space travel. In 1954, von Braun had proposed to his military chiefs the idea of launching a satellite. Von Braun's new missile, the Redstone, was intended to be a weapon of war, but it was the first missile ever to enter production, in 1955, that had the energy to send a satellite into orbit in accordance with Newton's laws.

In the Soviet Union, meanwhile, Sergei Korolev was developing an Inter-Continental Ballistic Missile (ICBM), much larger and more powerful than von Braun's Redstone, which had the reach to bomb American cities. Korolev also realized that this missile had the energy to launch a satellite into orbit around the Earth. In 1955, Korolev wrote an article about a Soviet scientific committee studying the space environment, prompting the Americans to jump to the conclusion that the Soviets were developing a satellite.

In response to Korolev's article, the White House announced that the US would launch a scientific satellite during the upcoming International Geophysical Year, from July 1, 1957, to December 31, 1958. Eisenhower wanted a *peaceful* scientific satellite. He was somewhat surreptitiously trying to hide the fact that the United States had already begun development of a military reconnaissance satellite in 1955 called Corona. He told von Braun's US Army team to stand down.

In response to the US announcement, Korolev wrote a letter to the Soviet Premier asking for support of a satellite. Korolev's strategy worked; he was directed to initiate work on a Soviet scientific satellite. A Soviet satellite was not a surprise. In 1955, the US Central Intelligence Agency (CIA) warned that the Soviets would be ready to launch a satellite before 1958, adding, "there is little doubt that the nation that first successfully launches the earth satellite, and thereby introduces the age of space travel, will gain incalculable international prestige and recognition."

Korolev was testing his ICBM by early 1957. Several failed, but two succeeded. On October 4, the third ICBM launched the first Earth satellite, which came to be called Sputnik. Only a month later, Sputnik 2 carried the first living thing, the dog Laika, into orbit. This Sputnik, nicknamed *Muttnik*, weighed 500 kg (1,100 lb).

By comparison, the first planned US orbital rocket, called Vanguard, was intended to launch a grapefruit sized satellite weighing 3 pounds. A month after Sputnik 2, the US was ready to try. But the rocket had never been tested, and it exploded on the launch pad. The public grew concerned, but on January 31, 1958, von Braun's modified Redstone missile carried the Explorer 1 satellite successfully into orbit.

MANNED SPACE VEHICLES

Work on the first US manned spacecraft began in the mid-1950s. Two systems were being considered; a winged rocket plane that could launch on a ballistic missile and fly around the Earth, or a *capsule* with a flat heatshield, based on nuclear weapon atmospheric reentry systems. The US military was developing both systems. In the wake of Sputnik a new civilian space agency, the National Aeronautics and Space Administration (NASA), was signed into existence on July 29, 1958, and given responsibility for the manned spacecraft that would be called Mercury.

After the worldwide recognition of Sputnik, Korolev asked for and received permission to develop a larger military reconnaissance satellite that could have its cameras and film replaced by a container for *biological materials*, a euphemistic code for a human. So by 1958 the race was on to place the first human into space.

MOONS, PLANETS, AND STARS

Before humans could truly dream of travel into space, they had to come to an understanding of what was in the sky. People on Earth noticed that some stars were not fixed in place. While the patterns of mythological constellations remained in place for generations, five points of light shone more steadily and changed position, traveling between the twinkling stars, usually in one direction, but then looping backwards and regressing for a time before continuing their forward motion. Study of these wandering stars, *planetes* in Greek, was the focus of superstition. Astronomers called these by the names of their gods, but we know them as planets.

As we saw in Chapter 2, the discovery of the difference between the planets and the stars began with an explanation of the planets' motion. Johannes Kepler was a sixteenth/seventeenth-century astrologer, mathematician, and astronomer. He worked for Tycho Brahe, one of the greatest observational astronomers before the invention of the telescope. Kepler himself was not a great observational astronomer because his eyesight was poor, but when Brahe died prematurely and left Kepler decades of data, Kepler used his knowledge of mathematics to simplify the prediction of planetary motion through geometry. He explained that the planets move in ellipses and not in perfect circles. Kepler could explain their rates of motion toward the Sun. Galileo, a contemporary of Kepler, studied the motion of stones dropping, noting that speed increased as they fell. Newton explained that the acceleration of a falling stone was due to a force; an equation showed the relationship between the force, the acceleration, and the mass. The same equation and the

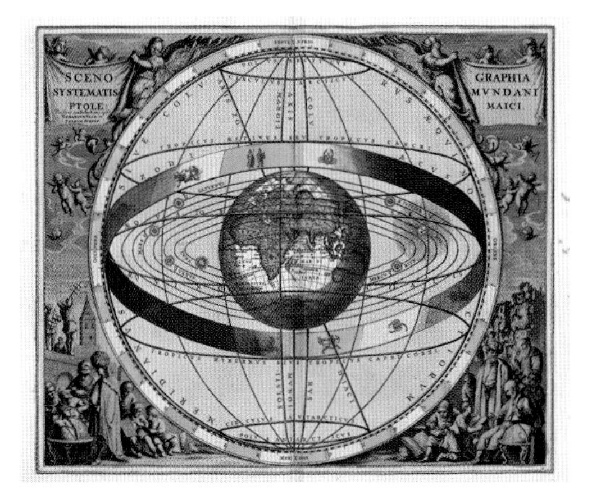

Ancient peoples believed the Earth to be at the center of the universe and everything revolved around this spherical planet. They traced the patterns of stars' movements in the sky..

The stars were formed into imaginary patterns representing mythological animals, heroes, or gods. Ancient Greeks recognized 48 constellations, but today 88 are officially recognized.

The ancients built astronomical observatories in many parts of the civilized world. Stonehenge was erected in England 4,000 to 5,000 years ago to track time and predict astronomical events.

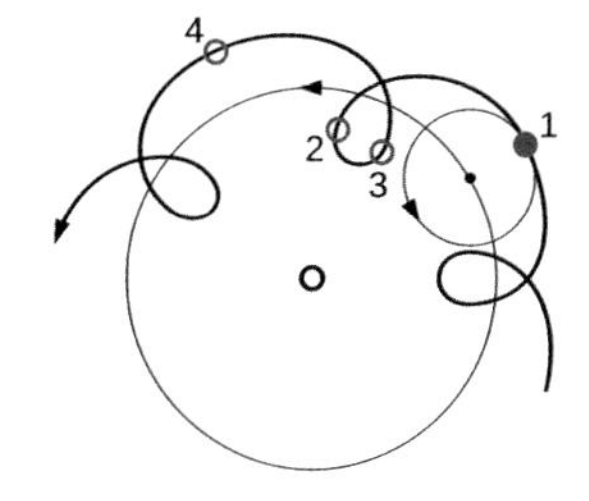

The last of the ancient Greek astronomers, Ptolemy, described the movements of planets and stars around the Earth. This cosmology was accepted for the next 1,500 years.

In the Ptolemaic model of astronomical movement, planets moved in perfect circles. In order to account for occasional backwards motion, each planet moved on a small cycle or epicycle that moved on a larger orbit called a deferent.

same force—gravity—applied to the planets. If there were no force, the planet would move in a straight line. A planet's motion, therefore, could be explained by gravity, and it explained the change in motion of the planets around the Sun (the strength of the gravitational force depends on the mass of the planet and its distance from the Sun). Thus, the scientific revolution revealed the planets to be other worlds. Observations of celestial objects also demonstrated that some objects traveled through space in linear fashion. Kepler wrote in 1630 that one day men would build ships to travel to the planets, and when that day came, sailors would come forward to sail them.

THE SPACE ENVIRONMENT

In addition to his work on gravity, Galileo changed the theory of heaven's "crystal spheres." One of Galileo's students, Evangelista Torricelli, invented the barometer and demonstrated the existence of a vacuum. In the seventeenth century, Pascal then simplified Torricelli's concepts and described how atmospheric pressure would be lower at higher altitudes. This theory in turn led him to postulate that high above Earth's air was the vacuum of outer space.

①

②

"WHEN SHIPS TO SAIL THE VOID BETWEEN THE STARS HAVE BEEN BUILT, THERE WILL STEP FORTH MEN TO SAIL THESE SHIPS."

JOHANNES KEPLER

Pascal's ideas were not just consigned to the realms of theory. In fact, the decrease of atmospheric pressure with the increase in altitude was something humans had experienced for themselves, in that people climbing high mountains experienced the symptoms of altitude sickness. In 1783, the French Montgolfier Brothers invented the hot air balloon. Using such a balloon, a Harvard-educated doctor, John Jeffries, in 1784, studied the decreasing temperature and pressure at higher altitudes. By the early 1800s balloonists were experiencing the symptoms of higher altitudes: increased pulse, mental and physical apathy, swollen lips and head. They suffered frostbite, vomiting, and loss of consciousness. In 1875, three balloonists attempted a world altitude record—two died from lack of oxygen. In World War I, military airships and early aircraft exposed their pilots and crews to high-altitude conditions. Thus, by the 1930s, aviation engineers had developed the first pressurized aircraft cabins. Early aircraft life-support systems were based on those in submarines.

COSMIC RAYS

As people flew to higher altitudes, they discovered there was also more radiation. High-altitude stratospheric balloons and later rockets and satellites studied cosmic, high-energy charged atomic particles that came from outer space. Most seem to originate in exploding stars and some from our Sun, but whatever their origins, cosmic rays pass quite readily through our bodies. Furthermore, they intensify in volume with altitude—there are more in a skyscraper or in a high-altitude aircraft than on the Earth, and even more in space, where they are an actual hazard to astronauts. During solar prominences, when the Sun actively spews charged particles through the solar system in the form of a solar wind, astronauts will periodically see light flashes with their eyes closed, caused by charged particles traveling through the astronaut's eyes and brain. The particles can damage the astronaut's DNA and cause cancer, cataracts, cognitive dysfunction, and physiological disorders. This is

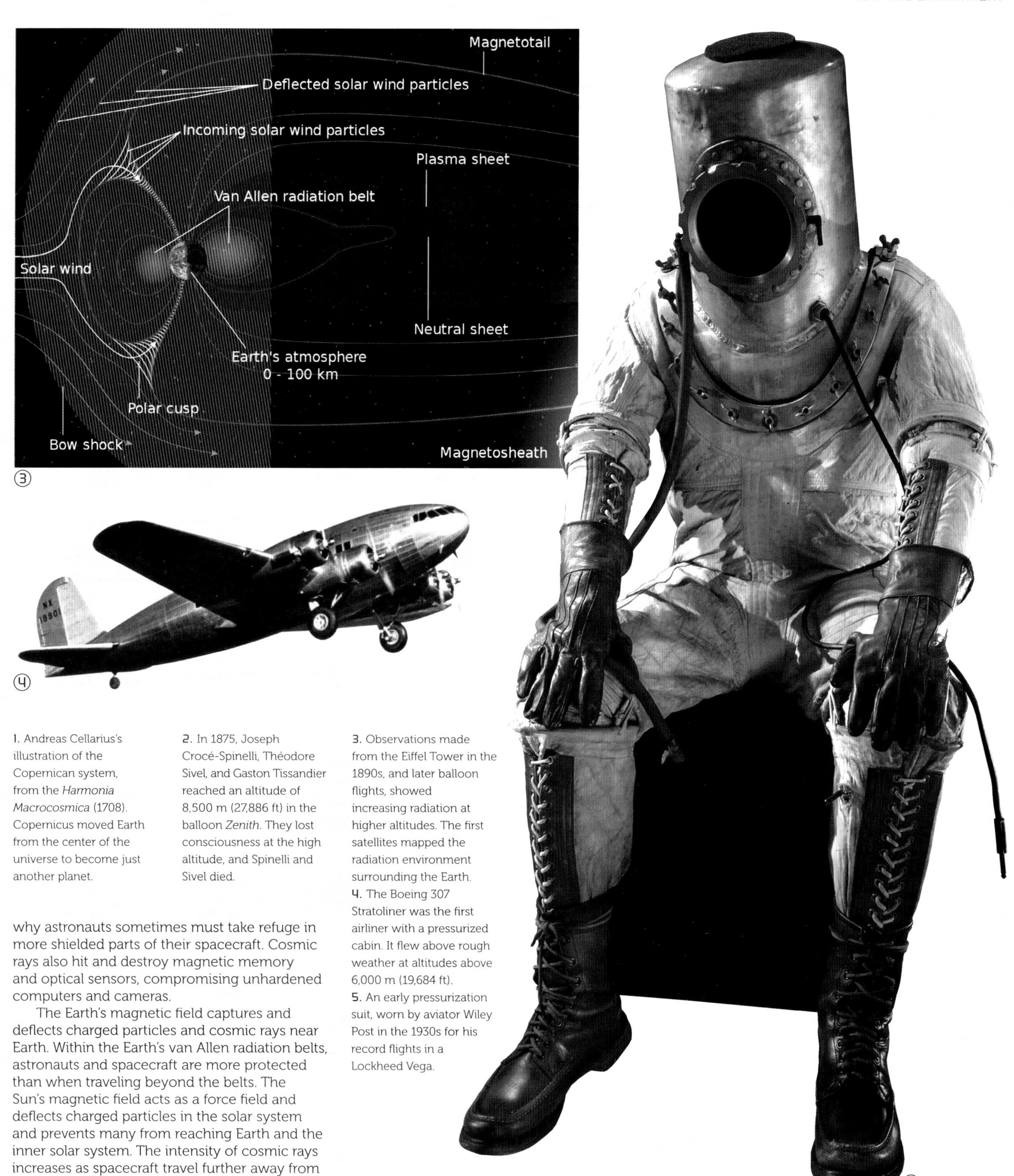

③

④

⑤

1. Andreas Cellarius's illustration of the Copernican system, from the *Harmonia Macrocosmica* (1708). Copernicus moved Earth from the center of the universe to become just another planet.

2. In 1875, Joseph Crocé-Spinelli, Théodore Sivel, and Gaston Tissandier reached an altitude of 8,500 m (27,886 ft) in the balloon *Zenith*. They lost consciousness at the high altitude, and Spinelli and Sivel died.

3. Observations made from the Eiffel Tower in the 1890s, and later balloon flights, showed increasing radiation at higher altitudes. The first satellites mapped the radiation environment surrounding the Earth.

4. The Boeing 307 Stratoliner was the first airliner with a pressurized cabin. It flew above rough weather at altitudes above 6,000 m (19,684 ft).

5. An early pressurization suit, worn by aviator Wiley Post in the 1930s for his record flights in a Lockheed Vega.

why astronauts sometimes must take refuge in more shielded parts of their spacecraft. Cosmic rays also hit and destroy magnetic memory and optical sensors, compromising unhardened computers and cameras.

The Earth's magnetic field captures and deflects charged particles and cosmic rays near Earth. Within the Earth's van Allen radiation belts, astronauts and spacecraft are more protected than when traveling beyond the belts. The Sun's magnetic field acts as a force field and deflects charged particles in the solar system and prevents many from reaching Earth and the inner solar system. The intensity of cosmic rays increases as spacecraft travel further away from the Sun.

HAYDEN PLANETARIUM CONFERENCE

After World War II, as governments were cutting military budgets, Wernher von Braun decided he would appeal directly to the American people to support space flight. In 1947, he began with talks to Rotary Clubs. He described the space wheel concept described in Chapter 2 and spoke of trips to the Moon and planets. He wrote a book about an expedition to the planet Mars, which he tried to sell to movie studios to produce a full-length motion picture. Von Braun's community talks gained him some notoriety, and the International Astronautical Federation (IAF) asked von Braun to write and present a paper based on his work. In October 1951, the First Annual Symposium on Space Travel was held in New York City at the American Museum of Natural History's Hayden Planetarium. In the audience were reporters for *Collier's* magazine, who decided to publish a series of articles covering the ideas discussed at the symposium. Correspondent Cornelius Ryan and artist Chesley Bonestell were assigned to the articles, but Ryan was not enthusiastic about his assignment at first. Yet three weeks after the symposium, Ryan attended a conference about Medicine and the Upper Atmosphere, where he met von Braun and other space scientists. They convinced Ryan of the viability of space travel and its potential for advancing technology, so Ryan became an enthusiastic supporter.

Heinz Haber (left), specialist in human physiology in space, von Braun, and Willy Ley, organizer of 1950s space conferences, discuss a "bottle suit" model, a type of one-man spacecraft used for the assembly of space stations.

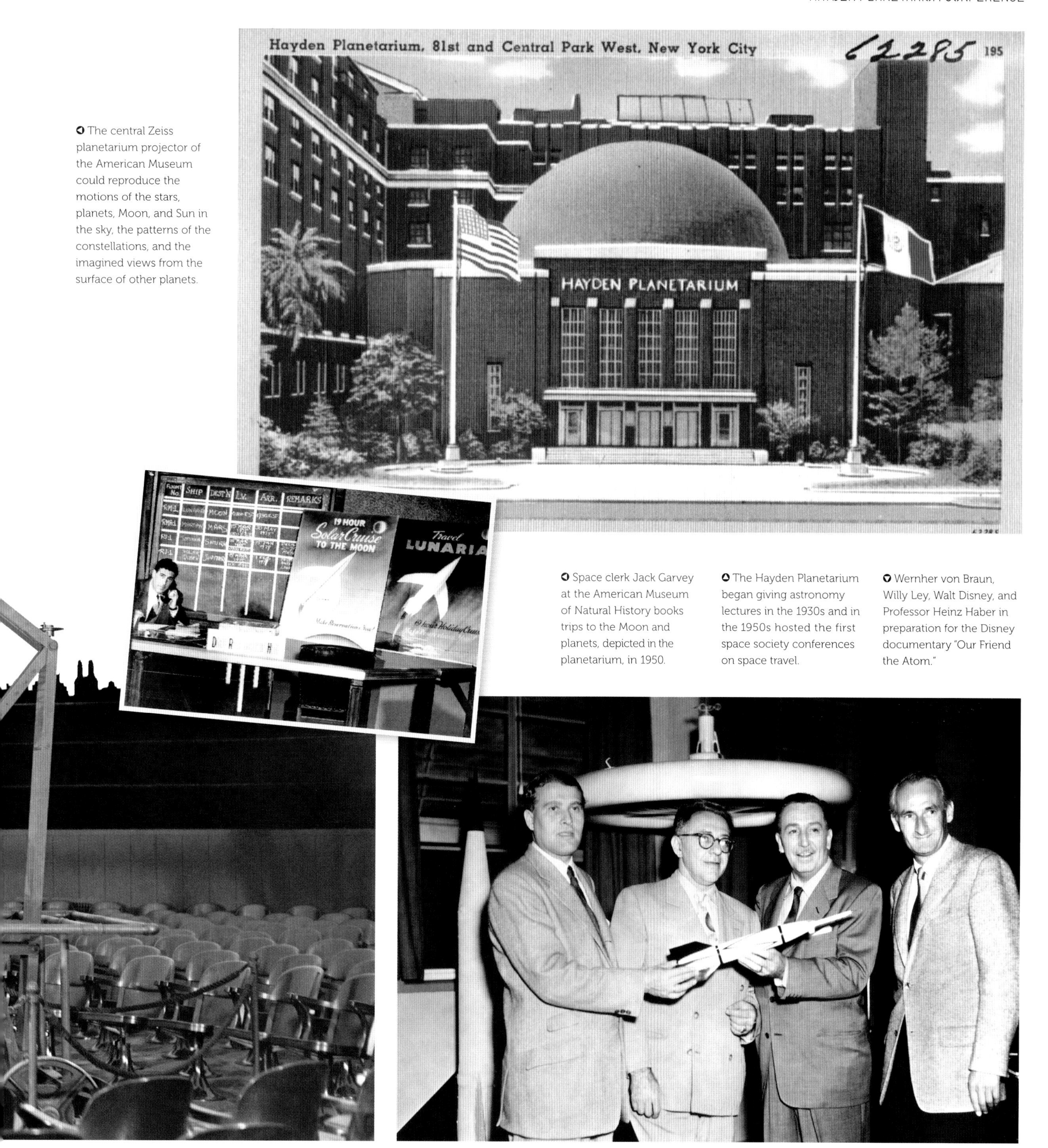

The central Zeiss planetarium projector of the American Museum could reproduce the motions of the stars, planets, Moon, and Sun in the sky, the patterns of the constellations, and the imagined views from the surface of other planets.

Space clerk Jack Garvey at the American Museum of Natural History books trips to the Moon and planets, depicted in the planetarium, in 1950.

The Hayden Planetarium began giving astronomy lectures in the 1930s and in the 1950s hosted the first space society conferences on space travel.

Wernher von Braun, Willy Ley, Walt Disney, and Professor Heinz Haber in preparation for the Disney documentary "Our Friend the Atom."

COLLIER'S AND DISNEY

By World War II, the most famous American scientist was the physics professor Robert Goddard, nicknamed the "Moon Man," who had invented the liquid-fuel rocket. At the same time, rocketry enthusiasts formed American, British, Russian, and German rocket societies. Two of the enthusiasts were the aforementioned Wernher von Braun, in Germany, and Sergei Korolev, in the Soviet Union—they would eventually lead their respective nations' rocketry and space programs. In World War II, von Braun led the development of the first ballistic missile; the V-2 was also the first vehicle to reach space. When the war ended, von Braun and other German rocket experts surrendered to the United States.

1. "Tomorrowland" at Disneyland in Los Angeles in California opened in 1955, to much fanfare and public enthusiasm. The TWA Moonliner, from Disney's the *Man in Space* TV program, sought to illustrate the future and was the tallest structure in the park.

2. Walt Disney (seen here on the left) sought to describe and shape a powerful vision of the future in space. He enlisted the support of von Braun (right) to help to define it. A film at the entrance showed America from space, as seen from the X-1 space station.

Von Braun dreamed of a world in which people would explore space, the Moon, and the planets. In the early 1950s, *Collier's* magazine invited him to share his vision for space exploration. Collier's was one of the most popular US periodicals in a time before the internet and when TV had not yet reached most people's homes. Over two years, millions of *Collier's* readers followed the eight installments highlighting von Braun's dream of space exploration.

INFLATABLE CRAFT

Von Braun challenged Americans to ensure their nation established space superiority. In the first *Collier's* story, von Braun wrote about the Earth-orbiting space station. Governments could use the space station as a tool for peace, because no nation could undertake preparations for war without the certain knowledge they were being observed from space. The station would circle around Earth every two hours, orbiting at 25,491 kmh (15,840 mph) at an altitude of 1,729 km (1,075 mi). Most of the station would be inflated rubberized fabric, pressurized to Earth-like conditions. It would be an outpost in the sky, providing all of the human's needs: 1.4 kg (3 lb) of air each day for each person, moisture absorption, air cooling and heating, and artificial gravity. In design, the station was a 76 m (250 ft) diameter wheel, rotating slowly, using centrifugal force to produce a feeling of artificial gravity. Power would be provided by reflecting sunlight into a tube of mercury that, when heated, would flow to drive a turbo-generator to create electricity. Technicians inside would use special telescopes connected to large optical screens, radarscopes, and cameras. "Nothing will go unobserved."

INTO ORBIT

To put the station in orbit would require a rocket capable of carrying the crew and 27,272–36,363 kg (60,000–80,000 lb) of cargo. It would accelerate to a fantastic speed, 28,163 kmh (17,500 mph), and at this speed, its path would match the curvature of the Earth—it would be in free-fall, but never fall back to the ground.

3. Wernher. von Braun, despite his controversial Nazi past, began his work with Walt Disney in the 1950s, when the rocket scientist appeared in three Disney television productions. The relationship was long-lived—here von Braun shows Disney and his associates around the Marshall Space Flight Center (MSFC) in 1965.

4. Wernher von Braun was a prolific genius. Here he stands surrounded by images of his many spacecraft concepts, including his wheel space station (bottom left).

Space artists depicted the station in orbit, the station interior, and dozens of scientists engaged in scientific research and Earth observations. Von Braun wrote of the military potential: "There will also be another possible use of the space station, a most terrifying one. It can be converted into a terribly effective atomic bomb carrier. . . . missiles with atomic warheads could be launched from the station. . . . atom bombing techniques would offer satellite builders the most important tactical and strategic advantage in military history."

"SCIENTISTS TODAY ARE OPENING THE DOORS OF THE SPACE AGE TO ACHIEVEMENTS THAT WILL BENEFIT GENERATIONS TO COME."

WALT DISNEY

The *Collier's* series ended on an issue about Mars, publilshed on April 30, 1954. Walt Disney was planning a new television program, *The Wonderful World of Disney*, and in the midst of building his first theme park, Disneyland, which would open in mid-1955. Disney felt spaceflight could help communicate the world of "Tomorrowland," which was one of the park's themes and a regular part of Disney's TV series. He sought out von Braun and the *Collier's* editors to produce a series of space-themed television programs. In the first TV program, *Man in Space*, aired on March 9, 1955, von Braun described the construction of a space station in Earth orbit. He described it as a way station for humankind's first expedition to the Moon.

EISENHOWER'S SPACE PROGRAM

Dwight D. Eisenhower learned the value of aerial photography in estimating enemy strength first as the Supreme Allied Commander in Europe in World War II, and later as the head of NATO forces in 1950. In 1953 he became US President, as the Cold War was escalating: communists had taken over China in 1949 and an invasion by North Korean communists began the Korean War in 1950. In response, US and Soviet militaries introduced thermonuclear weapons into their arsenals. In 1954, therefore, Eisenhower needed aerial photography to assess the Soviet threat, but the United States had been leading a public international discussion about free overflights of foreign countries. The question was whether the air and space above a country were like the territorial waters surrounding a country, and whether a country could legally board, confiscate, or destroy vehicles that encroached on their territory. US aircraft had been overflying Soviet

The Corona program, benignly called "Discover" to hide its military intentions, returned exposed reconnaissance film in reentry capsules that would be recovered in midair by Air Force cargo aircraft.

A cutaway presentation of the structure and components of the early Corona satellites. Reconnaissance cameras were located at the left. After film was exposed, it would be rolled into the recovery pod on the right.

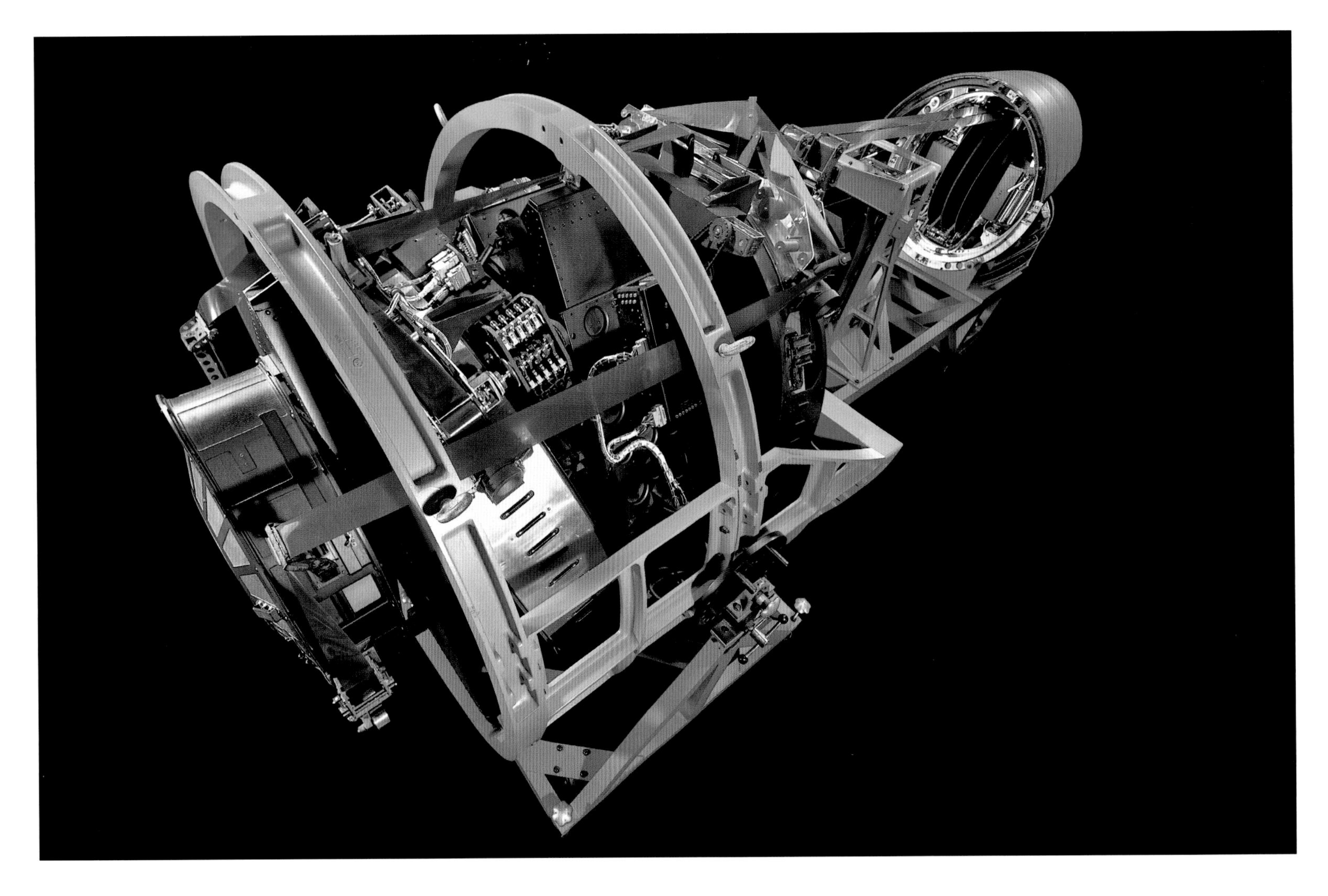

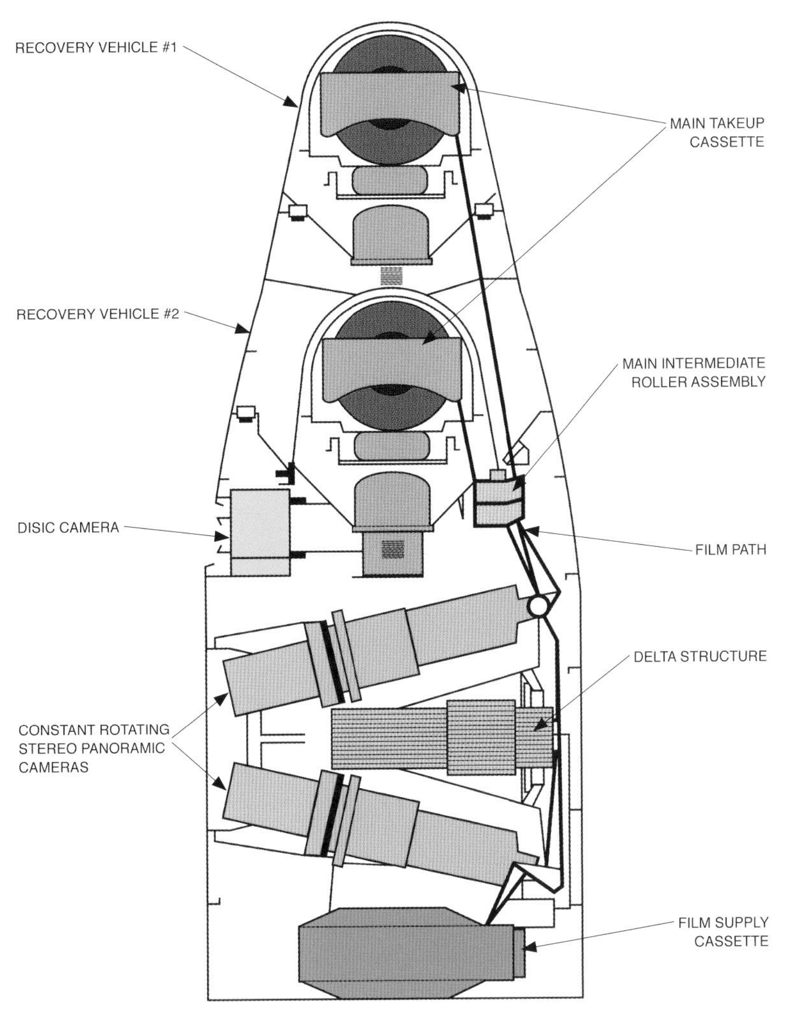

A diagram illustrating the layout and film transport path of the Corona reconnaissance satellite. The two-camera system enabled the production of a stereoscopic image.

President Eisenhower examines the nose cone of a Corona satellite, which sits on display in his office. Corona satellites were in operation between 1959 and 1972.

The high altitude U-2 spyplane. Essentially a powered glider, its light weight and long wings allowed it to fly at altitudes beyond the reach of missiles or interceptor aircraft, at least initially.

and Chinese territory surreptitiously since 1950, and as the Soviet weapons systems improved, US aircraft and their crews were being lost.

Eisenhower initiated the development of a top-secret spy plane, the U-2. It would fly at a high enough altitude that Soviet weapons would be unable to reach it, but this was only a temporary solution as Soviet surface-to-air missiles (SAMs) improved. The influential RAND ("Research ANd Development") Corporation had been recommending since the 1940s that Earth-orbiting satellites be developed for military reconnaissance. Eisenhower approved the Defense Department's Corona reconnaissance satellite system in 1956, hiding its military nature through the title "Discoverer Program." A separate program to build a scientific satellite, called Vanguard, began the same year. In response to Eisenhower's proposed "freedom of space," that space vehicles be permitted to freely overfly other nations, the Soviets responded negatively. Such overflights would, they said, be for reconnaissance purposes and they would not permit them.

SOVIET ADVANCES AND THE US RESPONSE

As Eisenhower developed his militarized space and aviation program, in the Soviet Union Sergei Korolev, the chief Soviet rocket engineer and spacecraft designer, was building a series of ever-larger military missiles. As noted in the introduction to this chapter, he also realized that an ICBM could be used to put a satellite into Earth orbit. The Soviets began studies aimed towards the eventual design of a satellite in the early 1950s. In 1955, Korolev's article about the formation of a scientific committee to study conditions in space galvanized the US authorities into action, and in July 1955 the United States announced that an intended US goal during the IGY 1957–58 would be the launch of a satellite. The satellite would be called Vanguard and would be used strictly for peaceful scientific purposes. It would be developed based on atmospheric sounding rocket technology, by the Naval Research Laboratory.

At this time, von Braun was completing development of the US Army Redstone missile. Von Braun's team completed the first test launch in 1953 and Chrysler was building production missiles by 1956. Von Braun proposed in 1955 that he be permitted to use the Redstone to launch a satellite. A variant for testing warheads had the capacity and power to launch a satellite and von Braun kept several of these in reserve at Cape Canaveral in Florida. A year later, when the Vanguard announcement was made, von Braun, much to his disappointment, was told to stand down on his satellite project. He nevertheless continued work on development of the Redstone and in 1956 had several of the more powerful variants set aside. After the US satellite announcement in July, 1956, Sergei Korolev wrote a letter to the Soviet Premier asking that he be permitted to develop a satellite in response. In August, the authorities directed Korolev to initiate work on the Soviet satellite.

Spacecraft designer Sergei Korolev (left) was the mastermind behind development of the early Soviet missiles, the first satellites, and the first man in space, Yuri Gagarin, seen here on the right.

The Soviet R-7 missiles featured the same booster upgraded with different upper stages. They carried the first satellite, first lunar probes, and manned Vostok, Voskhod, and Soyuz spacecraft.

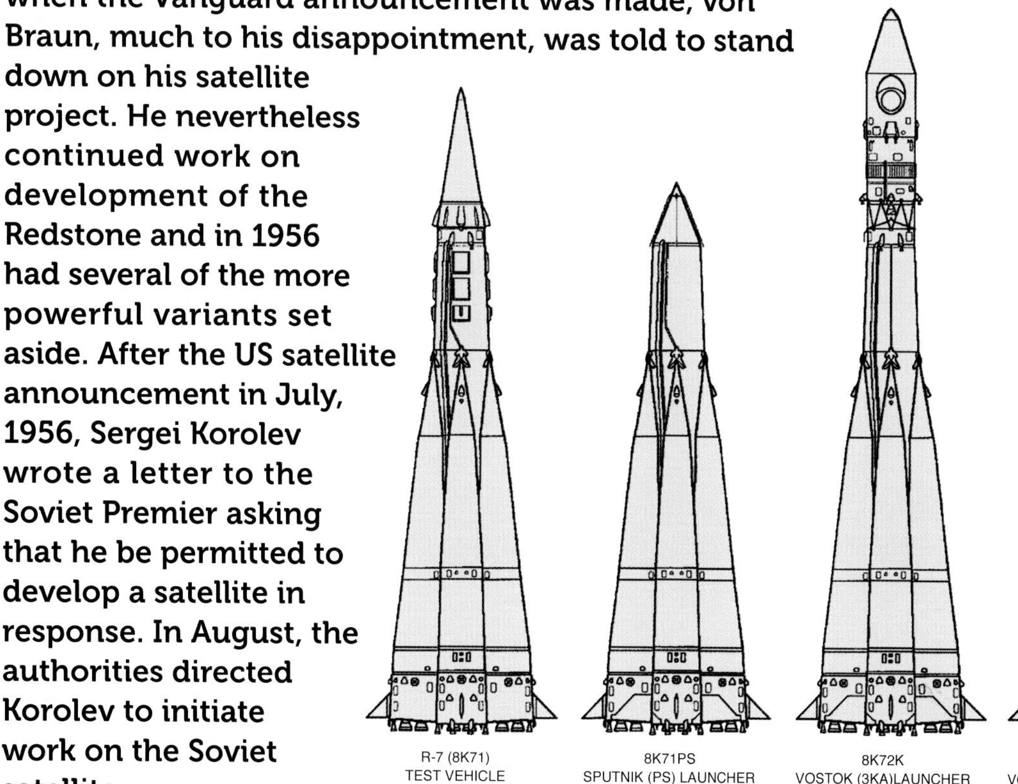
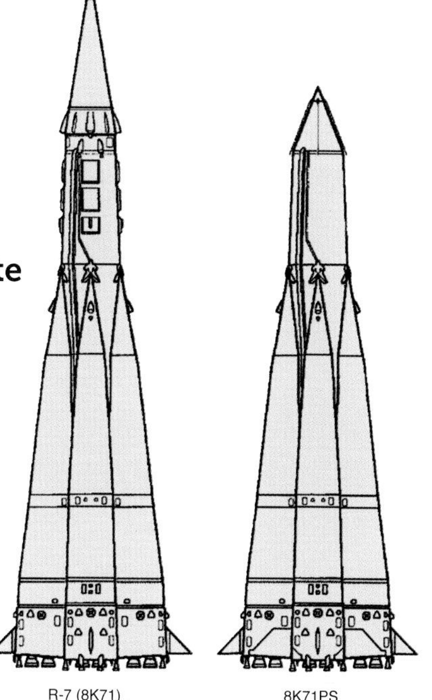

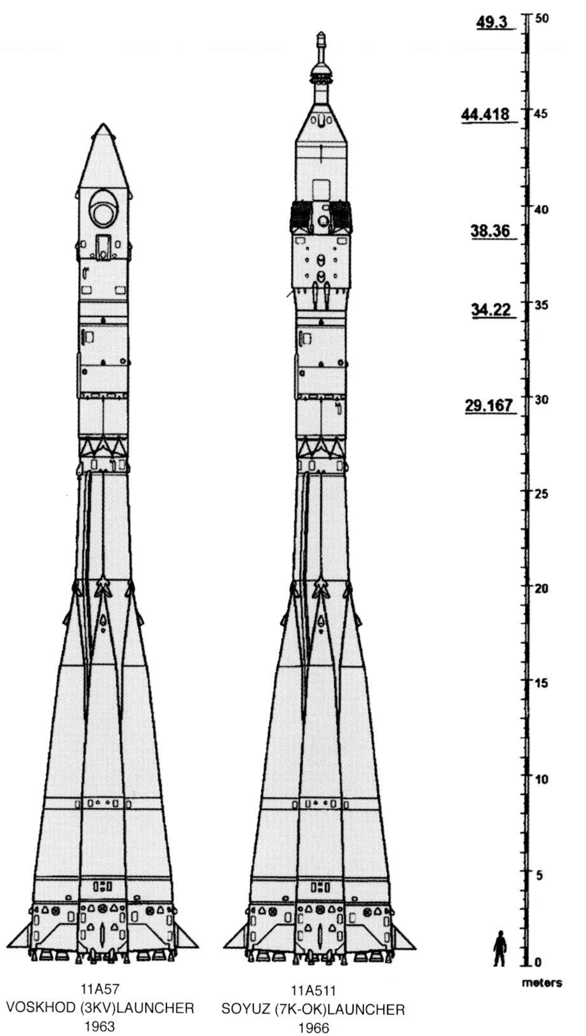

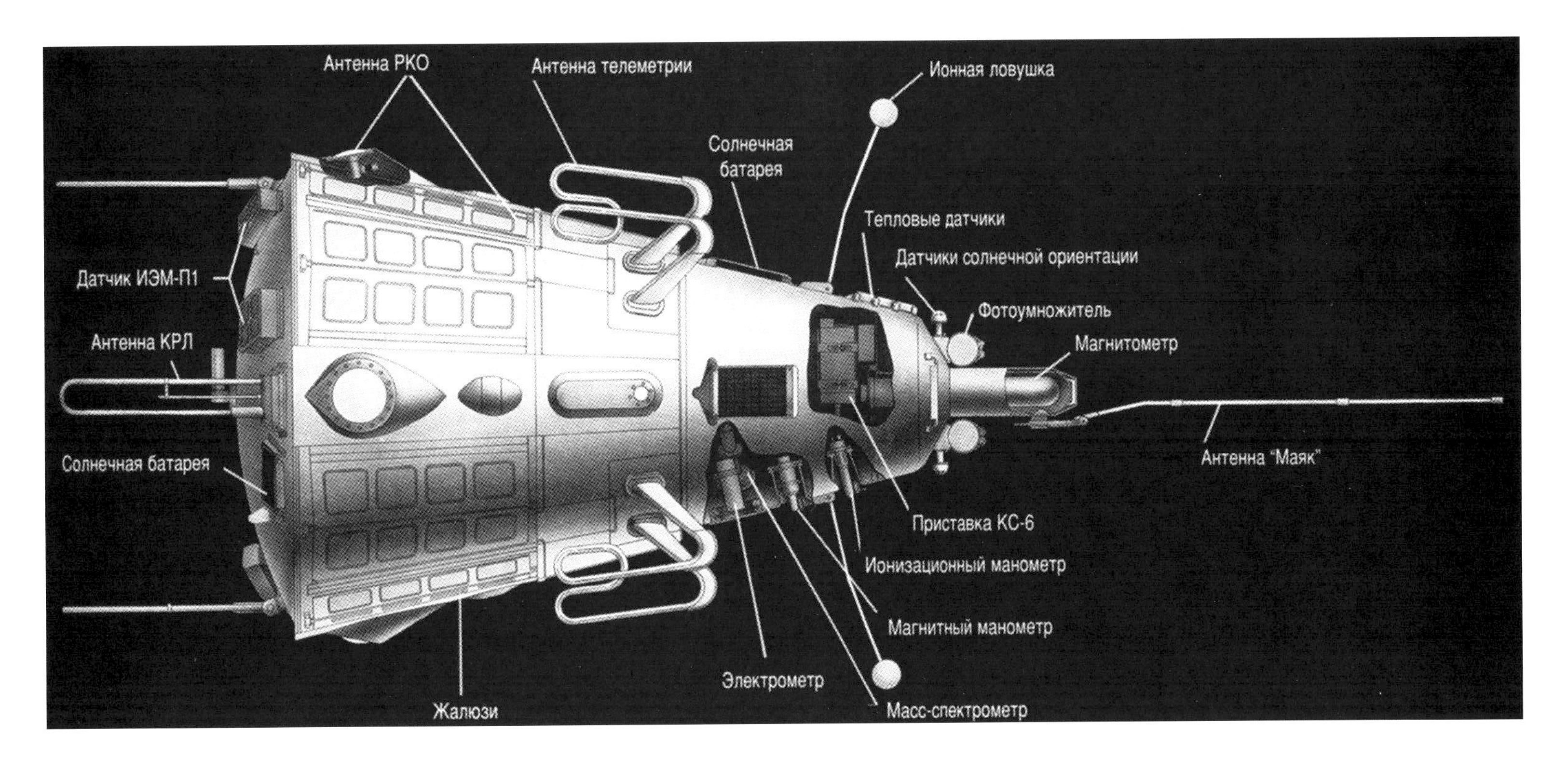

"Object D" was intended as the first Soviet satellite. Its launch was deferred until May 15, 1958, when it became the third Soviet satellite. It was massive by comparison with all prior satellites.

NASA launched Vanguard III (SLV-7) from Cape Canaveral, Florida on September 18, 1959. It was used to study Earth's magnetic field and radiation belt.

Test firing of a Redstone Missile at Redstone Test Stand in the early 1950s. The Redstone was a liquid-propelled missile, and was the first major rocket program in the United States.

SPUTNIK

Korolev now had a mission. At the end of World War II, the United States enjoyed a significant advantage over the Soviet Union, on account of its larger air force with strategic bombers. Soviet territory was easily reached from allied territory in Western Europe. Jet bombers were introduced and larger ballistic missiles were in development, so a decision was needed whether to focus efforts mainly on missiles or on jet aircraft. Because of the large number of pilots in the aftermath of the war, jet aircraft had a large constituency. Because of the novelty and concerns about the accuracy of long-range missiles, funding for missiles was questionable. Military spending was declining precipitously, so choices would need to be made.

1

2

3

The first US ICBM program started in 1946, but was cancelled after a year. The US military initiated a second ICBM program, called Atlas, in 1951, but received little funding. In the meantime, the size of the US thermonuclear weapon was getting smaller, and so missiles could shrink to accommodate the smaller, lighter bomb. Atlas, for example, started at 49 m (160 ft) tall and shrank to only 23 m (75 ft).

FROM ICBM TO SATELLITE

In the Soviet Union, Korolev was developing ballistic missiles based on the German V-2. Stalin died in March 1953 and was replaced by Nikita Khrushchev. In August that year, the first Soviet hydrogen bomb was tested. Shortly after, Korolev gave a briefing at which he was told to produce an ICBM that could deliver a 5-tonne (5.5-ton) payload over a distance of 8,046 km (5,000 mi). This was far larger than any previous missile and in order to design it, five similar stages were clustered together. On February 27, 1956, Khrushchev visited the production facility for the ICBM. Khrushchev was impressed by the missile's size. At the end of the presentation, Korolev told Khrushchev that one possible use for the missile was to launch scientific instruments, and that it would be easy and inexpensive to launch a satellite once work on the missile was complete. He added that the Americans were spending millions of dollars to build a rocket to launch a satellite, but that such funding would not be required with the Soviet missile already in development. Khrushchev said that if the main task, developing the ICBM, did not suffer, Korolev should launch the satellite.

The initial plan called for a complex laboratory as heavy as 1,400 kg (3,100 lb), carrying solar, cosmic ray, and environmental sensors and experiments from five Soviet ministries. The project was designated by the arcane name *Object D*. During the development of the design, one configuration could be modified to have a small compartment for carrying a dog into orbit. The animal compartment was based on similar cockpits used in Soviet atmospheric sounding rockets.

SLOW PROGRESS

By late 1956, the *Object D* project was falling behind. Test models remained unfinished. There were also problems with the ICBM main rocket motors, which were not producing the required thrust. The Soviets grew concerned. The Soviet Academy of Sciences stated that the goal must be to orbit a satellite before the Americans, and so a decision was made to develop a simple, minimum-mass satellite that could be available just as soon as the ICBM was ready to launch it. This had the name Simple Satellite 1 (PS-1) and consisted of a 580 mm (23 in), 83 kg (184 lb) sphere containing a signal generator, radio transmitter and batteries, without any scientific instruments.

Korolev tested his first ICBM on May 15, 1957, but it failed spectacularly when one of

1. A technician assembles the Soviet "Simple Satellite," which became known as Sputnik after its launch.
2. An early R-7, this one with an upper stage, lifts off from Launch Pad 1 at Baikonur Cosmodrome in Kazakhstan. The upper stages increased the mass that could be carried to orbit and was also used to send the first probes to the Moon and planets.
3. The dog Laika became the first animal to orbit the Earth, seen here in her compartment of the satellite Sputnik 2. She flew less than a month after the launch of first satellite.
4. *The New York Times* newspaper reports the launch of the first satellite in space history. While both the Americans and the Soviets had announced their intentions to launch satellites years before, it surprised many that the Soviets had been first.
5. An artist's depiction of Sputnik 1 in orbit, launched October 4, 1957. It carried a radio transmitter and battery to supply power inside of a 58 cm (23 in) polished metal sphere.

"All the News That's Fit to Print"

The New York Times.

LATE CITY EDITION

VOL. CVII..No. 36,414.

NEW YORK, SATURDAY, OCTOBER 5, 1957.

FIVE CENTS

SOVIET FIRES EARTH SATELLITE INTO SPACE;
IT IS CIRCLING THE GLOBE AT 18,000 M. P. H.;
SPHERE TRACKED IN 4 CROSSINGS OVER U. S.

HOFFA IS ELECTED TEAMSTERS' HEAD; WARNS OF BATTLE

Defeats Two Foes 3 to 1 —Says Union Will Fight 'With Every Ounce'

COURSE RECORDED

Navy Picks Up Radio Signals—4 Report Sighting Device

560 MILES HIGH

Visible With Simple Binoculars, Moscow Statement Says

Device Is 8 Times Heavier Than One Planned by U.S.

FAUBUS COMPARES HIS STAND TO LEE'S

Flu Widens in City; 10% Rate Predicted; 200,000 Pupils Out

ARGENTINA TAKES EMERGENCY STEPS

SATELLITE SIGNAL BROADCAST HERE

the boosters disintegrated after a fuel leak. Several others failed in the summer of 1957, but two functioned as intended. In preparation for the first satellite, a room-sized computer was installed at the Academy of Sciences Department of Applied Mathematics. It was the first high-speed computer in the Soviet Union and was used to compute the satellite's planned trajectory. The Ministry of Defense established a tracking, command, and telemetry network to track the satellite; it was the progenitor of the Command Measurement Complex (KIK) that would support every Soviet and Russian space launch to the present day. On October 4, the third successful ICBM launched the first Earth satellite, the PS. The Russian media called it *Sputnik*, which meant satellite.

LAIKA

The next day, Korolev returned home to Moscow, feeling ebullient after the launch. Khrushchev then called, wanting to know all the details about the launch and, most important, asking whether Korolev could launch another satellite in a month, for the November 7 fortieth anniversary of the Great October Socialist Revolution. Korolev said yes, and suggested they include a dog as a passenger. Khrushchev was ecstatic. They decided against using the biological specimen satellite already in development, but far from completed, and instead started to combine a suborbital sounding rocket compartment already in hand with the PS-2 duplicate of Sputnik 1. A transmitter was added to radio the vital signs of the animal and a slow scan TV was added to send back TV pictures.

Work on the new Sputnik began October 10, less than a week after Sputnik 1 entered orbit. Engineers added a regeneration unit to provide oxygen absorb carbon dioxide and water vapor exhaled by the dog. There would be no effort made to return the dog. The original PS-2 sphere, looking just like Sputnik 1, was mounted atop the animal capsule. Out of ten candidate dogs in training for sounding rocket atmospheric flights, Laika was chosen because of her quiet and placid temperament. A second dog, Albina, who already had flown twice, was backup animal.

On November 3, Sputnik 2 carried the first living thing into orbit, the dog Laika. Sputnik 2 weighed 500 kg (1,100 lb), but remained attached to the final stage of the ICBM throughout the mission and the mass of the satellite with the rocket stage was close to 6,000 kg (13,000 lb). Temperature in the dog's compartment was high, causing some discomfort. On the fourth day of the flight, the dog died from overheating.

By comparison, the first planned US orbital rocket, Vanguard, was intended to launch a grapefruit-sized satellite weighing 1 kg (3 lb). A month after Sputnik 2, Vanguard was ready to try. But the rocket had never before been tested, and it exploded on the launch pad.

VANGUARD AND EXPLORER

As the Soviets pushed ahead with their space program, the US National Academies began to encourage their government to launch a satellite as part of the IGY in 1952. President Eisenhower announced that the US would do so in July 1955, and the US Navy Research Laboratory was assigned the responsibility. The satellite was to be a 15 cm (6 in) sphere, featuring an early solar power panel as well as batteries. The Vanguard's booster rocket was based on sounding rockets used for atmospheric scientific observations and had never flown before.

Although a slower program of testing had been planned, when the Soviets successfully launched Sputniks 1 and 2 it was rushed along, and the first test launch was scheduled for December 6. At

The first US attempted satellite launch, Vanguard Test Vehicle 3 (TV-3) on December 6, 1957. The rocket rose just over a meter before settling back and exploding on the launch pad.

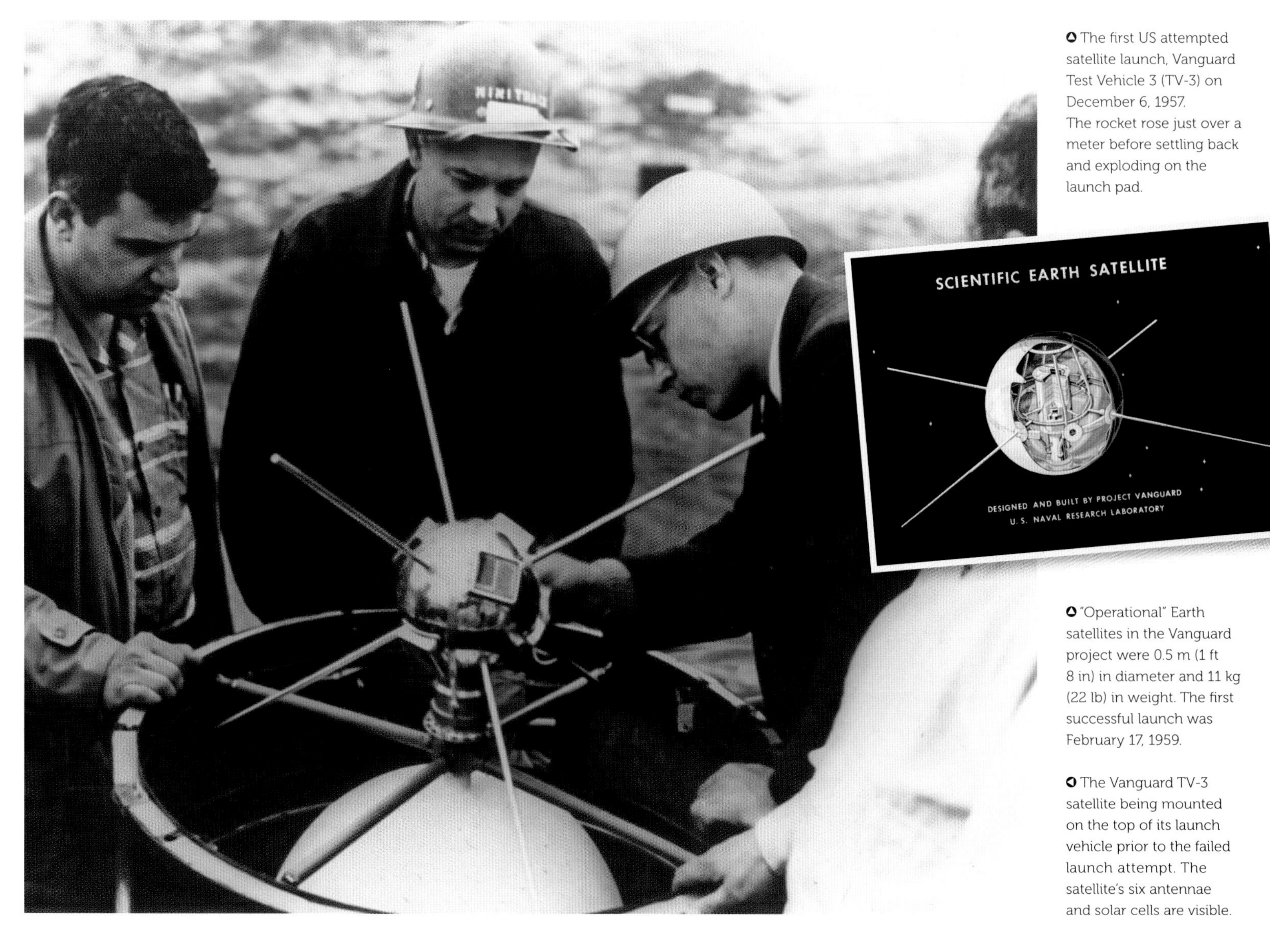

"Operational" Earth satellites in the Vanguard project were 0.5 m (1 ft 8 in) in diameter and 11 kg (22 lb) in weight. The first successful launch was February 17, 1959.

The Vanguard TV-3 satellite being mounted on the top of its launch vehicle prior to the failed launch attempt. The satellite's six antennae and solar cells are visible.

The Vanguard TV3 that failed on December 6, 1957, fell from the booster, but it continued transmitting. It was eventually recovered and went on to display at the Smithsonian Institution.

The first successful US satellite, Explorer 1, was launched from Cape Canaveral on January 31, 1958. The rocket was a modified US Army Redstone missile that had been held in storage for about two years.

the launch the rocket ascended about 1.3 m (4 ft), but then lost thrust—the rocket settled back to the launchpad and the fuel tanks exploded. The satellite was thrown clear and began transmitting as if in orbit.

Wernher von Braun had designed a somewhat more powerful rocket, the main part consisting of a modified, souped-up Redstone called the Juno 1 or Jupiter-C. For the satellite launch, two more stages were added, each composed of small solid rocket motors. The Explorer 1 satellite built by the Jet Propulsion Laboratory carried several instruments developed by Dr. James Van Allen of the University of Iowa. They measured acoustics, temperature, micrometeorite, and cosmic rays. The instruments were based on those used during US Army tests of captured V-2 rockets after World War II. Explorer 1 weighed 14 kg (31 lb) and was 205 cm (81 in) long. Permission to prepare Explorer 1 was given after the Soviet launch of Sputnik 2 in November 1957. After 84 days of preparation, the first US satellite was successfully launched January 31, 1958.

HUMANS IN SPACE

It was not only standard rockets that were propelling the space race. In World War II, the Germans and Russians both developed rocket planes, the first being the Russian BI-1 in 1942 and the German Me-163 in 1944. As the war ended in 1945, the United States created the Bell XS-1, later called the Bell X-1, which broke the sound barrier in level flight in 1947, piloted by Chuck Yeager.

H. R. 12575 [PUBLIC LAW 85-568]

Eighty-fifth Congress of the United States of America

AT THE SECOND SESSION

Begun and held at the City of Washington on Tuesday, the seventh day of January, one thousand nine hundred and fifty-eight

An Act

To provide for research into problems of flight within and outside the earth's atmosphere, and for other purposes.

Be it enacted by the Senate and House of Representatives of the United States of America in Congress assembled,

TITLE I—SHORT TITLE, DECLARATION OF POLICY, AND DEFINITIONS

SHORT TITLE

SEC. 101. This Act may be cited as the "National Aeronautics and Space Act of 1958".

DECLARATION OF POLICY AND PURPOSE

SEC. 102. (a) The Congress hereby declares that it is the policy of the United States that activities in space should be devoted to peaceful purposes for the benefit of all mankind.

(b) The Congress declares that the general welfare and security of the United States require that adequate provision be made for aeronautical and space activities. The Congress further declares that such activities shall be the responsibility of, and shall be directed by, a civilian agency exercising control over aeronautical and space activities sponsored by the United States, except that activities peculiar to or primarily associated with the development of weapons systems, military operations, or the defense of the United States [illegible]

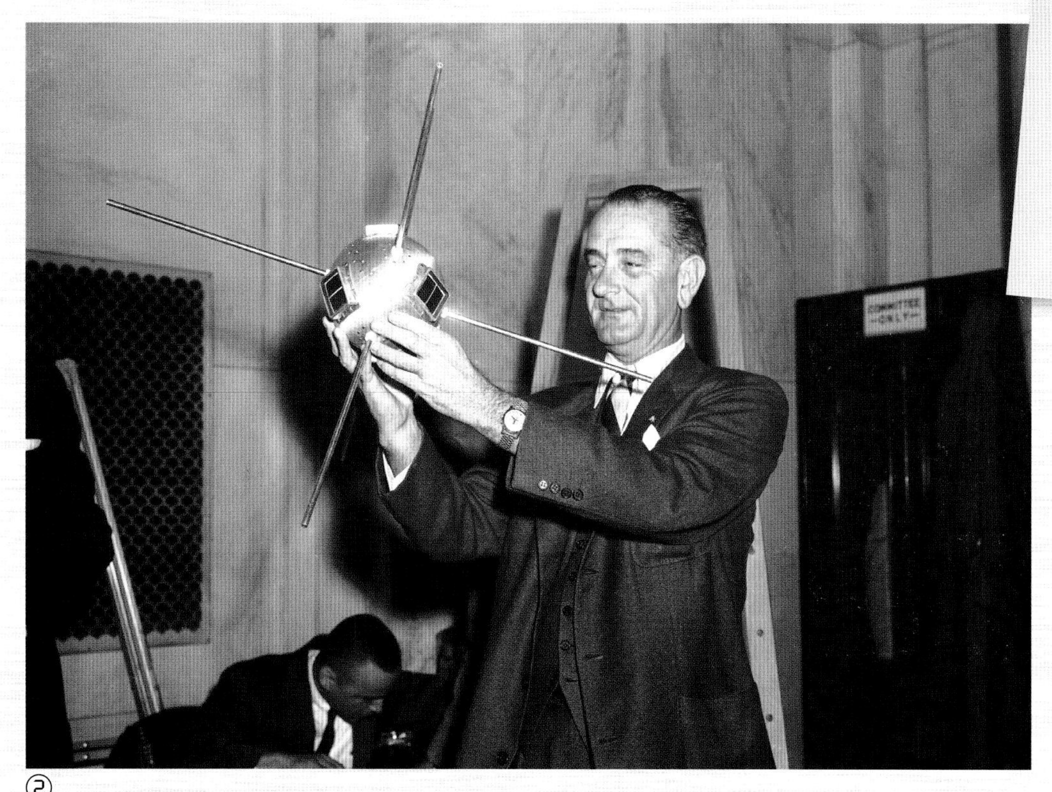

1. One US response to the perceived missile gap was the National Aeronautics and Space Act of 1958, which created NASA to pioneer and lead peaceful space exploration for the United States.
2. In 1958, US Senate majority leader Lyndon Johnson holds a replica of the Vanguard satellite. At this time, the perception was that the Soviets had developed an ICBM with much greater capacity than the United States.
3. Early NACA research aircraft on the lakebed at the High Speed Flight Research Station located on the South Base of Muroc Army Air Field in 1955: From left to right are the Bell X-1E, Douglas D-558-II Skyrocket, and the Bell X-1B, all capable of supersonic flight.

Several variations of the original US Army X-1—the US Navy D-558 and the US Air Force (USAF) Bell X-2—all followed. In 1954, the USAF began developing a hypersonic rocket plane, called the X-15, that could fly at five times the speed of sound and in space, at altitudes higher than 31,000 m (100,000 ft). This was an astonishing level of performance at this point in history, given that jet-powered aircraft were scarcely more than a decade old.

The USAF was also working on a successor aircraft to the X-15, in a program that was revealingly called Man in Space Soonest (MISS). The program could use either a spaceplane or a nose-cone-like spacecraft, although ballistic missiles would be launched both. In January 1958, an USAF conference looked at the alternatives. The National Advisory Committee for Aeronautics (NACA) and McDonnell Aircraft recommended a conical nose-cone-shaped vehicle launched by an Atlas ICBM. They concluded that a wingless, conical capsule would be easier to design, build, and fly.

BIOLOGICAL TRANSPORT

Sergei Korolev was also interested in sending a man into space and, beginning in 1955, was looking at alternative designs. The Defense Ministry funded Korolev's operations, thus he was told to focus on ICBMs first, and then on a reconnaissance satellite in which cameras and film in a special pod would eject and return to Earth. Yet Korolev developed an alternative pod for carrying *biological samples*, large enough to accommodate a human passenger. The return capsule was a sphere, weighted towards one side—it would passively stabilize itself during return from orbit. It had a plastic and asbestos thermal shield, but this was too thick, partly by mistake, and made the spacecraft too heavy to land softly under a parachute. The designers opted for an ejection seat so the passenger would land separately.

NASA IN COMPETITION

The launch of Sputnik kicked off a missile crisis in the United States, because the Soviet ICBM that carried it could reach any US city with a nuclear bomb. The missile gap was real, although temporary. Congress began hearings on space and missiles, and they decided that a new organization, NASA, would take over civilian air and space programs. They gave NASA the responsibility for spacecraft and the USAF retained spaceplanes.

NASA engineers laid out a plan: their spacecraft would carry one human into orbit. The basic design was ready in weeks, and the program and the spacecraft were named Mercury. McDonnell Aircraft would build the spacecraft, and by April 1959 test pilots were selected as the first astronauts.

③

Korolev, meanwhile, watched the establishment of NASA and the US Mercury program with interest. He convened a council of Soviet spacecraft designers and they unanimously recommended that orbiting a human be given top priority. In January 1959, work on the program, already underway for two years, was accelerated. It appeared that the Soviets had a well-organized program, meeting challenging goals in an orderly manner. But it was a misperception, mainly because the ample size of their ICBM enabled massive things to be placed in orbit. Internally the Soviet leaders were concerned that they had no organization and no plan for their program. When they chose passengers for their space missions, Korolev said that they did not need highly trained engineers or highly experienced pilots. Their spacecraft, he argued, was highly automated, unlike the US spacecraft, which would require a pilot to guide it during its mission stages. The Soviet candidates simply had to be intelligent, comfortable with high-stress situations, and physically fit.

> "NOTHING WILL STOP US. THE ROAD TO THE STARS IS STEEP AND DANGEROUS. BUT WE'RE NOT AFRAID . . . . SPACE FLIGHTS CAN'T BE STOPPED."
>
> YURI GAGARIN

For the Soviets, it was imperative that the first manned orbital flight be carried out before the Americans could accomplish the same feat. A Soviet document directed that the first human pilot be in orbit before the end of 1960. On May 15, 1960, the first Vostok (East) spacecraft prototype was placed in orbit.

MERCURY LAUNCHES

On July 29, 1960, the US first production Mercury capsule was to be launched on a sub-orbital flight on an Atlas ICBM. The mission was supposed to demonstrate many of the requirements of an orbital flight. A minute into the launch, the Atlas rocket exploded and the capsule was lost. On the Soviet side, two dogs and a number of other living specimens were placed into orbit on the second Vostok spacecraft on August 19. The spherical capsule successfully came back from orbit, only the second time a spacecraft had ever returned from space.

Before orbital missions, an unmanned sub-orbital test of the US Mercury was to be carried

Feature Index

The Huntsville Times

HUNTSVILLE, ALABAMA, WEDNESDAY, APR. 12, 1961

Man Enters Space

'So Close, Yet So Far,' Sighs Cape

U.S. Had Hoped For Own Launch

Soviet Officer Orbits Globe In 5-Ton Ship

Maximum Height Reached Reported As 188 Miles

Hobbs Admits 1944 Slaying

Praise Is Heaped On Major Gagarin

First Man To Enter Space Is 27, Married, Father Of Two

'Worker' Stands By Story

Reds Deny Spacemen Have Died

VON BRAUN'S REACTION:

'To Keep Up, U.S.A. Must Run Like Hell'

WERNHER VON BRAUN

No Astronaut Signal Received At Ft. Monmouth

①

"TO BE THE FIRST TO ENTER THE COSMOS, TO ENGAGE IN AN UNPRECEDENTED DUEL WITH NATURE, COULD ONE DREAM OF ANYTHING MORE?"

YURI GAGARIN

out using a smaller booster, von Braun's more tested and reliable Redstone, instead of the newer and failure-prone Atlas. On November 21, 1960, the Mercury Redstone was ready for launch. The Redstone's engine roared to life, but then shut down; the Mercury's escape rocket, mounted at the top of the capsule was released and launched, but neither the rocket nor the capsule followed. Then on December 19, 1960, a Mercury Redstone was successfully launched and for the first time a Mercury spacecraft reached space, though not orbit.

YURI GAGARIN

In January 1959, a Soviet state commission, which included military representatives, met to discuss the criteria for selecting human subjects for their spacecraft. They decided that Soviet Air Force training gave candidates the appropriate instruction and experience. During the summer of 1959, the records of more than 3,000 pilots were reviewed. A number were eliminated based on physical issues. By September, candidates were being interviewed, though they had no idea why, and more than 200 individuals were selected to continue testing in October. Many dropped out as they were exposed to highly strenuous physical and mental tests. By January 1960, however, the final 20 candidates were selected and their identities were announced on February 25. Unlike the Americans selected in April 1959, none of the Soviets were test pilots. Only one had flown a high-performance jet aircraft and many had few flight hours. Yuri Gagarin was one of these—he had only 230 hours of flight time.

A simulator for the spacecraft cockpit was developed. It was the first ground-based flight simulator built in the Soviet Union, and because there was only one, only six candidates were selected to train for the first flight. Some were eliminated based on their physical attributes of height and size and others based on medical conditions. They were called the Vanguard 6. Four of the six would fly the first spacecraft, by this time named Vostok; the other two were injured during training.

In January 1961, the candidate crew members took a series of oral and written tests and practice sessions in the simulator. Yuri Gagarin was the favorite and had done well on the tests. Gagarin came from a working-class family and was focused, likable, and well adjusted. He had been in the Air Force for five years and he seemed to be Korolev's personal favorite. All 20 of the cosmonauts were polled about who should be first, and Gagarin was selected by 17.

The Soviets were concerned with the reliability of their rocket and spacecraft and with the safety of their crewman. The same type of rocket with the same type of third stage that was required to place Gagarin's Vostok in orbit had been launched 16 times previously. Of the 16 launches, six had failed because of malfunctions in the ICBM and another two failed because of faults with the upper stage, so its success record was 50 percent. Of seven attempted Vostok missions, two had failed to reach orbit, and two were unable to complete their missions after orbit was reached. But the Soviets decided they had done all they could do, and on April 12, 1961, at 9:06 a.m., Vostok lifted off with a 27-year-old passenger, Yuri Alekseyevich Gagarin. His words at launch were "We're off!"

The launch proceeded as planned. After two minutes, the side rockets were turned off

1. Von Braun asked for one extra test of his Redstone booster with a Mercury capsule. The test gave the Soviets the opportunity to place the first human in space only weeks before the US and spurred Kennedy's goal of a man on the Moon by the end of the decade.
2. The Mercury space capsule undergoing tests in the Full Scale Wind Tunnel, January 1959.
3. A Bell X-2 (46-674) on the ramp at Edwards Air Force Base, California. Behind the X-2 are ground support personnel, the B-50 launch aircraft and crew, chase planes, and support vehicles.
4. The D-558-II was the first aircraft to fly faster than twice the speed of sound, Mach 2 or 2,400 kmh (1,491 mph). The flight took place on November 20, 1953, at the hands of Scott Crossfield, the NACA pilot.
5. NASA tested the Mercury 1-man spacecraft in suborbital flight using the Redstone rocket. This would enable testing the spacecraft without subjecting it to the full effects of orbital flight.

and jettisoned, then after five minutes, the main ICBM core stage finished its firing and was jettisoned and the upper stage began firing. The g level increased and Gagarin's pulse increased to 150 beats per minute, but he continued a steady stream of reporting on the mission events. At 11 minutes and 16 seconds, orbit was attained; the rocket had fired too long and the orbit was 70 km (43 mi) higher than planned.

Gagarin reported he felt fine, and spoke about the view out the porthole for much of his mission. The entire 98-minute long flight was intended to be automated, although there was a control for the manual orientation system. In order to access the system, the operator would have to type in a code, which was sealed in an envelope given to the cosmonaut—the cosmonaut was not to open the envelope unless the automatic system did not work. Less than an hour later, while Gagarin was in space, in a choreographed series of announcements to ensure that in the case of an off-target landing the cosmonaut would be rescued, Radio Moscow reported that a Soviet Air Force pilot was in space, in orbit, in the spaceship Vostok and that his name was Gagarin.

The Vostok's automated systems worked and, thankfully for Gagarin, the spacecraft was oriented properly. One hour and 19 minutes after lifting off, the retrorocket automatically fired for 40 seconds, slowing the spacecraft. Then the spacecraft's instrument section with the retrorocket engine and extra breathing air and power systems was supposed to jettison. Gagarin, who was out of communications with the ground, felt a jolt, then the spacecraft began to tumble rapidly, at one rotation every 12 seconds. He could see the African continent whiz past his porthole, then the horizon, then space, then the Sun, and then it would repeat. The spacecraft was tumbling out of control and the instrument section had failed to separate. The reentry heating began, but instead of his back facing the direction of flight with the thickest part of the heatshield protecting him, he was coming in face first, forced against his restraints, the manned spherical capsule heavier and therefore leading the still connected lighter instrument section, with the thinnest and least protective part of the heatshield taking the brunt of the frictional heating. He knew the situation was not as planned, but there was nothing he could do about it. After about ten minutes, reentry frictional heating had melted the cables and straps that connected the two parts of the spaceship, then the instrument section separated, and the weighted, spherical capsule righted itself. He could see the heatshield melting away and hear it crackling. Temperature in the cabin began to rise. G level increased to about 10 g. Gagarin's vision blurred and he felt faint, but he exerted himself and recovered. The spacecraft oscillated slowly.

At 7,000 m (21,000 ft), the main parachute opened. The hatch blew off, and immediately after this event his ejection seat rocket fired, shooting Gagarin out of the spacecraft. Then the seat separated, his personal parachute opened, and he immediately recognized the terrain, the Volga River, and the Saratov region, the village of Smelovka and the town of Engels. He landed gently. The air valve on his helmet was jammed, and it took Gagarin six minutes to open the valve and breathe fresh air.

The Soviet achievement was a major milestone. It was the first time that a human had left the Earth and flown in space, experiencing weightlessness for more than just 30 seconds.

NASA

If it were to compete with the Soviets, the United States needed world-class astronautical organizations. NACA had been formed in 1915, for although the United States had led heavier-than-air flight in the early twentieth century, it quickly gave up technological leadership to Europe and fell behind in aeronautical research. NACA was intended to centralize such research and information distribution. NACA was a relatively minor player in missiles, principally doing technology research and development (R&D) and participating in the rocket plane programs of the 1940s and 1950s, like Yeager's sound-barrier breaking Bell X-1 flight. NACA also did R&D on wing shapes, propulsion, instruments, and control systems, plus provided some test pilots. Much of its work was in report preparation and publication. The military usually provided the leadership and supported much of the cost burden for developing the new aircraft. In the 1950s, however, President Eisenhower became interested in separating civilian spaceflight from military efforts. He stipulated that the satellite for the International Geophysical Year (IGY) was to be civilian, but there was no US civilian organization equipped to design and develop it. When the

NASA Chief Research Pilot Joseph Walker was the first NASA pilot to fly the X-15 rocket plane and he achieved its fastest speed: Mach 5.92 or 7,060 kmh (4,387 mph) in 1962. He would also achieve a high-altitude record, 108 km (67 miles), in 1963.

President Eisenhower swears in Dr. T. Keith Glennan (left) as first NASA Administrator and Dr. Hugh Dryden (right) as Deputy Administrator on August 19, 1958.

Von Braun and Major General August Schomburg officiate the transfer of the Army Ballistic Missile Agency to the NASA George C. Marshall Space Flight Center on July 1, 1960.

Sputnik crisis hit, a new civilian space agency was needed and Congress and the president converted NACA into NASA, which would be responsible for air and space research of a peaceful nature. The National Aeronautics and Space Act of 1958 directed that US activities in space be "devoted to peaceful purposes." NASA acquired responsibility for non-military satellites and the manned space program.

At the time the Americans created NASA, there were Military Man in Space programs; the USAF also had three such programs in 1958. The X-15, latest in a series of rocket planes, would be capable of reaching space, and the X-20 was intended to be launched on a Titan missile with the goal of flying in space and in orbit. Man in

The X-15 rocket plane reached an altitude of more than 100 km (62 mi). Its research focused on hypersonic flight and human interface control systems for advanced air and spacecraft.

The 4 m (13 ft) diameter Beacon and Shotput inflatable satellites were launched in 1959–60 and served as prototypes for the 30 m (98 ft) diameter Echo passive communication satellites.

A prototype Mercury spacecraft on top of a Little Joe I solid rocket booster. The Little Joe was used to test the Mercury escape rocket and parachute recovery system.

Space Soonest (MISS) looked at the use of a nose-cone-shaped spacecraft, using a blunt heatshield for flights in orbit. This program was specifically intended to place a US man into space before the Soviets.

At the time of NASA's establishment, the MISS program was transferred to NASA, while the winged rocket planes were retained by the USAF. Part of the organization responsible for the US Army moonbase program was von Braun's Redstone Arsenal group, which was transferred to NASA. Information developed for the moonbase concept was factored into the Apollo program.

A technician makes some fine adjustments to a rocket motor during the attachment of the escape tower to the Mercury capsule prior to assembly with the Little Joe launcher, August 20, 1959. The capsules were produced in-house by Langley Research Center engineers.

Hangar 4802 at the NASA Flight Research Center in 1966. Aircraft on the left side include (left to right): HL-10, M2-F2, M2-F1, F-4A, F5D-1, F-104 (barely visible), and C-47. Aircraft on the right side (left to right) include: X-15-1 (56-6670), X-15-3 (56-6672), and X-15-2 (56-6671).

Key NASA figures, 1966 (left to right): Dr. C. A. Berry, MSC Medical Director; Dr. D. K. Slayton, Director of Flight Crew Operations; E. F. Kranz, Flight Director; C. W. Mathews, Gemini Program Manager, MSC; W. C. Schneider, Gemini Mission Manager, NASA HQ; General L. Davis; Dr. R. R. Gilruth, MSC Director; Dr. G. E. Mueller, Associate Administrator for Manned Space Flight, NASA.

VOSTOK

NASA would have to compete with an energetic Soviet space program, for by 1960 Vostok spacecraft were in the final stages of development. There was no formal plan for Vostok—Korolev simply wanted each successive flight to achieve more than the last. After the first one-orbit flight by Gagarin, the next cosmonaut, Gherman Titov, flew for a full 24 hours, experiencing space-sickness throughout his flight. He had the same reentry problem as Gagarin when the two modules failed to separate, but like Gagarin he landed safely.

The Soviets launched the next two missions together, Vostok 3 with Andrian Nikolaev on August 11, 1962, and Vostok 4 with Pavel Popovich, on August 12, flew in nearby orbits but never approached closer than 5 km (3 miles). They landed six minutes apart after three- and four-day flights. The next mission of two Vostoks would include the first woman in space, Valentina Tereshkova. Vostok 5, with Valery Bykovsky, launched June 14 and was supposed to orbit for eight days, but due to a low orbit had to return after only five days. Vostok 6, carrying Tereshkova, launched two days after Vostok 5. Unlike the previous pair, which were launched into very similar orbits, the orbital planes of Vostoks 5 and 6 were 30° apart, so the two spacecraft would briefly approach close to one another twice each orbit.

Tereshkova became ill during the mission. It started in her first few orbits, and the Soviets considered ending the mission early, but when they discussed the situation with Tereshkova, she said she felt better and asked for the mission not to be interrupted. Although telemetry indicated that her biometrics were normal, TV pictures showed her to be tired and weak. Later she reported that at one point she threw up, though she blamed it on the poor quality of the food. Her flight plan called for her to take manual control of the spacecraft on the second day, but she was unable to orient the vehicle. This caused serious concern on the ground, because if she was unable to assume manual control, she could be in danger if the automatic systems failed. On her third flight day, June 19, Tereshkova successfully assumed manual control for 20 minutes, which reassured the ground control.

Bykovsky, by comparison, was thoroughly enjoying his flight. He felt excellent the entire time and floated freely in the cabin, one time for a complete orbit of the Earth. Yet the authorities made the decision to end Bykovsky's flight early, because his altitude was

Valentina Tereshkova became the first woman in space, taking this honor aboard Vostok 6 when she orbited Earth on a three-day mission. Tereshkova was also the first non-pilot to fly into space and was selected on the basis of her experience as a parachutist, with a total of 126 jumps.

A Vostok launch on an R-7 Semyorka rocket. Although it was developed as an ICBM weapon, it was actually never deployed operationally in that capacity. It has been in use continually from 1957 to the present day and has more launches than all other satellite launchers combined.

Technicians prepare a Vostok for flight. The spherical reentry module houses the astronaut. Its mass is off center, which maintains the sphere's orientation automatically during reentry. The entire spacecraft was automated and designed not to require the intervention of the pilot.

An R-7 Semyorka on its launch pad. Atop a lattice structure is the rocket's upper stage, which places the Vostok into orbit. The Vostok spherical capsule is hidden beneath its nose cone shroud.

lower than required and the spacecraft might reenter on its own in an unplanned part of the world unless the early return was planned.

On Tereshkova's third day, the vehicle was to automatically line itself up and perform its retrofire maneuver, but the ground staff grew concerned because while the automatic operation took place as expected, Tershkova failed to report her status throughout the retrofire and reentry. She parachuted to a safe landing after her nearly three-day flight, yet Tereshkova's performance was seen as less than adequate. In addition to her failure to take control of her spacecraft when planned, she also failed to record her medical condition, including her intake of food. Then after her landing, she gave her remaining food to villagers, which meant there was inadequate data on her food consumption during the flight. Reports written after her flight by the medical group and the cosmonaut training center chastised her performance. Bykovsky's return suffered the same problems as Gagarin's and Titov's, two years earlier, when the equipment module failed to separate and the spacecraft reentered in a tumble and then backwards. But after a nearly five-day flight, he landed without incident.

MERCURY

The US Mercury program ran at the same time as the Soviet Vostok program. The objectives of Mercury were simple: place a manned spacecraft in orbit; investigate human ability to function in space; and recover the man and the spacecraft safely. The first two Mercury missions flew on a Redstone, but it was not powerful enough to place the Mercury in orbit. Instead the flights were suborbital lobs lasting 15 minutes. Everything worked well, and Alan Shepard became America's first man in space.

The second Mercury flight, manned by astronaut Virgil "Gus" Grissom, followed the same pattern. The flight went perfectly, but not the recovery, for while Grissom waited for a helicopter to latch onto his capsule, the hatch blew off prematurely. Water rushed in around Grissom, who had already removed his helmet. Grissom got out but nearly drowned, and his capsule was lost. Grissom's

Launch of the Mercury spacecraft Friendship 7, carrying John Glenn, on its Atlas ICBM booster, on February 20, 1962. Glenn made three orbits of the Earth in a journey lasting about five hours.

The first American astronauts were known as the "Mercury 7." They were selected in 1959, first flew in space in 1961, and the last, John Glenn, flew the last spaceflight in 1998.

McDonnell Aircraft technicians build a one-man Mercury spacecraft, 1960. The capsule had a wide base acting as a heatshield to dissipate reentry heat.

Launch of the first American astronaut, Alan Shepard, in the Mercury spacecraft Freedom 7, on its Redstone booster rocket, May 5, 1961.

John Glenn enters his Mercury spacecraft, Friendship 7, through a small hatch for the first US manned orbital flight. The spacecraft was small, about 3 m (10 ft) long and 2 m (6 ft 6in) wide.

spacecraft had worked fine, however, so Mercury proceeded to an orbital flight. By the time John Glenn flew on February 20, 1962, the Russians had already placed two men in orbit, one for a full 24 hours. Glenn's mission was only five hours, but it was successful and he became the first American to orbit the Earth. The next flight repeated Glenn's. The astronaut, M. Scott Carpenter, however, fell behind on planned tasks; he used excessive fuel and ran out during reentry, opening his parachute early to maintain control. Due to a late and out-of-position retrorocket firing, the splashdown was 400 km (250 miles) off course. In contrast, the next mission—a six-orbit flight—was considered a textbook mission in which all objectives were met. The final flight of the program was a 22-orbit mission lasting 34 hours; many systems had failed by the end, requiring the astronaut to take manual control, but all mission objectives were ultimately achieved.

VOSKHOD

After Vostok 6, the Soviet military wanted more Vostoks for military experiments. Korolev was ordered to place a three-man crew in orbit before the two-man US Gemini. The mission was named Voskhod (Sunrise), but used the same Vostok spacecraft. The Vostok engineers objected to three people in a Vostok; they felt that it was too dangerous. Korolev said one of the engineers would fly as a crew member, and this gave the design engineers a more positive attitude. Konstantin Feoktistov was a lead Vostok designer, and he proposed flying three men without spacesuits or ejection seats—after all, the cabin had to hold air and there was no way to get off a failed rocket. Two rockets were added; one for use in orbit as a backup retrorocket to ensure the spacecraft would return on schedule. On Vostok they would allow the spacecraft's orbit to naturally decay and reenter and the crews were always provided with ten days of food and oxygen. A second rocket would slow the final impact on Earth, although shock-absorbing couches were also added. Besides the tight fit inside the spacecraft, there was also a concern that the environmental control system, sized for just one person, would overheat with three on board.

The vehicle was completed in August 1964, but in September a test parachute drop failed and the test capsule was destroyed. Considerable disagreement over the composition of the crew members also ensued until just weeks prior to launch. Korolev made the final decision—he selected Feoktistov because the man probably knew more about the design of the spacecraft than anyone else, but he

The actual Voskhod 2 spacecraft, as displayed at the Energia Museum near Moscow, Russia. The inflatable airlock system and Alexei Leonov's spacesuit are both backup versions.

The first multiple-person crew in space, the crew of Voskhod 1, Feoktistov, Komarov, and Yegorov. They wore only jackets, since they could not fit into the spacecraft with their pressure suits.

A Soviet stamp showing the spacewalk during Voskhod 2. Although the Soviets revealed the design of the one-man Vostok at this time, this illustration looks nothing like the Voskhod spacecraft.

⊙⊙ The actual Voskhod 2 spacecraft with a backup inflatable airlock mounted on top. Surrounding the airlock are high pressure gas bottles containing has for the airlocks inflation and for re-pressurization of the airlock after the spacewalk.

⊙ More Voskhod multi-man spaceflights were planned but not flown. This is the Voskhod 3 spacecraft prepared for flight. Most of the exterior configuration is identical to the one-man Vostok, except for the cylindrical backup retrorocket mounted at the top.

⊙ The Voskhod 1 spacecraft is assembled on its R-7 booster rocket at Baikonur Cosmodrome. At the far left is its backup retrorocket, required to ensure a return to Earth within one day, as the spacecraft only carried supplies for a one day of flight.

was seen by many as a difficult and unsociable personality with medical conditions such as ulcers and nearsightedness. Doctors said that he was unfit for the mission, but Korolev overruled them.

On October 12, Voskhod 1 went into orbit for a day-long, three-man mission, including engineer Feoktistov, a medical doctor, Boris Yegorov, and a pilot, Vladimir Komarov. The next mission would include a spacewalk, preempting the Americans. The mission was called *Vykhod* (Exit), though it was renamed in the days before launch to Voskhod 2. The two-man crew was launched March 18, 1965. For the spacewalk, a collapsible, inflatable, rubber and fabric airlock would be deployed in orbit. The airlock was inflated and Alexei Leonov floated outside the spacecraft. But at the end of 10-minute spacewalk, his spacesuit had ballooned and he could not fit back inside the airlock. It was a moment of crisis and potential disaster, but he acted by partially depressurizing the suit to be able to get back in.

At the end of the mission, the cosmonauts had to take control of the retrorocket firing. This was difficult with their seats reoriented, as there was not good access to the ship's controls or to a porthole. They consequently fired the retrorocket late and landed way off target—it took two days for rescue crews to find them.

GEMINI

The Apollo mission had to perform specific activities and functions for astronauts to be able to walk on the Moon, and Project Gemini was to test the necessary steps. Gemini's objectives were to show that men and spacecraft could operate in space for two weeks and that they could demonstrate rendezvous and docking, demonstrate precise orbital trajectory changes, perform useful work outside the spacecraft, and make pinpoint landings.

1. A cutaway of the Gemini spacecraft. To the right is the rendezvous radar and parachute compartment. Then the two-man crew sit in ejection seats. Behind them is a retrorocket compartment, and to the left the adapter section containing supplies and systems.
2. Gemini 9 is prepared for launch in 1966. The white adapter section attaches the spacecraft to the booster rocket and contains air, fuel, electrical, and environmental systems. The black command module houses the astronauts and is the only section that returns.
3. Nose of the Gemini 7 spacecraft in orbit, the photograph taken by the visiting Gemini 6. The nose of the Gemini houses a rendezvous radar transponder. Red outlines show the location of the two crew hatches.

The Gemini program began in December 1961, and in 1964 the first Gemini flew. In total, ten manned missions met every objective, while two unmanned missions tested the Gemini spacecraft and its booster rocket. The first manned Gemini mission was brief—only three orbits—and simply showed the functioning of the Gemini with men on board. On the spacecraft, astronauts fired thrusters to change each orbit.

Geminis 4, 5, and 7 were long-duration missions: Gemini 4 was four days, Gemini 5 doubled it to eight days, and Gemini 7 doubled it again to 14 days. The Gemini 7 astronauts said that living for two weeks in such a confined space was the most difficult thing they had ever done. Fuel cells were used to produce electrical power and drinking water, combining oxygen and hydrogen.

RENDEZVOUS AND DOCKING

Seven Gemini missions demonstrated rendezvous and docking. Gemini 5 used a small automated pod released from the spacecraft. Gemini 6 met 7 during the two-week mission. Gemini's 8, 10, 11, and 12 all used an unmanned Agena rocket stage, launched separately; the Gemini would actively maneuver to meet the Agena in orbit, both components linking together. On Gemini's 10 and 11, the Agena's large rocket engine was used to send astronauts to world record altitudes, reaching 1,126 km (700 miles). Geminis 4, 9, 10, 11, and 12 all included spacewalks.

Also during the Gemini missions, astronauts conducted 51 research experiments in human physiology, environment measurement, photography, astronomy, biology, engineering, and physics.

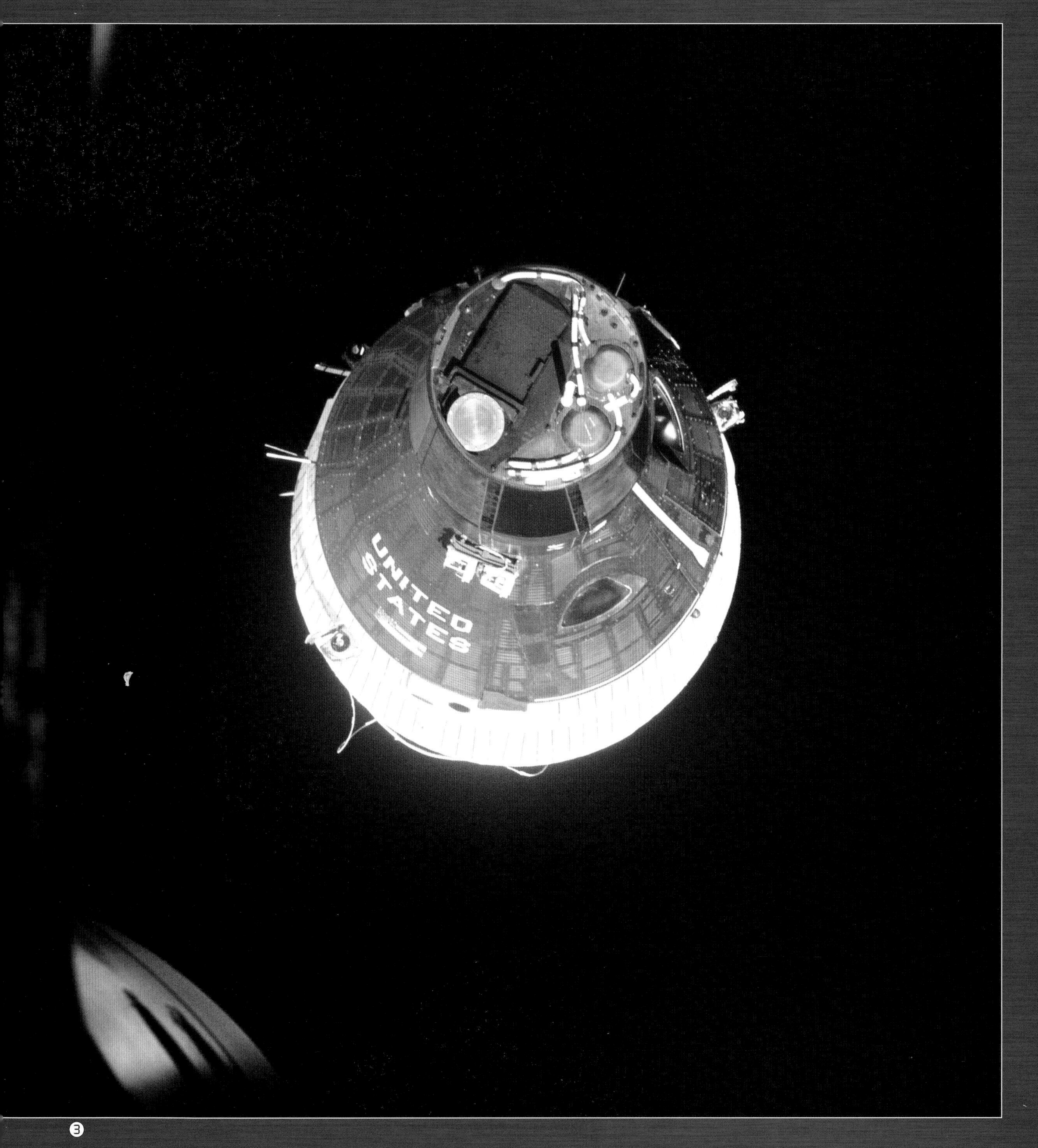

❸

❶

1. The aft adapter section of the Gemini 7 spacecraft. The white section houses maneuvering engines, retrorockets, air, fuel, and electrical and environmental systems. The gold is a lightweight aluminized mylar sun shield.

2. The crew of Gemini 4, Ed White (left) and James McDivitt, atop the Titan launch vehicle at Cape Kennedy, Florida. They are seen here just prior to launch on June 3, 1965, for a four-day stay in space.

3. When the Soviets made the first spacewalk, the NASA authorities set out to repeat the feat. They did so when Ed White left the cockpit of Gemini 4 on June 3, 1965, two months after Leonov's spacewalk. White stayed outside for 20 minutes and used a propulsion gun to maneuver.

2

3

Chapter Four

THE EARLY SPACE STATIONS

Scientists, engineers, and thinkers have imagined space stations since the late 1800s. They are designed to remain in space, housing long-duration crews for months or even years. Unlike other spacecraft, space stations have no systems that allow landing back on Earth or on other worlds. Throughout the twentieth century, and for the first years of the space age, numerous concepts for space stations were developed, before the first stations were orbited in the 1970s. Most were designed as observation platforms for military or civilian purposes, and later their uses were expanded to include scientific research.

(Left) Skylab, the first US space station, orbits the Earth in 1974. Skylab was launched by a Saturn V Moon rocket, making use of the Saturn's enormous lifting capacity. The main station module was larger than any since that time: 7 m (23 ft) in diameter, 9 m (30 ft) long, and weighing 75,000 kg (165,000 lb).

(Above) The first crew of Skylab—(left to right) Kerwin, Conrad, and Weitz—were also called upon to be the first space repairmen, fixing the power array and thermal shield to make Skylab habitable after a launch accident.

The Early Space Stations

HERMANN OBERTH COINED THE TERM "SPACE STATION" IN *THE ROCKET INTO INTERPLANETARY SPACE* (1923). OBERTH, UNLIKE MANY OTHER EARLY THINKERS ABOUT SPACECRAFT, WOULD LIVE TO SEE HIS IDEAS PUT INTO ACTION, WITH MISSILES AND ROCKETS CUTTING THROUGH THE HEAVENS, HUMANS IN SPACE AND WALKING ON THE MOON, AND ALSO THE FIRST SPACE STATIONS IN THE 1970S. IT WAS A REMARKABLE SCIENTIFIC AND TECHNOLOGICAL EVOLUTION IN JUST HALF A CENTURY.

As we have seen, Herman Potočnik, writing under the pseudonym Hermann Nordung, elaborated on Oberth's ideas in his book the *Problems of Rocket Travel* (1929). Oberth and Potočnik originated the concept of the torus or doughnut-shape space station that rotated to produce artificial gravity. The idea would later be popularized by Wernher von Braun—an 18-year-old von Braun wrote a story in 1930 about the Oberth/Potočnik station, which he named "Lunetta" or Little Moon. It would be used to make observations of the Earth and atmosphere. Although von Braun diverted his attentions to serve the Nazi military in World War II, he would return to the idea of "Lunetta" almost immediately after the war, while working for the United States.

EXPERIMENTAL DESIGNS

Most of the pre-space age conceptual space stations were so large that they would have to be assembled in orbit; von Braun described spacewalking men performing the assembly while floating freely. But until the mid-1950s, no one had developed a successful full-pressure spacesuit. Von Braun described "The Bottle Suit," which was a hard shell, miniature spacecraft with robotic arms.

Some of these early conceptual stations were made of metal cylinders, others of inflatable rubber and fabric. Some were to be powered by solar energy and others by atomic energy. All the scientists working on such ideas understood the free-fall of orbital flight and anticipated the feeling of weightlessness, and thus all of them believed that artificial or synthetic gravity would be needed inside their space stations to permit people to live and work. They created artificial gravity by rotating their stations—Newton's first law said that an object in motion would continue in motion in a straight line. In a circular structure, like a space station, this caused centrifugal force. The person on the inside would tend to continue in motion, but the curved floor of the space station would keep the person from flying off.

In 1959, the London Ideal Home Exhibition approached Saturn's rocket builder, Douglas Aircraft, and specifically asked for them to design a space station to exhibit at the show.

> "SINCE THE BEGINNING OF TIME, MANKIND HAS CONSIDERED IT AN EXPRESSION OF ITS EARTHLY WEAKNESS AND INADEQUACY TO BE BOUND TO THE EARTH."
>
> HERMANN NOORDUNG

TECHNICAL CHALLENGES

At the 1951 International Astronautical Federation Congress, German aeronautical engineer Heinz-Hermann Koelle described the problems of developing a space station. He included the technical issues, like assembly in zero-gravity and the political problems of long-term national and international support and funding. But the problems did not deter the global community of scientists. By the late 1950s, the Soviet engineer Korolev described the need for Earth-orbiting stations to provide a base for testing systems before a long planetary mission. One of the concepts—the SKB station—was small enough that it could be launched by the ICBM. In 1960, a larger 3–5-man military

The unmanned Skylab 1/Saturn V space vehicle is launched from the Kennedy Space Center on May 14, 1973, to place the Skylab space station cluster in Earth orbit.

About a minute into the launch of the Skylab space station, a thermal shield was ripped from the Skylab's main body and this in turn ripped off a solar power and jammed a second array.

space station was part of the 1960s Soviet space plan. First there would be only a single module. Later versions could have two or more modules. An even larger station design was started in 1963. Multiple Soviet space station program plans continued in parallel, designed by different groups, but no modules were actually launched. Later and larger Soviet stations required a rocket the size of the American Saturn V called the N-1; the original requirement for the N-1 was large space modules. Later it was intended for use in the Russian Moon-landing program.

FROM EXHIBITION TO DESIGN

Each year in the United Kingdom, the popular newspaper *Daily Mail* sponsored the Ideal Home Exhibition in London. New ideas in home décor and design were exhibited. In 1960, however, the theme was "A Home in Space." A space station mock-up was built inside the second stage of a Saturn rocket. Ten years later many of the design ideas were actually used in the Skylab orbital workshop.

The exhibition catalog described the exhibit:

. . . for the first time in history visitors can see a full-sized replica of a space ship of the future. A crew of 4 ascend in a nose cone in which they also return to Earth. Solar batteries provide 5 kw of power. The atmosphere is oxygen and nitrogen pressurized to simulate 10,000 feet [3,000 m] altitude. CO_2 breathed out by the crew is absorbed in special containers. There is no "floor" or "ceiling." The crew can work equally easily in any position. One of the crew is working on a telescope, in a space suit, outside the vehicle. In a gravity-free condition he uses "restraint" straps to prevent "drifting."

The crew returns to Earth fastened into special seats. A small rocket motor is fired to reduce the speed of their nose cone. They sink into the upper atmosphere and into the Earth's gravitational pull. They lose speed and a large parachute opens and the capsule floats to the ground.

Following the exhibition, NASA defined the need for a station, and in 1962 NASA awarded a contract to study alternative designs. The study began with the Ideal Home Exhibition station.

The design used the empty upper stage of a Saturn rocket. Most of the stage would carry oxygen and hydrogen propellants, and the payload would be a docking adapter, airlock, and centrifuge. Astronauts could be launched to the station in either Gemini two-man spacecraft or Apollo three-man spacecraft. Enough provisions would be provided to support four crew members for 100 days. A centrifuge inside the stage would subject the crew to partial gravity to readapt them before returning to Earth.

After the crew arrived, they would dump as much remaining hydrogen as possible. They would open a hatch in the docking adapter/airlock and enter the now largely emptied upper stage in their spacesuits. The astronauts would spend an early part of the mission cleaning the interior. While the stage provided a large amount of open pressurized volume, the logistics of cleaning, and then delivering and outfitting the empty shell, would use a lot of the crew's time.

The study was modified over its four years. In the new thinking, an Apollo spacecraft would carry supplies. A mission module launched with the Apollo would provide living quarters and a laboratory. Payloads included a side-looking radar to produce detailed maps of the Earth and a large astronomical telescope; additional modules with specialized functions like materials science or life sciences could be added. Crews would remain for six months to a year.

It was the intention to produce several stations. Early stations would be placed in a low Earth orbit, while later stations would be launched by the larger Saturn V rocket into high Earth orbits or to the Moon to map its surface. An even later generation could be sent on a trajectory to fly by Mars. These ideas were taken seriously—by the late 1960s technology funding was being spent to develop the prototype regenerative life support systems for the space station. Human beings were beginning to see the potential reality of what it would mean to live and work in space.

MILITARY MAN IN SPACE

In November 1945, with World War II over, General Henry H. "Hap" Arnold, commanding general of the US Army Air Forces (USAAF), predicted that "true spaceships, capable of operating outside of the Earth's atmosphere" would be launched "within the foreseeable future." As the Cold War set in, the US military services contended with one another for control of missiles and access to space. The US Army had responsibility for military missiles with a range of less than 322 km (200 mi), because they were a form of artillery, while the US Air Force (USAF) had ICBMs that supplemented strategic aircraft. Sea-launched missiles belonged to the US Navy.

In the 1950s there were several "Military Man in Space" programs. The US Army's program under Wernher von Braun was studying space stations and Moon bases to store, launch, and control nuclear weapons. The USAF had rocket planes to take men into space. The X-15 would fly in 1959; it promised to reach space by flying at altitudes of over 80 km (50 mi). The X-20 would be launched on a missile into orbit and dip into the atmosphere to perform reconnaissance; it also had to solve technological problems faced by craft beyond the speed or altitude of the X-15. Man in Space Soonest (MISS) looked at the use of a nose-cone-shaped spacecraft with a blunt convex heatshield, to place men into space before the Soviets. The USAF retained responsibility for military space programs, arguing that air and space were indivisible. But as interest in space stations grew, the USAF established the Manned Orbiting Laboratory (MOL) Program. MOL was designed to put astronauts in orbit for 30 days to show the value of the military man in space.

The X-20 Dyna-Soar aerodynamic model being tested in the Langley Research Center full-scale wind tunnel. The "Dyna-Soar" name is an abbreviated of the phrase "Dynamic Soaring."

The X-20 Dyna-Soar was studied from 1956 as the "Manned Glider Rocket Research System." It was this effort, begun in parallel with "The Manned Ballistic Rocket Research System," which led directly to NASA's Project Mercury.

A Dyna-Soar vehicle lifts off atop an Air Force Titan II launch vehicle, in this 1961 artist's concept. The Dyna-Soar would be launched by a Titan I or II rocket into a suborbital trajectory, and by a Titan III into orbital flight.

While the X-20 program was cancelled in 1964, several vehicles with similar design characteristics followed. These were the "Lifting Bodies." The X-24b was the last of these and flew 36 flights, including 24 rocket-powered flights to 23 km (14 mi) altitude.

The X-15 was the most successful of the rocket-powered research aircraft, flying to record altitudes and velocities and testing a number of systems adapted for use on crewed spacecraft.

EARLY STATION CONCEPTS

Following the establishment of NASA in October 1958, Congress mandated that NASA should define the US civilian human space flight program. The Eisenhower government was fighting against any new and large human space flight programs until a specific purpose could be defined. NASA administrators, however, in testimony before the Senate Committee on Aeronautical and Space Sciences, described the long-range objective of the nation's space program to include a multi-man, permanent orbiting space station.

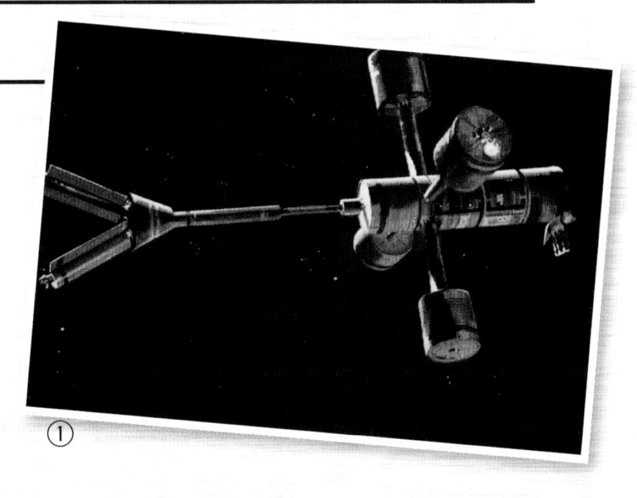

①

②

③

In 1959 the NASA Research Steering Committee on Manned Spaceflight concluded that the follow-on to the one-man Mercury Program should include a two-man spacecraft. For missions longer than three days, they recommended a cylindrical module that would be attached to the two-man spacecraft. They concluded that an Earth-orbiting space station was a more logical follow-on to the Mercury Program than a Moon-landing program. However, they were concerned that any space program needed a well-defined, clearly stated objective that would inspire policy makers, industry, and the public. The committee said that the lunar landing goal had a more clearly defined goal and that the space station, by contrast, did not.

Later, in July 1959, NASA convened an important conference at the Langley Research Center to better define the goals of a space station and to identify the issues a space station could resolve. The primary goal identified was to study physical and psychological reactions of humans in the space environment for extended periods of time and to better understand human capabilities and usefulness during long-duration missions. A second goal for the future space station would be to study materials, structures, and systems for extended-duration space vehicles, and specifically systems for communication, orbit control, and rendezvous. A third goal would be to evaluate techniques for astronomical and terrestrial observations and to understand better how a human's abilities could be used most optimally in spaceflight.

APOLLO-BASED SPACE LABORATORY

In 1961, as the first human space flights were being undertaken by Yuri Gagarin in Vostok and Alan Shepard in Mercury, NASA developed plans for a space laboratory based on an Apollo spacecraft. The laboratory could support experiments related to crewed spacecraft design and scientific observations, including monitoring of the Sun, testing the ability for men to work outside the spacecraft, and micrometeoroid impacts.

The RAND Corporation, a US military think tank, was established in 1948 by the USAF to help research and analyze public policy issues. In April 1960, RAND and NASA hosted a symposium to discuss space station planning and feasibility. While RAND and other groups were focused specifically on space station concepts and missions, there was additional work going on in NASA which identified the types of required orbital operations and laid plans for developing the required capabilities. In May 1961, NASA headquarters published a report on a Program for Orbital Operations. It included studies of the space environment, investigation into operational requirements for mission dependency on ground support versus

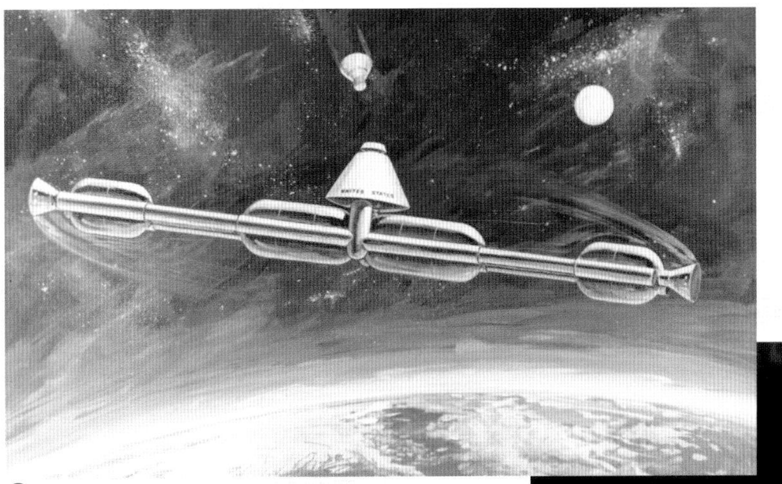

(4)

"A SPACE STATION IS A BASE IN SPACE. THE BASE IS A CENTRAL LOCATION FOR POWER, VOLUME, LOGISTICS, EXPERIMENTAL EQUIPMENT, COMMUNICATIONS AND DATA REDUCTION."

ROBERT GILRUTH

(5)

1. In the late 1960s, NASA was interested in large Earth orbital bases which would house as many as 50 to 100 people. NASA was proposing that portions of these bases would rotate for artificial gravity. Nuclear power, separate from the habitat, would provide unlimited electricity.
2. It was originally envisioned that space stations would rotate to provide artificial gravity for its crew. Of these concepts from the 1950s–60s, only the one at lower left was envisioned to permit the crew to live in zero-gravity.
3. A 25-crew nuclear-powered station proposed by NASA in 1966 would be used to manufacture new materials in space. It could also serve as a staging and departure point for missions to the Moon and planets, launched by two Saturn V rockets.
4. A 1963 Grumman proposal for a space station with 7 m (23 ft) diameter and 12 m (39 ft) modules connected by a 100 m (328 ft) long tunnel. The entire station would spin to create artificial gravity. Gemini spacecraft would carry crews to and from the station.
5. Owen Maynard and Will Taub of NASA patented this concept of a space station in 1962. It was made of several modules linked together in orbit.

flight crew independence, ground tracking and communication networks, logistical support of an orbiting base, and operation of space stations when not occupied by crews.

Prior to May 1961 and the Kennedy Moon-landing decision, the direction in which the NASA program would go was undecided. NASA's primary mission might be a lunar landing, but it could just as likely be a space station. An Apollo Program Office was established to design and build a three-man Apollo spacecraft, but no one was entirely certain what the spacecraft would be used for.

Some recommended a phased approach to a space station. During the initial phase, the focus would be on design, manufacture, and initial orbital operations of a space station. During this phase, engineers would study the basics of supporting and operating a space station: the rotation of flight crews, logistical support of station supplies, and the addition and replacement of space station modules. A second phase would focus on technology and operational demonstrations: inflatable modules and structures; the creation of artificial gravity conditions; and rotation of the entire complex to maintain gyroscopic stability. For a rotating spacecraft, additional studies would be required into structural and dynamic stability, materials and fabrication techniques, life support, thermal control, and power-generation systems. New logistics and resupply vehicles, including winged fly-back gliders, could be developed and their use explored.

REFINING THE CONCEPT

There were lively discussions and disagreements about what a space station should look like and how it would be used. By 1962, as the first Mercury astronauts were orbiting the Earth, research was ongoing about the potential uses and experiments that would be done on a space station. NASA had decided by this time that a space station was technologically feasible but wanted to justify the need for such a program. Questions arose whether the space station was uniquely required to meet the science and technology goals, or whether the goals could be achieved using other spacecraft, like the Mercury, Gemini, and Apollo already in development. Generally, NASA headquarters stated that the objective ultimately had to contribute to a crewed planetary mission. If the need could be justified, then a station might be in orbit in five years.

A number of studies tried to define the design and characteristics of a space station so that a political decision could be made about how to proceed. Some stations were based simply on the upper stages of rockets already in development, and others added components such as inflatable arms or additional modules to build rotating space stations.

MANNED ORBITING LABORATORY

In 1946, the RAND Corporation recommended that the United States consider outer space as a vantage point for military reconnaissance. The USAF was already using high-altitude spy planes, but as Soviet air defenses improved more than two dozen US aircraft were shot down, with most of the crews lost. In 1956, a year before the first satellite was launched, President Eisenhower approved a reconnaissance satellite program. Its military code name was Corona, although its public name was the more scientific Discoverer. It would take 13 unsuccessful attempts but in 1960, on the fourteenth attempt, Corona returned the first man-made object from space, a canister carrying exposed film from its reconnaissance cameras. Corona would ultimately prove reliable enough for gathering images of the Soviet Union in place of spy planes, and later generations became much more sophisticated and capable. The early unmanned reconnaissance satellites, however, could not always be relied upon. Clouds often obscured ground targets. Systems were not yet sophisticated enough to respond to targeting changes, operating on pre-programmed instructions. Technical malfunctions could not be fixed.

Manned Orbiting Laboratory (MOL) was a military space station that would use the NASA-developed Gemini spacecraft. Military personnel would be placed in orbit to conduct military experiments. They would manually collect strategic reconnaissance, overcoming the limitations of unmanned satellites. It was thought that the manually operated system could return better photographs with a better resolution than the unmanned systems already in use. MOL astronauts could focus on crisis areas and identify ground activities using their eyes and brains. The USAF provided the spacecraft and crew while the National Reconnaissance Office provided cameras and supporting systems.

MOL was announced in 1963. By this time, the longest spaceflight had been a Vostok five-day mission, and the longest US mission was only 34 hours, but MOL was planned to orbit astronauts for 30 days. The program faced criticism that it

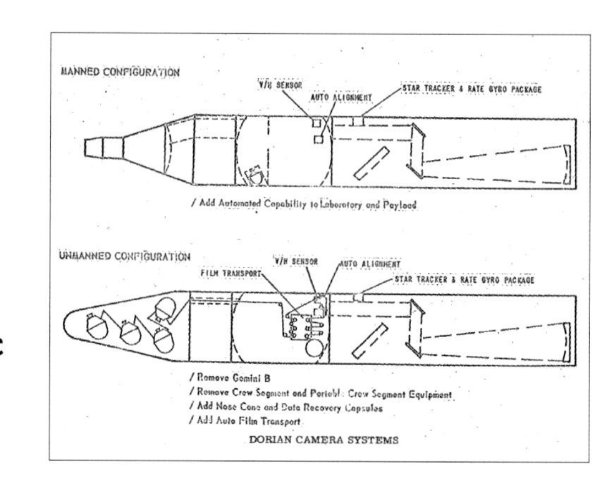

▲ The Manner Orbiting Laboratory (MOL) was selected after comparing human-operated and unmanned systems for reconnaissance. The unmanned system would return film in reentry pods. The human system could rely on orbiting astronauts to review and identify critical features.

▼ A modified Gemini spacecraft was used to transport returning astronauts to Earth. The Gemini had a hatch located in its rear heatshield to allow astronauts entry from the cylindrical laboratory. The hatch was tested on November 3, 1966, and proved safe for flight.

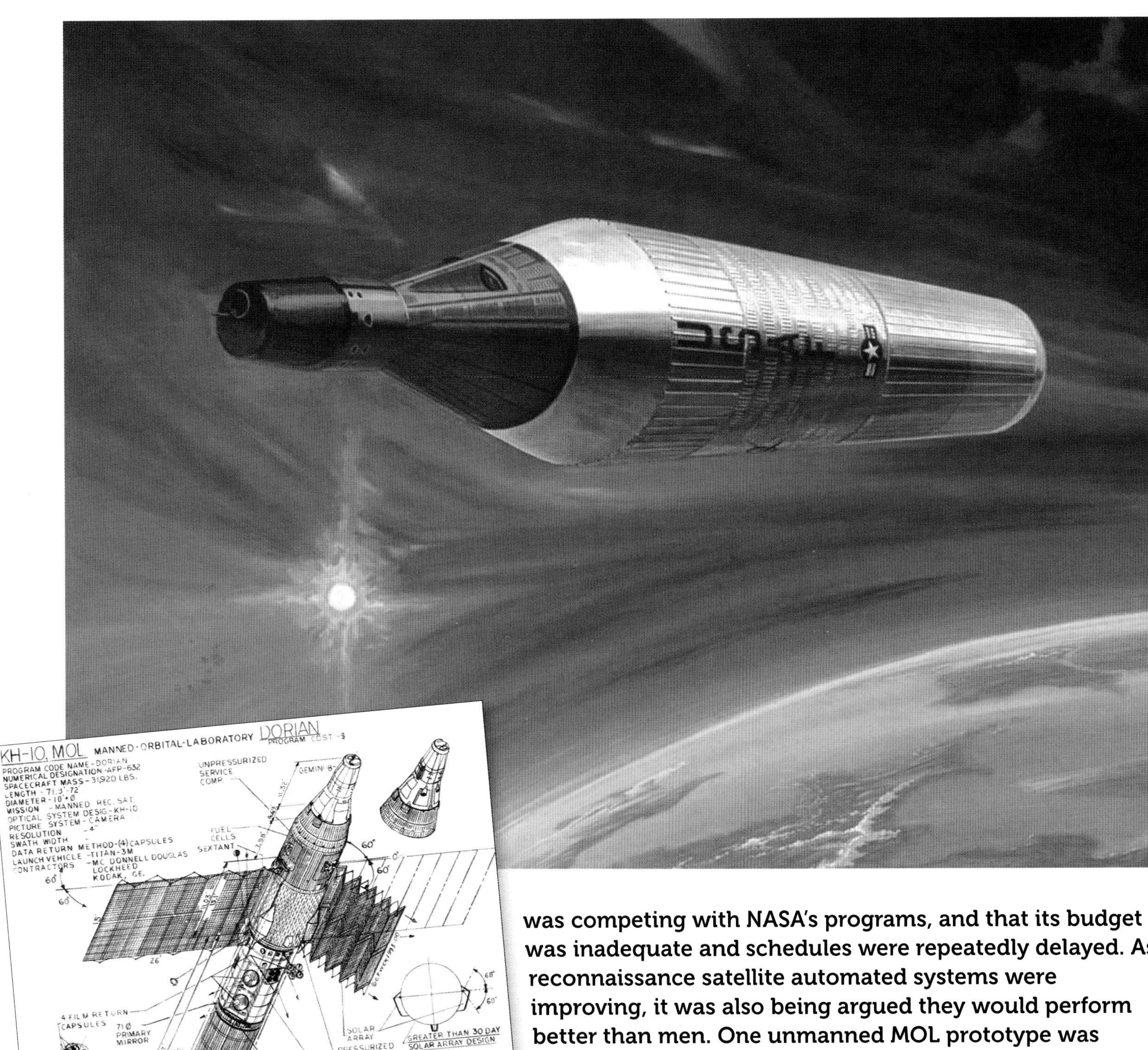

was competing with NASA's programs, and that its budget was inadequate and schedules were repeatedly delayed. As reconnaissance satellite automated systems were improving, it was also being argued they would perform better than men. One unmanned MOL prototype was launched carrying experiments and testing the return of a modified Gemini command module. The Gemini had a hatch cut into its heat shield, and the test verified that this posed no technical problems for the spacecraft or the returning crew. As schedules were extended, the costs of the program also increased. Simultaneously the program's budget was being cut. With the cost reductions, flights and hardware production were curtailed. In June 1969 President Nixon cancelled the MOL program. although some of its experiments and test hardware were transferred to NASA's Skylab Program.

MOL was advertised to conduct 30-day missions, but later information indicated a mission duration of up to 90 days. In early designs power was provided by fuel cells.

Later versions of MOL were powered by solar cells. There were compartments for habitation, work, exercise, a centrifuge and a telescope and camera.

MERCURY-BASED SPACE STATION

In 1960 the Mercury spacecraft was being tested in preparation for its first crewed launch. Mercury prime contractor McDonnell Aircraft proposed a One-Man Space Station. For normal Mercury missions, the spacecraft was launched by the Atlas ICBM, but for the space station mission, the Mercury would be launched by an Atlas with an RM-81 Agena upper stage. The Agena would stay attached to the Mercury, used as a service module to provide propulsion. The laboratory, 3.3 m (10 ft) long and 2 m (6 ft) in diameter, would be launched by its own Atlas Agena. The Mercury-Agena would actively maneuver and then dock with the laboratory.

The Mercury spacecraft was tiny and its normal mission length was typically no more than a day, but the space station provided a little more room for the astronaut and extended the mission length to 14 days. The station had an interior volume of just 8 m³ (282

The Agena was developed as an upper stage and automated "bus" for maneuvering unmanned payloads in orbit. It was boosted into space by different first stages; here an Atlas rocket lifts an Agena off the launch pad.

Early Mercury prototype capsules are built by NASA personnel at NASA's Langley Center in Virginia. Once design was determined, manufacturing was turned over to McDonnell Aircraft.

The one-man station would be launched on an Atlas-Agena followed by a Mercury launch with an astronaut on a second Atlas-Agena. The two would rendezvous and link together for missions.

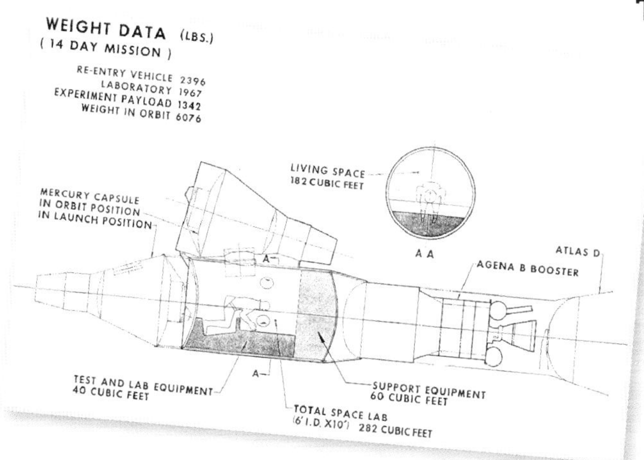

The Mercury one-man space station in orbit. The station was to be 3 m (10 ft) long by 2m (6 ft 6 in) in diameter, not including the Mercury spacecraft. The station's power would be provided by solar arrays.

cubic ft). Open living space would account for about half of this volume and about half by payload and support systems. The astronaut could remove his pressure suit to work in shirtsleeves.

Two variants were considered. In one an inflatable tunnel would have been extended between the Mercury's side hatch to a hatch on the station cylinder. An alternative was to hinge the Mercury so that it would swing back, lining its hatch up with the hatch on the station cylinder. At one end of the station, a systems module would provide life support, communications, and power systems. Fuel cells would deliver 1,500 W power and augment the water supply that could be used for drinking, cleaning, and cooling. Clothing and food storage would be provided for two weeks in orbit. McDonnell recommended that a series of one-man Mercury stations could carry out specialized missions in astronomical research, Earth observations, or communications. About 1 m^3 (40 cubic feet) of volume was specifically allocated for the experiment/research payload, and 13 kg (28 lb) was allocated for the return of experiment results.

KRAFFT EHRICKE CONCEPTS

Krafft Ehricke emigrated to the United States from Germany with Wernher von Braun. Initially he worked with the German engineers at White Sands, New Mexico, and in Huntsville, Alabama, but in 1954 he moved to the Convair Astronautics Division of General Dynamics, where the US Atlas ICBM was in development. Ehricke worked under Karel Bossart, chief designer of the Atlas ICBM. Bossart's design of the Atlas was unique because the missile was essentially a stainless steel balloon made of thin metal that contained the fuel. The metal retained its shape through pressurization. If for any reason the pressurization was lost, the missile collapsed. Bossart had Ehricke develop concepts for advanced space projects based on the Atlas, including ideas for space stations made from converted Atlas rocket tanks.

In 1954 Ehricke described a small four-man space station. Other pre–Space Age stations like von Braun's were large with crews of up to 50 or 100. Ehricke argued that the larger the station, the more logistics and maintenance and reboost fuel would be required, so size was therefore not desirable. Ehricke also suggested that several stations operating in different orbits would permit more wide-ranging research.

Krafft Ehricke emigrated to the United States after World War II. He went to work for the Atlas Project and developed ideas for using the Atlas as more than an ICBM and satellite launch vehicle.

A series of Atlas launches would place in orbit all of the components necessary to assemble the station. A living quarters would be inserted inside the Atlas fuel tank and inflated to support a crew of four people.

A nuclear generator would supply electrical power to the Atlas crewed space station. To keep the station operational, fly-back triangular shaped spacecraft would carry two crew members between Earth and orbit.

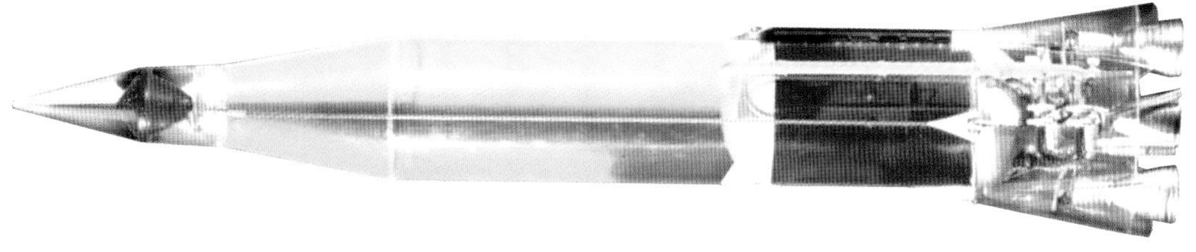

Atlas was a thin stainless steel shell with no internal structure. Rigidity was maintained by pressurizing the interior. Without such pressure, the rocket would collapse at ground level.

Ehricke shows the interior of the station's living quarters. The Atlas nose would house a hygiene facility, galley, sleeping quarters, and a laboratory.

In 1958 Ehricke had further developed the concept of the station based on the Atlas. It would also require vehicles to carry four crew members at a time, plus supply ships. The astronauts's living quarters were in an inflatable cell in the forward end of the Atlas, divided into four levels. The Atlas shell would be launched first, followed by launch of the cargo ship carrying logistics, and then the launch of the astronauts, who would integrate the logistics to bring the station to operational status. Additional logistics vehicles would bring up further supplies, while a final cargo vehicle would arrive carrying a nuclear power supply. In later versions the crew ferry vehicles were small diamond-shaped lifting bodies that would glide back to runway landings. The conceptual station was named Outpost II.

Ehricke proposed a series of Atlas-based station designs, some requiring the linking together of several spent Atlas rocket stages in orbit.

PROJECT HORIZON

Planners working under Wernher von Braun initiated Project Horizon while he was still working for the US Army, although the study was completed after his Army Ballistic Missiles group had been transferred to NASA. The project proposed to establish a military crewed intelligence observatory on the Moon. In addition to the lunar base, the study identified other systems required, such as crew vehicles for flight to and from Earth orbit and to the Moon. Von Braun also proposed using empty Saturn stages left in orbit to form an orbiting "Space Dock." Astronauts would empty the stages entirely of fuel, clean them, and convert them into living quarters. Two or three stations would be placed into equatorial orbit around Earth to support the initial lunar base. Several more would be added as the build-up of the lunar base continued in the late 1960s. They would serve as locations for consolidating, assembling, staging supplies, and resupplying fuel to the Moon-bound freighters. Each orbiting base would house approximately 10 people and they would rotate between orbit and Earth's surface by three-person shuttles every few months. It was thought that personnel going to the Moon would stay longer on the Moon than those only going to the orbiting stations.

The plan called for 12 men to live and work on the Moon by 1965. Further build-up of the base would continue for several years after that. When Project Horizon started, work had begun on the

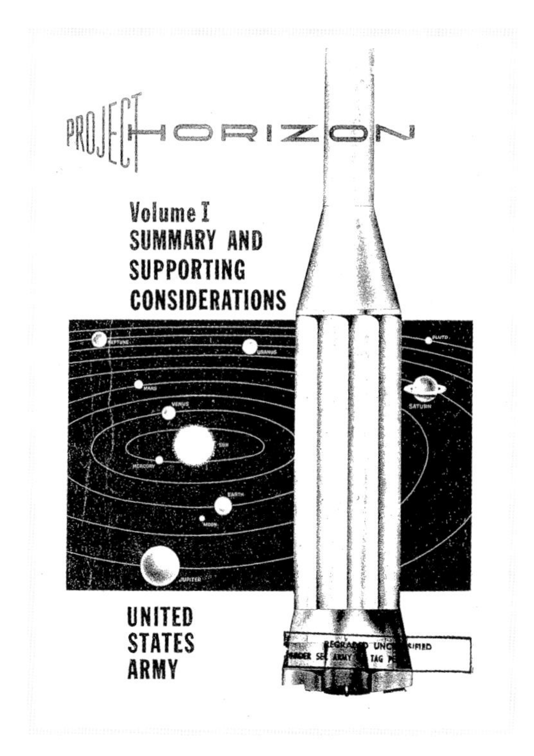

The cover of the Project Horizon report illustrates the Saturn 1 rocket that would be used to develop the first space station in Earth orbit and a military base on the Moon.

Horizon Earth orbital stations would be assembled using the upper stages of Saturn 1 rockets. Later this would be called the "wet workshop" concept.

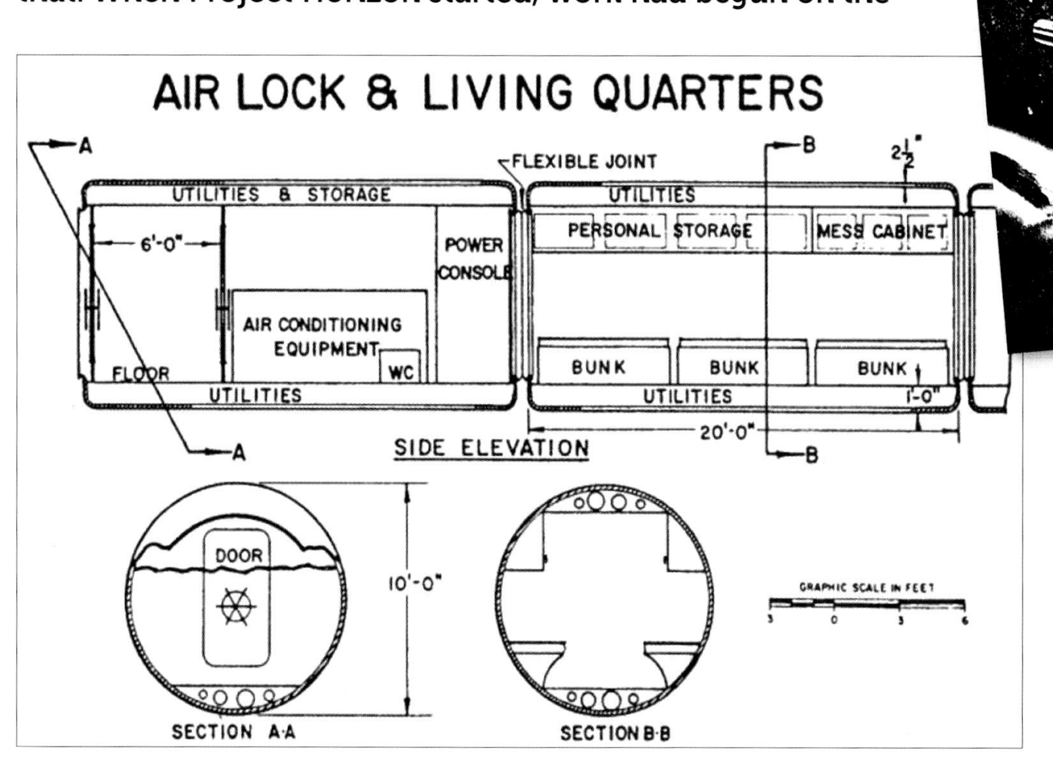

The "wet workshop" concept involved astronauts living in used (empty) Saturn 1 rocket upper stages. The Horizon stations would serve as test beds to prove the critical environmental control equipment on which crews would depend on the Moon.

Horizon also proposed using spent Saturn stages as living quarters on the Moon. The stages would be buried beneath lunar soil in order to protect crews from radiation.

Concepts pioneered in Project Horizon have been adopted in other lunar base studies. This inflatable base concept dates from the Space Exploration Initiative of 1989.

Saturn 1 and Saturn 1b rockets and on some elements of the Saturn V Moon rocket. However, this was one to two years before the Kennedy Moon-landing decision, and the Saturn V design was not finalized until 1962. Engineers would develop the moonbase using "smaller" Saturn 1 rockets, which were then in development at Redstone Arsenal. These were still larger and more powerful than anything the United States or Soviets had ever developed previously. At least 140 Saturn rockets would be needed to launch the elements for Project Horizon.

MANNED ORBITING RESEARCH LABORATORY (MORL)

In 1962, Douglas Aircraft began to study a space station that could be based on a Titan II or Atlas rocket and use Gemini spacecraft to carry astronauts. The goal was a space station that would be in use by 1965. The early station modules would be launched "wet" with fuel, the fuel obviously being required to propel the stage and their payload into orbit. Once in orbit, they would have little fuel remaining, so astronauts would then need to remove any remaining fuel and clean the interior to prepare them for habitation. An elaborate docking and crew transfer system would require the crews to transfer from their Gemini spacecraft to the station by spacewalking.

NASA subsequently adopted the study and issued contracts to expand upon the early ideas. In the later parts of the study, a "dry" space station module would be launched by a Saturn 1b rocket, ready for crew habitation. The station would include an airlock, docking adapter, and a centrifuge for re-adaptation of crew members before return to the Earth. The crew would be launched in Apollo spacecraft carrying additional mission modules designed for their particular functions. Modules for propulsion, science experiments, communications, and supplies were all identified. Several contractors provided hardware requirements for each function.

The last stage of the study also examined the kinds of space research the station would conduct. Stanford coordinated studies of oceanography, cartography, and photogrammetry. A large astronomical telescope, the same size as the much larger Hubble Space Telescope, would be launched by a separate Saturn 1b and

MORL was to be launched by a Saturn 1b rocket. Its crew of up to six astronauts would fly to and from the station in either Gemini (shown here) or Apollo spacecraft.

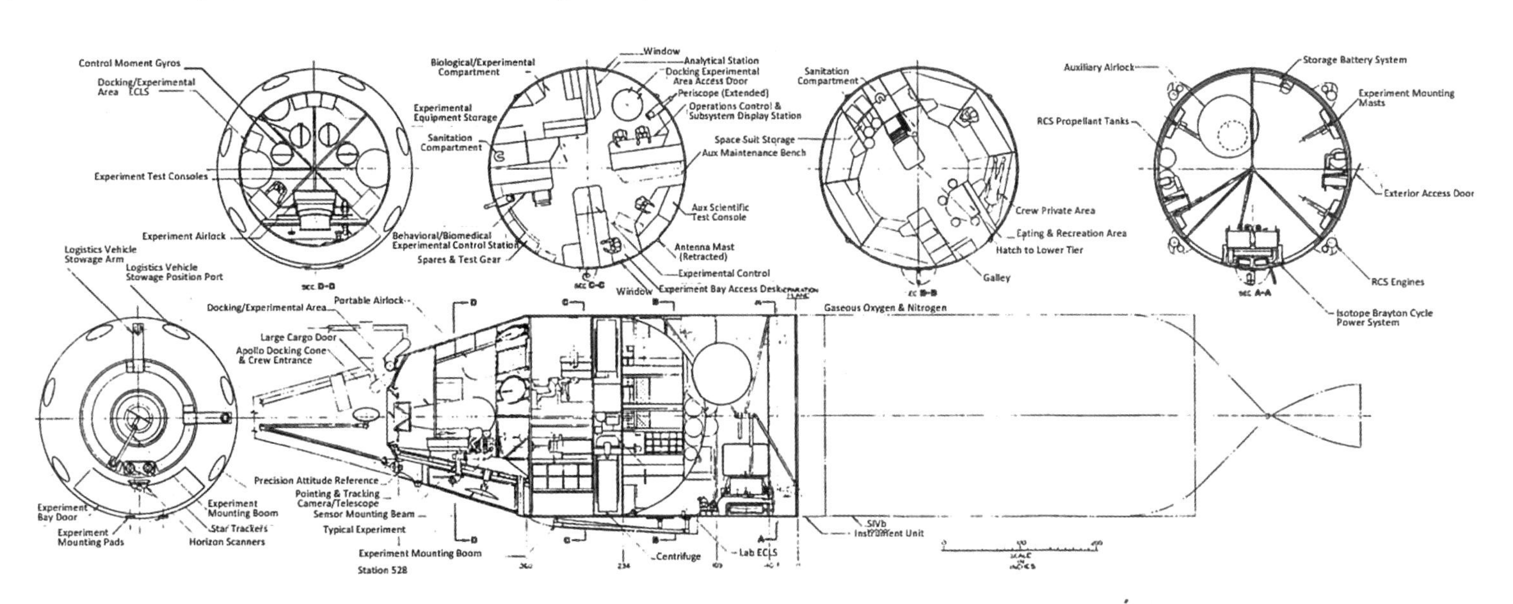

A schematic of the 1963 MORL. A crew would fly to/from the station in either Gemini (shown here) or Apollo spacecraft; one astronaut would attempt a year's stay in orbit.

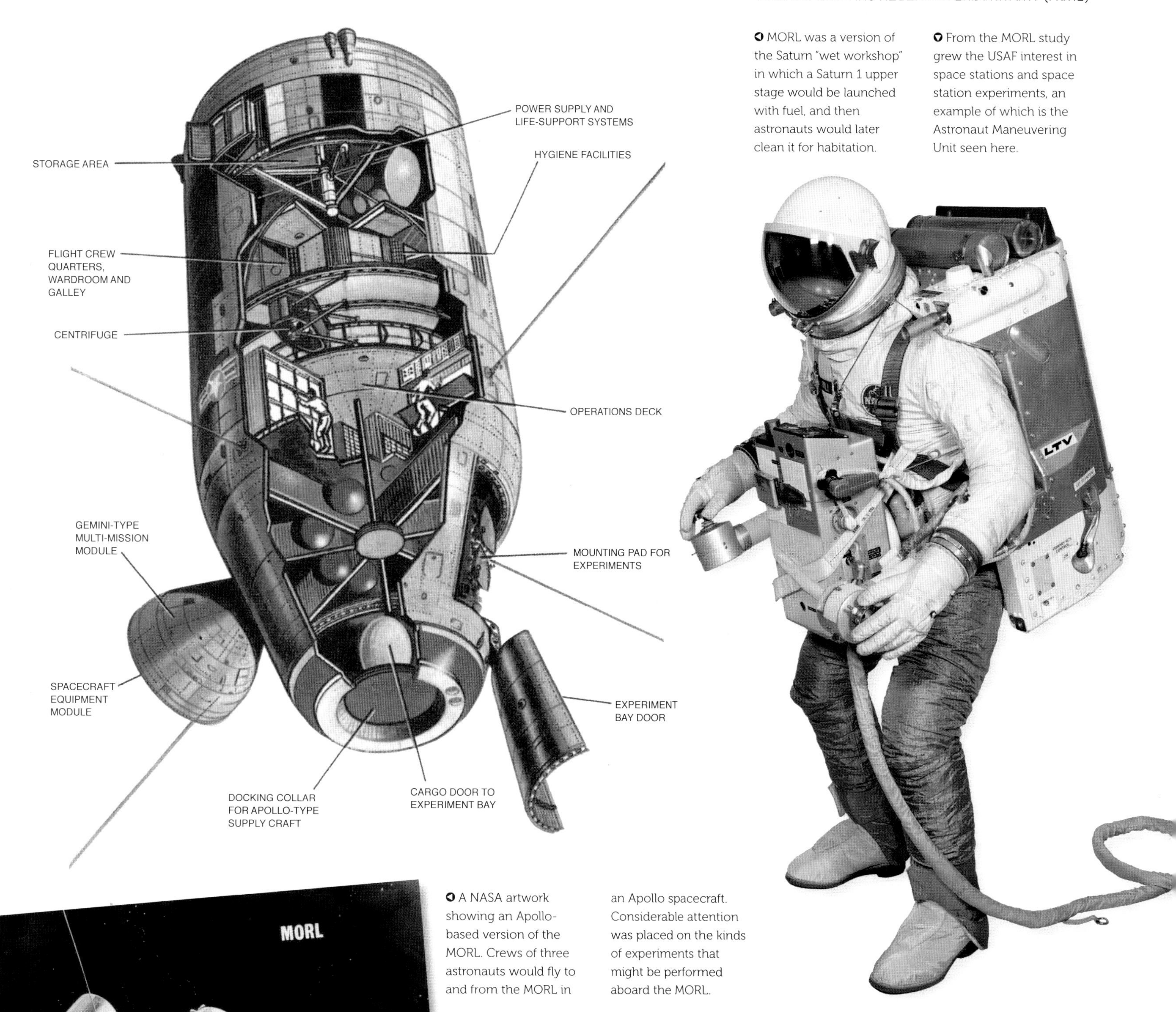

MORL was a version of the Saturn "wet workshop" in which a Saturn 1 upper stage would be launched with fuel, and then astronauts would later clean it for habitation.

From the MORL study grew the USAF interest in space stations and space station experiments, an example of which is the Astronaut Maneuvering Unit seen here.

A NASA artwork showing an Apollo-based version of the MORL. Crews of three astronauts would fly to and from the MORL in an Apollo spacecraft. Considerable attention was placed on the kinds of experiments that might be performed aboard the MORL.

dock to the station for operation by the crew. The crew would conduct EVAs to retrieve film from the telescope's cameras.

In 1965 a mission was planned that used a MORL module for a crewed Mars fly-by mission. The mission module would be unique for the Mars mission, and in total it was intended that six to eight crew would make the trip. The Mars vehicle would be launched by an advanced, lengthened Saturn V rocket. Many of the MORL concepts were used by the Apollo Applications and Skylab programs later in the 1960s and 1970s.

ALMAZ TKS

As the US MOL began in 1963, in the Soviet Union Premier Khrushchev wanted to keep close watch over the movements of US nuclear-powered aircraft carriers. A small space station called Almaz (Diamond) was proposed for the job. Almaz, like the US MOL, would be used for reconnaissance and radar imaging. Almaz was also an interceptor and would carry a cannon. A Proton rocket would launch the station; Proton was more powerful than the ICBM used to launch Vostok. Each Almaz station was designed to operate for up to two years. Originally, Almaz was to have an Orbital Piloted Station (OPS) and a crewed Transport Supply Ship (Russian abbreviation: TKS)—the latter looked like a US Apollo. Like Gemini, the crewed capsule had a hatch in its heatshield. After the first Soyuz spacecraft flew in 1968, the crewed capsule of the TKS was cancelled, however. Soyuz and Almaz were products of two competing design bureaus. Almaz was supposed to have been in operation in 1967, but by 1970 two spacecraft were in construction and still years away from flight.

The US won the Moon race in 1969 and during the same summer announced that the first US space station, Skylab, would orbit in 1973. Responding, the Soviets set a goal to launch a station before the US Skylab. They developed a plan to use Soyuz

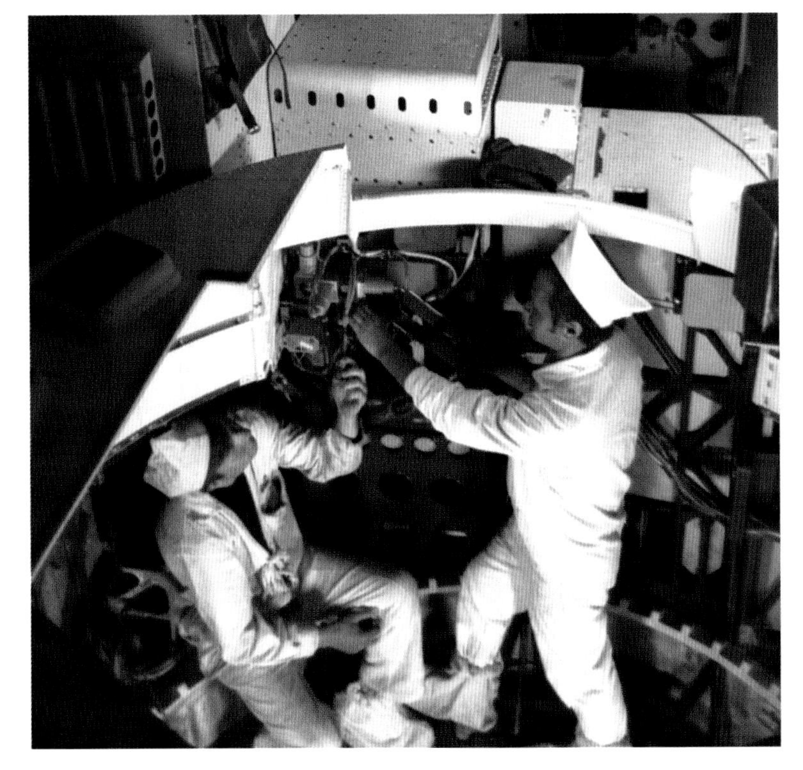

Russian technicians assemble an Almaz module. Almaz and DOS development began in the late 1950s and were to be used for space stations and planetary fly-bys.

The TKS Manned Transport Supply Ship, as originally planned for use in conjunction with the Russian space stations. Astronauts were to ride in the cone-shaped capsule.

Control station of the early Salyut station. The systems of each Salyut were upgraded and were fully computerized by the time the ISS Service Module flew in 2000.

This unique station was composed of parts of the Soyuz spacecraft plus components of the Almaz stations. It was the DOS, renamed Salyut 1, and first crewed when the Soyuz 11 crew visited in June 1971.

The first crew of a space station, namely Georgy Dobrovolsky, Vladislav Volkov, and Viktor Patsayev (front to back). They launched on Soyuz 11 and spent three weeks on the Salyut 1 space station. During the return to Earth, however, the air leaked from their Soyuz and all three died of suffocation.

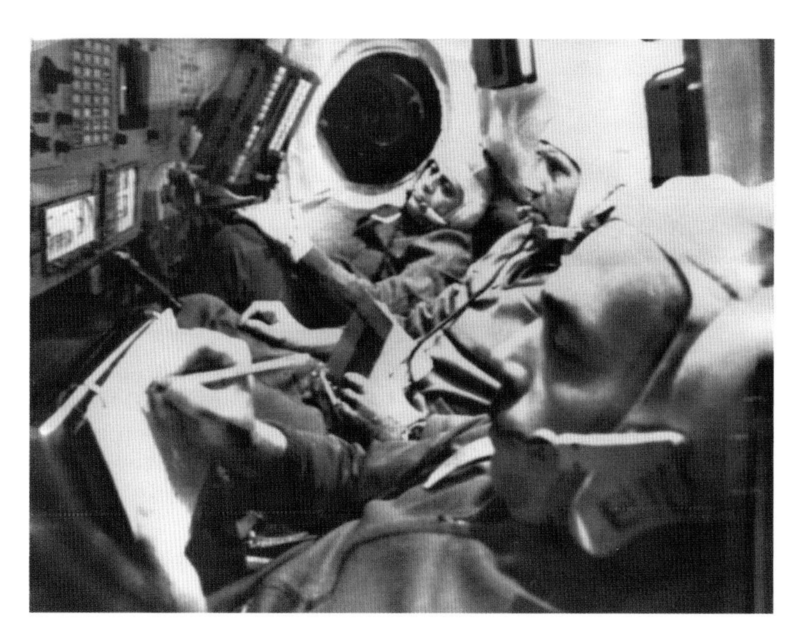

systems on an Almaz military space station shell within a year. The new station would be orbited unmanned, then a Soyuz 11 craft would ferry cosmonauts aboard. The Almaz with Soyuz systems was named the Long Duration Orbital Station (DOS). In orbit it was termed Salyut 1.

Salyut 1 launched April 19, 1971, and a crew launched on June 7 and lived on-board for 23 days. Each night they transmitted images of their work on-board and these were broadcast nationwide. On June 30, they entered the Soyuz for the return journey and separated from the station. The retrorocket in the Soyuz service module fired properly, but then as the service module separated from the reentry capsule, an air valve opened inadvertently. The crew wore no spacesuits. They could feel in their eardrums, and probably hear, the air escaping, but in the confined volume of the return capsule they could not reach the malfunctioning valve. One of the cosmonauts was near the valve so may have been trying to block the air loss, but was not able to do so. No communications were received from the spacecraft, but otherwise the reentry and parachute landing appeared to go well and the spacecraft landed normally without any apparent damage, but when the hatch was opened the three cosmonauts were found lifeless. Cardiopulmonary resuscitation was attempted, but without success—the cause of death was identified as suffocation. The very first stay of a crew on a space station had resulted in their fatalities; the only time a crew has ever died in space.

Later military Almaz missions flew under the names Salyut 3 and 5. The Soviets launched five Almaz (OPS), which resupplied Salyut 6 and 7. Later variants were used for the Quant II, Kristall (Crystal), Spektr (Spectrum), and Priroda (Nature) modules of the Mir (World) station, and for the Functional Cargo Block (FGB) of the ISS.

SALYUT (DOS)

After the first successful Apollo Moon landings, the Soviets developed a civilian Long Duration Orbital Station (DOS). The civilian DOS and military TKS/Almaz were products of competing design bureaus as they were being developed in the late 1960s. In order to beat the US Skylab schedule, the two bureaus began to work together to prepare Salyut 1.

1. Salyut 7 was launched April 19, 1982. It was a second-generation Salyut and could host visiting spacecraft at either end. A Soyuz is shown docked in this view.
2. Outside Salyut 7, Svetlana Savitskaya became the first woman to perform a spacewalk, on July 25, 1984. The EVA lasted for a total of three hours and 35 minutes.
3. Inside Salyut 6, Georgy Grechko tests an Orlan spacesuit in December, 1977. The test was to verify the operation of the spacesuit prior to the first spacewalk conducted outside a Salyut space station.

Korolyev died in 1966, and Vasily Mishin took his place in the Design Bureau responsible for Sputnik, Vostok, and Soyuz. Mishin was interested in a large space station to be called the Multirole Orbital Complex (Russian abbreviation: MOK). MOK would be a large-diameter space station powered by a nuclear energy source, and was to be involved in a variety of science and applications work, including astrophysics and mapping Earth resources. The station would also support military functions. A goal was to consolidate Earth-to-orbit logistics in order to save on costs and produce more results for less investment. MOK would be adaptable for multiple purposes. A crucial requirement to launch the large MOK station was the Saturn V class N-1 rocket.

TRAGEDY

The crew of Soyuz 11 was killed returning from 23 days on the Salyut 1 station in 1971. They were without spacesuits when their Soyuz spacecraft depressurized. In the wake of the tragedy, urgent direction was given to complete development and begin using the Sokol lightweight pressure suit inside Soyuz and then to orbit another DOS station before the American Skylab, in early 1973. A DOS station was to precede the Almaz military station. A second DOS station was launched a year later, July 29, 1972, but failed to reach orbit.

The DOS station was seen as a forerunner to the MOK, as DOS performed similar functions on a smaller scale. In 1974, after the last launch failure of an N-1 rocket, the N-1 and the MOK were both cancelled. Later DOS stations after Salyut 1 were upgraded with advanced orientation, navigation, and thermal control systems and with pointable solar arrays.

The later Salyut (DOS) modules were used as the core modules of the Salyut 6 and 7 (DOS- 5 and 6) and Mir (DOS-7) orbital stations and later as the Service Module (DOS-8) of the collaborative ISS.

3

1

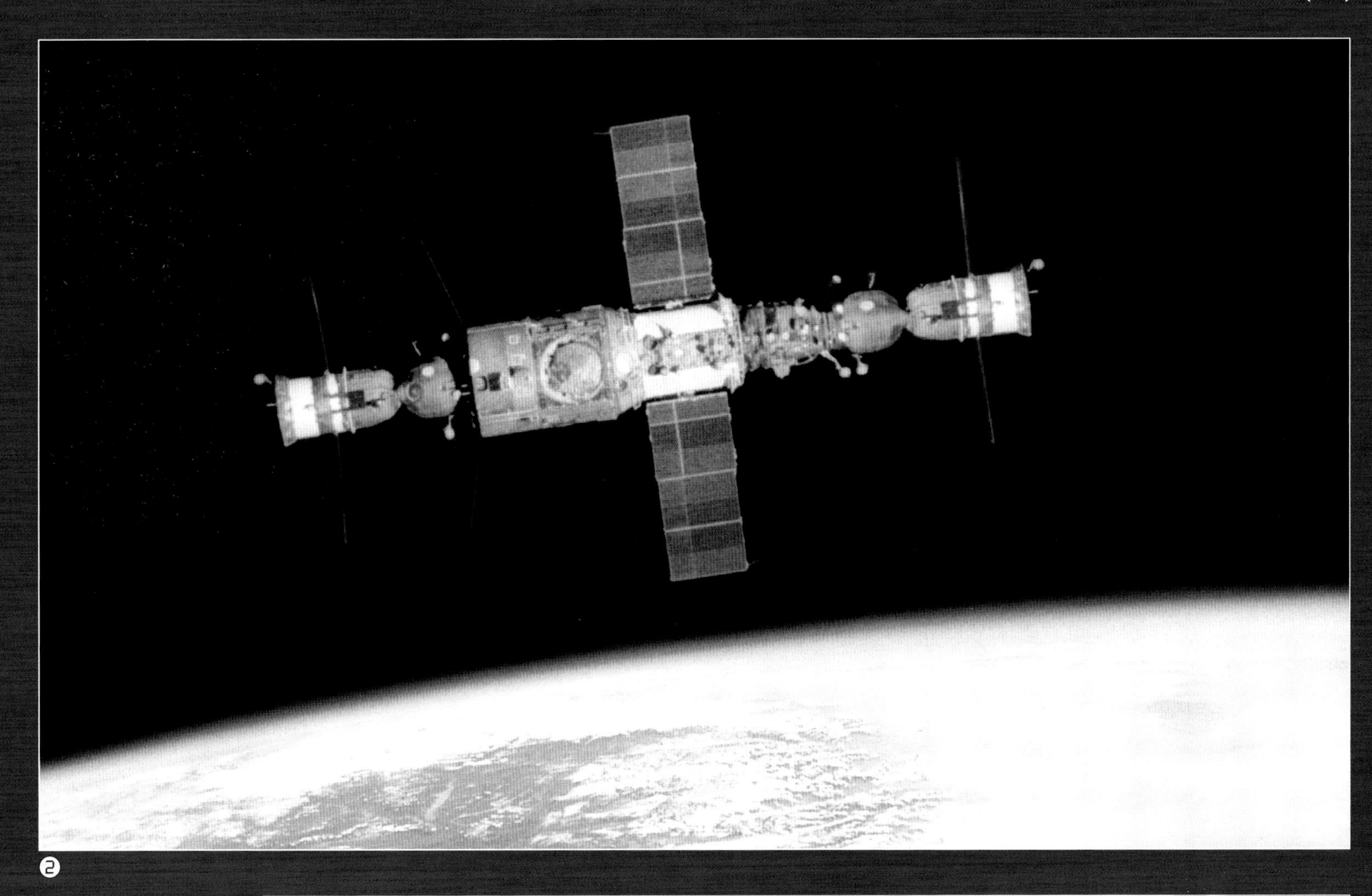

1. August 1978, at the galley table of Salyut 6, the first flight in space of a German citizen, Sigmund Jahn (left), is celebrated with the Soviet and East German flags on the table. Vladimir Kovalyonok (right) and Aleksandr Ivanchenkov hosted the visitng crew of Jahn and Valery Bykovsky.
2. Salyut 6 in orbit. This was the fourth Soviet station, launched in 1977. It hosted five long-duration crews and 11 short-duration. It was the first second-generation station with docking ports at both ends. Two Soyuz spacecraft are docked in this view.
3. On Salyut 6 (left to right): Georgy Grechko, Vladimir Dzhanibekov, and Oleg Makarov during a handover of the Soyuz spacecraft.

SKYLAB

In the Project Horizon study, Wernher von Braun outlined a plan to use Saturn rocket second stages, repurposed as living quarters for astronauts in Earth orbit. In 1960, the Ideal Home Exhibition in London built a space station mock-up based on the same rocket stage, and this became a model for Skylab. Skylab was the first NASA space station to reach orbit, in 1973, a ground-breaking feat of engineering and vision.

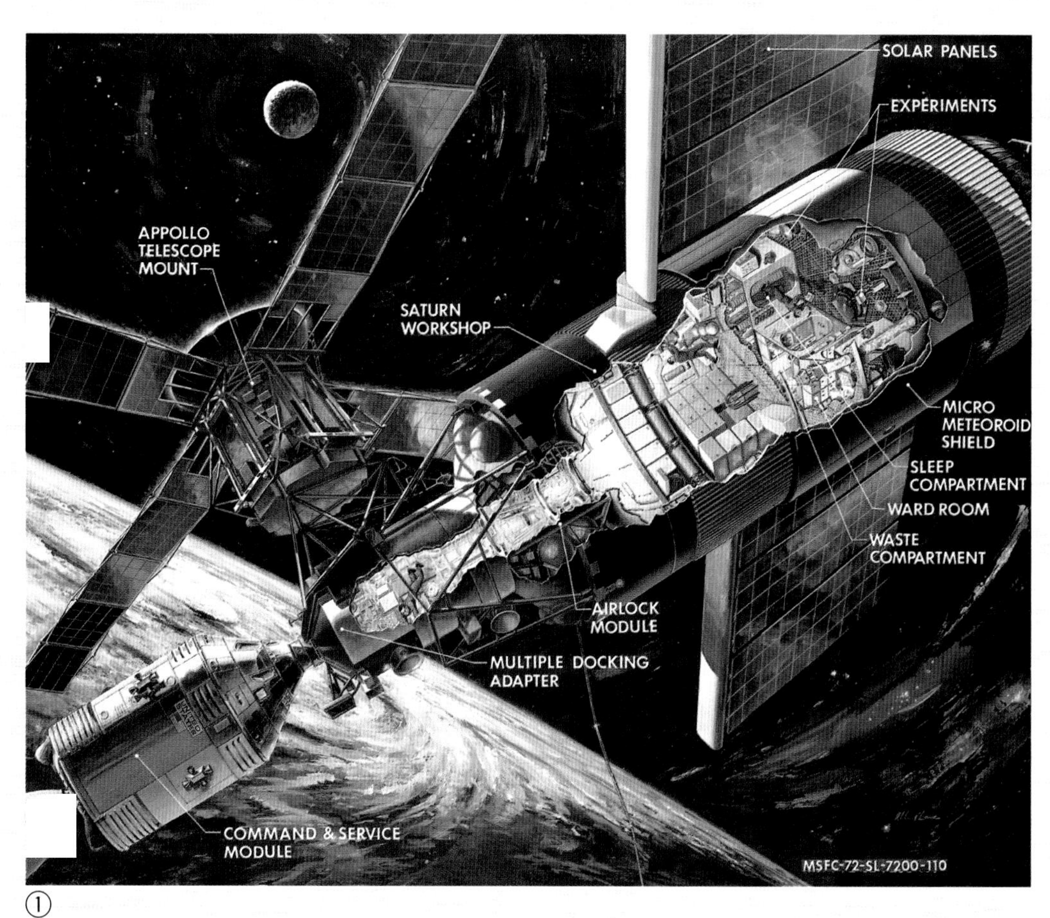

①

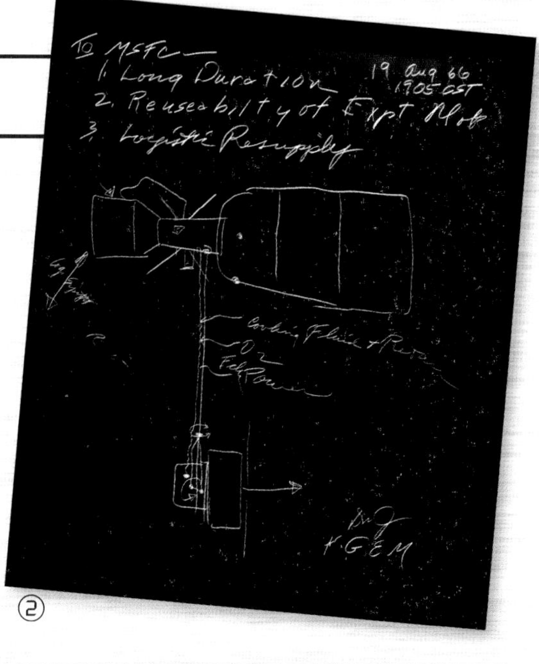

②

"SKYLAB WAS CALLED A WORKSHOP, BUT INDEED IT WAS A SPACE STATION, A TRUE SPACE STATION. A LOT OF PEOPLE WANTED TO EXTEND SKYLAB, BUT I THINK THAT WAS LUDICROUS. WE DESIGNED THE PROGRAM TO ACCOMPLISH CERTAIN EXPERIMENTS, AND WE DID THAT AND THEN IT WAS OVER."

KENNETH "KENNY" KLEINKNECHT
NASA PROGRAM MANAGER

In 1966, NASA began studying the use of Apollo hardware for programs that would come after the Moon missions. One proposal was to test von Braun's idea of using a rocket stage emptied of fuel as a habitat; the astronauts would enter the spent stage in spacesuits. First they would assess the condition of the inside of the stage. Then they would clean the interior and deploy equipment to convert the rocket stage to living quarters. This was called the "wet workshop."

To refine the process, NASA Manned Spaceflight chief George Mueller and von Braun practiced the spacewalk in a Marshall Space Center water tank in 1968. Mueller determined it was too difficult and risky to rely on the astronauts' ability to clean the stage in order to proceed with an entire space program.

When several Moon-landing missions were cancelled, one of the Saturn V rockets was reserved for the orbital workshop project. Using the Saturn V Moon rocket instead of the smaller Saturn booster meant that a two-stage Saturn V rocket would launch a third stage that had already been converted to a fully equipped space station before launch. In 1969 the program was formally approved and was given the name Skylab.

LAUNCH PROBLEMS

NASA sponsored several science teams, who developed sophisticated instrument packages that would study Earth, the Sun, and the effects of the astronauts' exposure to the periods of weightlessness. For the Earth observation program, aircraft and ground crews performed comparative analyses verifying conditions of crops and plant life, while the aircraft and Skylab would take multispectral images above.

Skylab was launched May 14, 1973, by a Saturn V rocket. This would be the last launch of a Saturn V and its missions was to orbit the Skylab space station. The entire Skylab program rested on the success of this launch. About one minute after the lift-off, the rocket was traveling through the speed of sound and experiencing the maximum amount of atmospheric turbulence. At this point telemetry signals went awry,

③

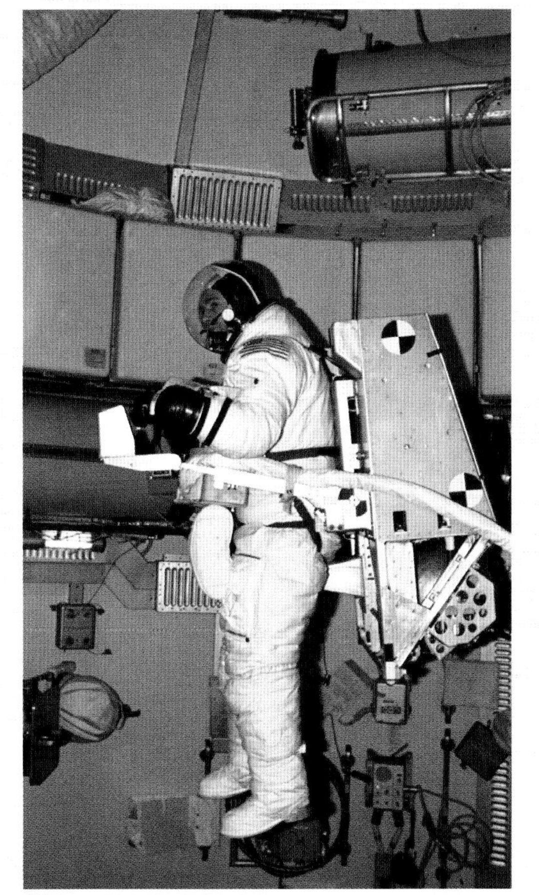

④

⑤

1. A cutaway drawing of the Skylab as it was to appear in orbit in 1973 and in 1974.
2. The idea of using spent rocket stages as a habitat went back to at least the early 1950s. This early sketch shows how the station could be used for extended missions, though NASA was careful not to call it a space station.
3. On the last Skylab crew, Jerry Carr balances Bill Pogue on one finger. Carr and Pogue would serve as primary consultants for the later design of the ISS.
4. On the second Skylab crew, Alan Bean test "flies" a gas-powered maneuvering unit around the interior of the orbital workshop.
5. During launch on May 14, 1973, a thermal and meteorite shield was ripped from Skylab. Here two seamstresses prepare a replacement shield, which was launched with the first crew on May 25.

indicating that a thermal and meteorite shield and two solar panels, all mounted on the station, had deployed. This was clearly anomalous, but during powered flight there was nothing that could be done.

The rocket continued flying, accelerating and gaining altitude. Ten minutes after leaving the launch pad, the rocket reached 28,000 kmh (17,500 mph) and 430 km (270 mi) altitude, and released the station into orbital flight. During the release, the second stage, in order to pull away, fired small retrorockets. One of the retrorockets hit one of the station's partially deployed solar arrays, blowing it off. At this point, several deployments were to take place. A large telescope was to rotate 90 degrees—this proceeded as planned. The telescope also carried four large solar panels, which unfolded. Then the two main solar panels mounted on the station were to open; one of these was that just blown away by the impinging retrorocket and the other was jammed against the station's side, pinned by a metal strap—some of the debris from the meteoroid shield. Little electricity was being generated. The meteoroid shield was also to serve to reflect sunlight and cool the station. With the shield gone, temperatures in the station began to rise, reaching 165°C (325°F). More than 50 telemetry parameters were giving anomalous readings. After eight hours it became obvious the station was in serious trouble, and the launch of the first Skylab crew, planned for

①

1. Launch of the last Saturn V and the first US space station, Skylab, on May 14, 1973. NASA has hoped that the Saturn could be used for more station launches, but costs were prohibitive and production of Saturns was curtailed in 1966.
2. On the first crew, Pete Conrad exercises on a bicycle ergometer. On this, his fourth spaceflight, Conrad stayed for 28 days.
3. Owen Garriott, scientist on the second crew, "sits" at the Apollo Telescope Mount control panel. No seats were needed, the astronauts reported, and they removed them. The astronauts particularly enjoyed their duty at the telescope, where they would watch the Sun to try to capture the start and evolution of solar disturbances.
4. The "dinner table" had heated receptacles to hold food and drinks. Many of the foods were vacuum sealed and canned.
5. On the first crew, P. J. Weitz takes a shower. The collapsible shower was easy to use, but it was difficult to clean and maintain. Astronauts recommended only sponge baths for future weightless missions.

the following day, was postponed indefinitely until a plan could be worked out the save the space station.

The solar arrays on Skylab were not designed to point at the Sun itself. Orienting the station to gain maximum solar energy also placed the station in an orientation that increased temperatures to dangerous levels. By placing the station at an oblique angle, therefore, temperatures could be reduced, but as a consequence power generation dropped. The central problem became one of balancing temperature and power generation. Maintaining the accurate pointing caused further problems by using orientation propellant at a faster than planned rate.

Aside from the anomalous readings and the necessity of balancing temperature against power, telemetry signals did not show exactly what had gone wrong. On the second day after launch, a US spy satellite photographed the station and was able to reveal the extent of damage more clearly.

For the next five days, teams of engineers, designers, and seamstresses worked around the clock designing new solar shades for protecting against the Sun's heat and cutting tools for cutting the jammed solar array free. New food and supplies were prepared to replace the items spoiled because of high temperatures. Repair procedures were developed. Eleven days after the launch of the Station, Skylab 2, an Apollo spacecraft, carried the supplies and repair materials to orbit with its three-astronaut crew.

HUMAN INTERVENTION

One of the first things the crew did after entering the Skylab was to deploy an umbrella-like sunshade out of a small science-experiment airlock on the side facing the Sun. Within minutes of the shade's opening, temperatures started coming down so that the astronauts could remain inside and perform useful work. Two weeks later, astronauts Conrad and Kerwin performed a spacewalk trying to unjam the still jammed solar array. It required one of the astronauts holding a cable and a second astronaut "walking" on the hull of the Skylab,

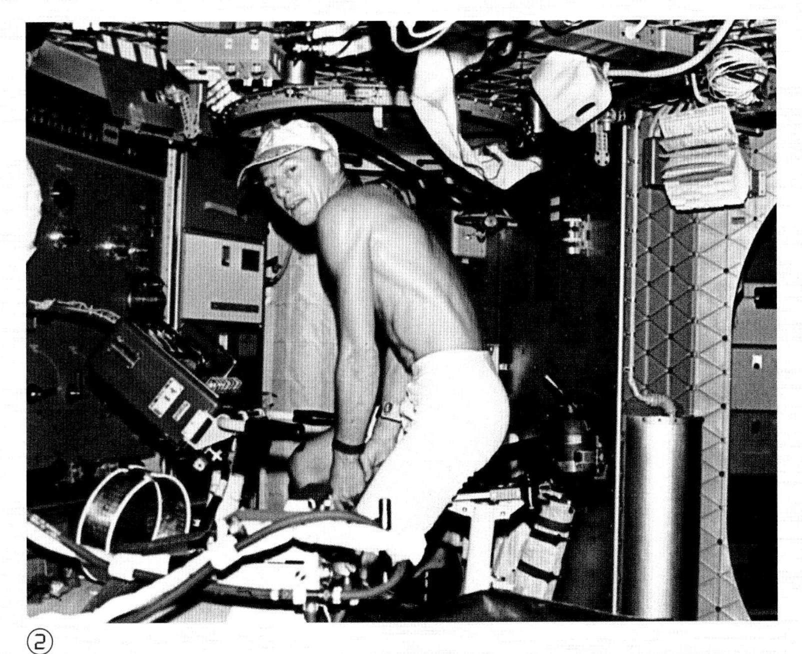

②

③

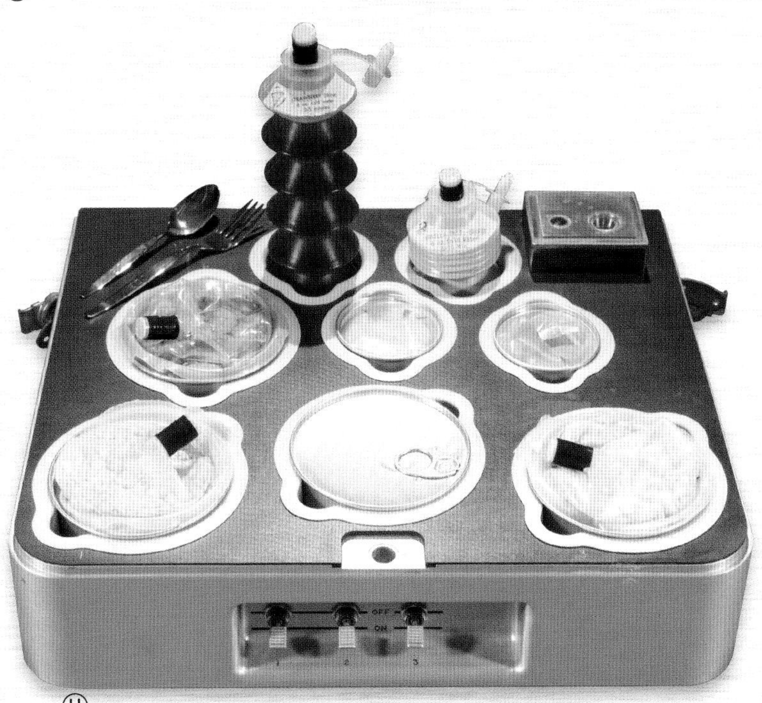

④

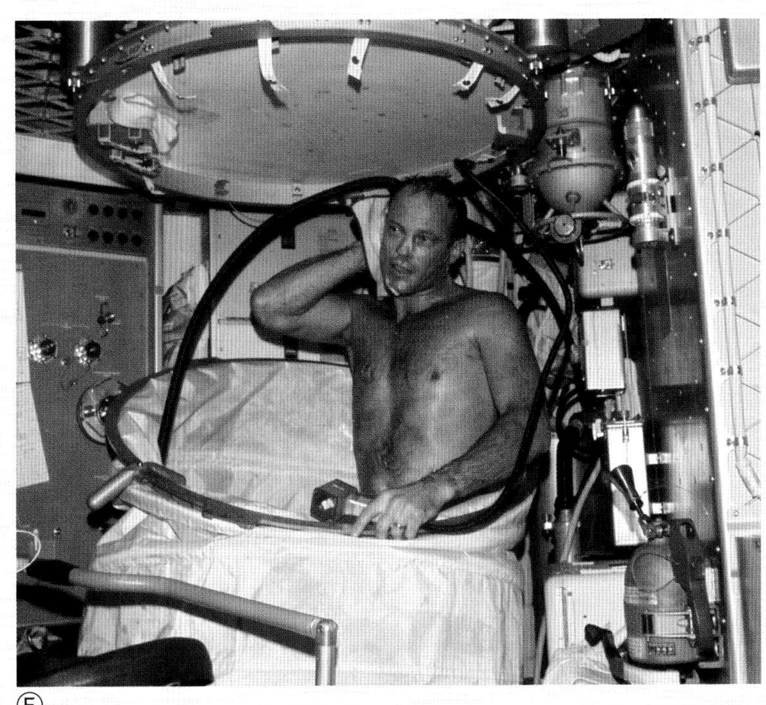

⑤

using a tree-pruner to cut the restraining strap. When the strap suddenly came free, the array popped open and the astronauts were catapulted, essentially like arrows, off the Skylab. Fortunately the restraint cables were secure and the astronauts were able to compete the spacewalk and then the rest of their mission. The freed solar array meant that much of the required power for the additional two planned Skylab missions could be carried out.

Each of the three crewed Skylab missions doubled the duration of the previous longest US mission durations. Medical investigations were performed in-flight and post-flight to trace each astronaut's health. Biology experiments studied a variety of living plants and animals including mice, spiders, and fruits flies. High-school students from across the US competed to have their experiments flown and performed on the Skylab. The studies also resulted in some of the best images of the Sun ever recorded. Earth observation recorded crop growth and seasonal changes and were compared with observations made on the ground and by aircraft flying similar types of instrumentation.

The final crew remained on board longer than planned in order to photograph a comet, *Kohoutek*, discovered during the course of the Skylab program. The last mission was extended to 84 days from the planned 56. By the end of their mission they had run out of food and were rationing granola bars.

"THERE WAS A CHALLENGE OF BEING UP THERE FOR A LONG PERIOD OF TIME, TO SEE IF YOU CAN ACTUALLY STAY UP THERE AND DO USEFUL WORK."

JACK LOUSMA, SKYLAB ASTRONAUT

①

"WHEN I LOOK AT WHAT WE'RE DOING ON THE ISS AND WHAT WE DID ON SKYLAB, WE ARE LEARNING HOW TO USE THE NEW ENVIRONMENT WITH EXPERTISE. WE ARE EXTENDING THE FRONTIER. WE ARE EXTENDING THE BOUNDARIES. . ."

ED GIBSON, SKYLAB ASTRONAUT

1. On the second crew, Owen Garriott spacewalks to the solar telescope to recover cassettes of exposed photographic film. 2. View of the fore end of the Skylab taken by an approaching crew. The black circle at lower right is their docking port. The solar telescope is to the upper right, with four large solar arrays marking an "X." The one remaining main solar array is to the left. An identical array that would have been on the right was ripped away during launch. 3. The last Skylab spacewalk, February 3, 1974. Scientist Ed Gibson emerges from an airlock. 4. A view of the top side of Skylab taken by the departing final crew. The main workshop is protected by an improvised golden solar thermal shield.

2

3

4

SATURN-LAUNCHED LARGE-DIAMETER STATIONS

The Manned Orbiting Research Laboratory (MORL) was another concept developed at the time of Skylab. It had a mass of 13,636 kg (30,000 lb), while the Apollo Applications (later Skylab) Orbital Workshop was 34,090 kg (75,000 lb), and they were both 6.7 m (22 ft) in diameter. But a Saturn V could launch a payload 10 m (33 ft) in diameter and 118,181 kg (260,000 lb) in weight into Earth orbit, so by combining several of these modules a large space station could be assembled. NASA called this the Large Orbiting Research Laboratory (LORL). The NASA Administrator, James Webb, said in 1965 that it was time to begin planning the next step after Apollo, and that this would be just such a large space station boosted into orbit by a Saturn V rocket. Lockheed did a study saying that it made sense to proceed with this space station because of the resources already in place for Mercury, Gemini, and Apollo, which would reduce development time, cost, and technical risk.

In 1969, NASA released contracts to study the design of space stations they hoped to pursue as a follow-on to the Apollo Moon missions. Several contractors were enlisted. McDonnell Douglas proposed a station in which two large 10 m (33 ft) diameter modules would be connected by a tunnel.

A large 10 m (33 ft) diameter multi-floor core module. The Saturn rocket had a 10 m (33 ft) diameter, so this would be near the maximum size of module the Saturn V would be able to launch. Ports would be provided near either end to attach additional smaller modules.

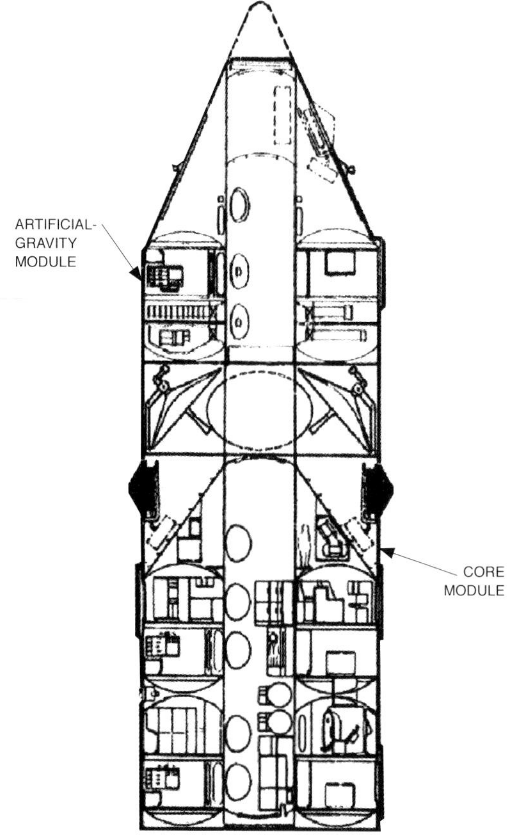

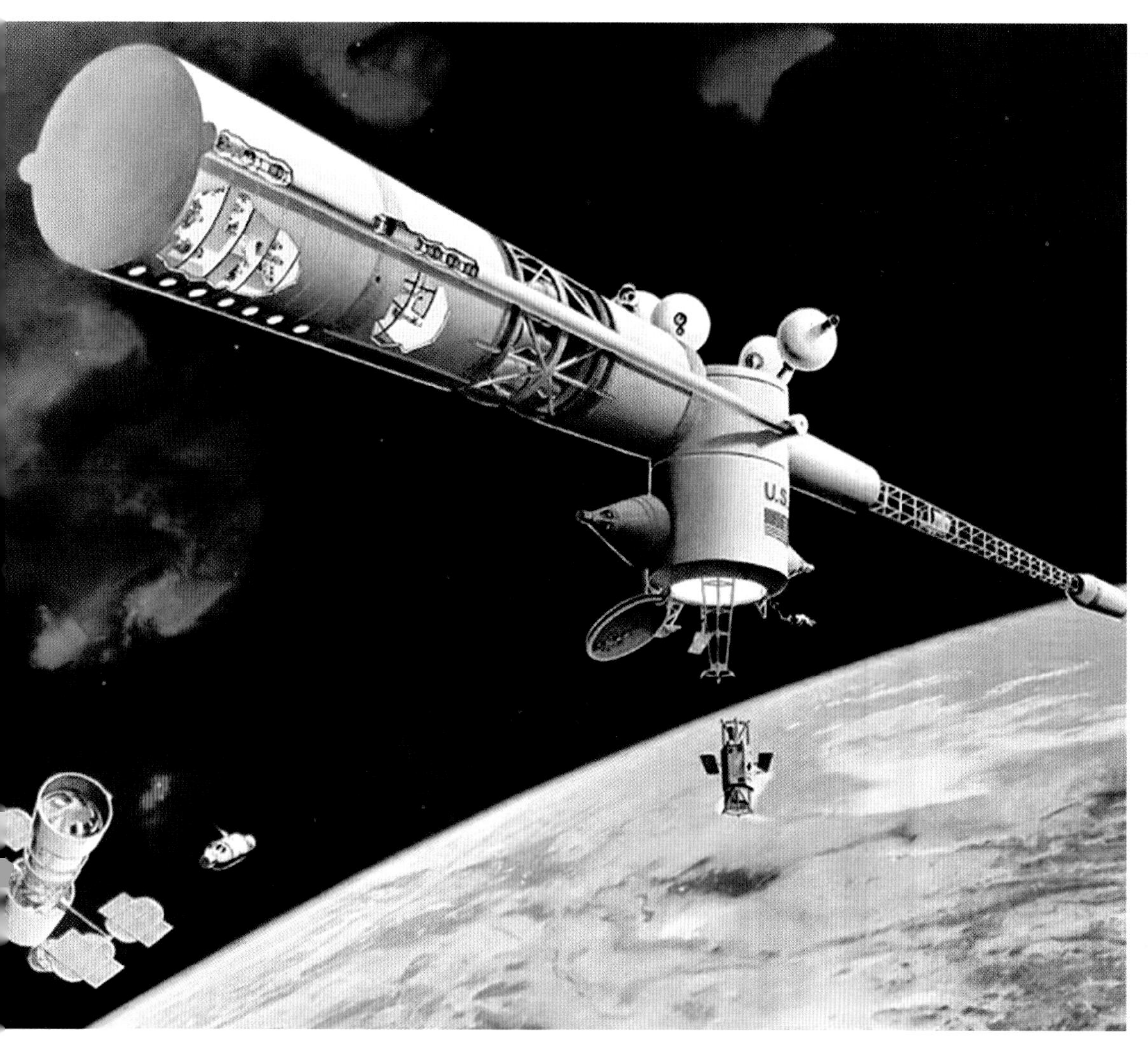

An illustration used by the Manned Spacecraft Center Director, Robert Gilruth, in 1968, as discussions first began about assembling a large space station in orbit, with some modules that would be boosted by Saturn rockets. This station would rotate around a central core module so artificial gravity would be produced towards either end. Crew and logistics vehicles would enter the center core. To the left is a depiction of the earlier Manned Orbiting Research Laboratory.

A technical study identified key parameters. A two-stage Saturn V would launch a 4.9 m (16 ft) or 6.7 m (22 ft) diameter core module. Other reports identified a 10 m (33 ft) diameter core module. Docking ports would allow the addition of more modules. The station would be launched unmanned, but could accommodate a crew of 24, who would stay between three months and a year, with resupply every 90 days. Six-man Apollo, or 12-passenger ballistic or winged transports, would rotate the crews. Cabin pressure could be varied in different modules between 3.5 to 14.7 psia (pounds per square inch absolute). Yet only a year later, in 1966, funding for additional Saturn rockets was not provided. Funding was also requested in 1967, but not provided. A total of 15 Saturn V rockets were built: 12 were used for test and lunar landing mission, one was used to launch Skylab, while two more remain as evocative museum pieces in Florida and Texas.

While the large central core of the space station would be launched by the Saturn V rocket, smaller modules, for specialized research or for logistics, could be launched by smaller booster rockets and berthed to ports on the core.

SHUTTLE-LAUNCHED MODULAR STATIONS

Now that funding for Saturn rocket production was terminated, before a Saturn rocket ever sent anyone to the Moon, there was no way to launch large-diameter space stations. NASA's last hope for Apollo and Saturn production was the new presidential administration that took office in 1969, but the White House supported terminating the program after the first few Moon landings. President Richard Nixon was clear that he did not favor terminating US human space flight, but he would only support a single new program.

Even before the Apollo Moon program had begun, a space station was determined to be the next logical step in human spaceflight. George Mueller, NASA's Associate Administrator for Manned Spaceflight, reiterated that the nation and the agency need a space station, and it needed a space shuttle to ferry the crews and supplies for the station. Without a heavy-lift rocket like the Saturns, a key factor in any shuttle would be a size adequate to transport modules in which humans could live and work.

But without a space shuttle, there was no way to launch modules of a space station or the astronauts who would be the crew. With only one choice to make, NASA recommended that the United States pursue development of a space shuttle large

Discussion about a space shuttle began in the late 1960s. The shuttle would become NASA's only large launch vehicle and would need to be able to carry astronauts and space station modules.

As ideas about a shuttle design became more hardened, the size of the shuttle and its payload bay were determined by the size of modules and spacecraft it would be required to carry. These designs became known as the Modular Space Station or MSS.

Design of the shuttle began in 1969 and the size and lift capacity of the shuttle would have implications for the assembly process and design of a station, as is evident here. A Vice Presidential committee recommended NASA proceed with a shuttle, a space station, and a Mars mission, all simultaneously.

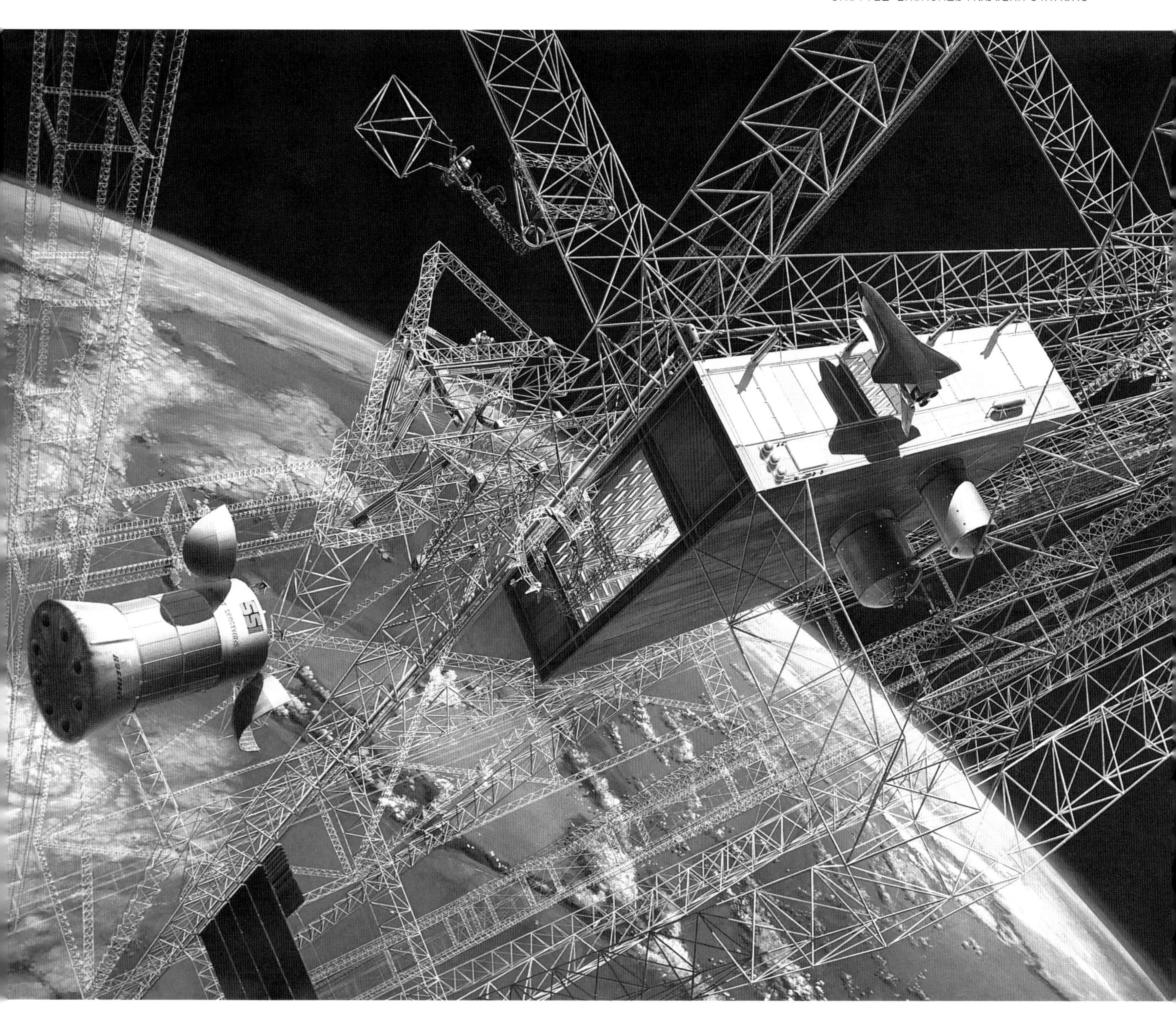

Space aficionados in the 1970s had grand plans for large space structures. These would be used for solar power satellites, bases at which to construct Mars vehicles, and even space colonies. NASA felt this could be done with an efficient transportation system for carrying payloads into orbit.

enough to bring up station modules later. Once development of the shuttle was complete, they could try again to sell a space station to a new administration. The shuttle as a launcher dictated that future space station elements would need to be shuttle transportable. NASA identified a minimum payload bay diameter of 3.6 m (12 ft) or a more optimal diameter of 4.5 m (15 ft) to accommodate modules that would be large enough for people to live and work inside.

SPACELAB

The space shuttle was approved in 1971, but the space station was deferred. NASA was concerned that they now had a transportation system, but the system for supporting experiments requiring extensive human interaction was not available. Almost immediately NASA's Marshall Spaceflight Center (MSFC) studied a pressurized module to be carried in the shuttle's payload bay. The European space community expressed interest in providing the module, now called Spacelab. NASA and the European Space Research Organization (ESRO) signed agreements in 1973, and the first astronauts for the missions were selected in 1978. Spacelab served as an intermittent space station for research conducted by scientists and astronauts in the 1980s and 1990s.

The Spacelab module was connected by a tunnel to the shuttle crew cabin. Interior racks housed subsystems, supplies, and experiments that varied from mission to mission. Unpressurized platforms could be carried behind the module in the payload bay. Crews worked around the clock in two shifts, with dozens of

With the US decision to build a shuttle and then a space station, NASA studied the idea of a science laboratory that could be flown inside the shuttle. The European Space Agency (ESA) partnered with NASA to build the Spacelab.

Spacelab-J was sponsored by Japan. Here Jan Davis conducts science experiments.

Spacelab missions were often leaned toward the physical sciences. Life science payloads often had equipment mounted in the modules's central corridor.

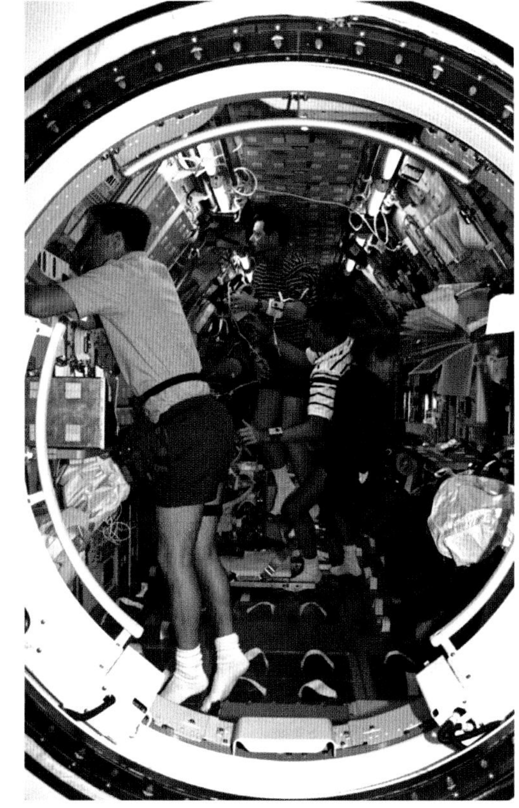

Spacelab's rear segment, seen here, was a place for conducting experiments. Equipment was mounted in standardized equipment racks.

The shuttle's payload bay was optimized to carry habitable modules. The Spacelab provided an opportune time to learn to design and build many of the elements of a station module.

In this photograph, shuttle STS-71 carries a Spacelab module in its payload bay and is just departing the Russian Mir orbital station on July 4, 1995.

scientific investigations performed during missions lasting ten to 16 days. Much of the research was performed by scientists from many nations; missions were often dedicated to one science discipline or another. Many of the researchers were selected to fly on the missions to perform their own experiments. Astronauts conducted research in life and space sciences, Earth observations, fluid and material sciences, astronomy, and atmospheric physics. Many of these functions were precursors to later experiments and research on space stations. Spacelab was used to support the Russian Mir Orbital Station on the first Shuttle–Mir mission in 1996.

Spacelab was a versatile and effective extra-terrestrial laboratory that established many of the operating principles used later on the ISS. In total, the Spacelab module itself flew on 16 missions, while other Spacelab components flew on a total of 32 missions.

MANNED ORBITING FACILITY

While it developed the space shuttle in the mid-1970s, NASA carried out a number of space station studies in the expectation that once shuttle development was completed direction would be given to start on a space station. NASA's MSFC worked with McDonnell-Douglas to develop the Manned Orbiting Facility (MOF) concept starting in 1974.

The MSFC approach was evolutionary, beginning with existing systems and expanded upon their capabilities. NASA based the design of the MOF on the Spacelab module and unmanned platforms then in development, for use as a sortie laboratory on the shuttle. It would start out as a man-tended facility, launched into orbit by the shuttle. It had its own power supply and support systems and the shuttle would leave it in orbit. Experiments could then be conducted even though the MOF operated unmanned, through pre-programming or by remote control from the ground.

NASA worked closely with the ESA developing Spacelab. They proposed the free-flying science platforms based on the Spacelab modules and pallets.

Over time, engineers could add modules to the space station platforms. Humans could live and work in these modules while the shuttles remained attached.

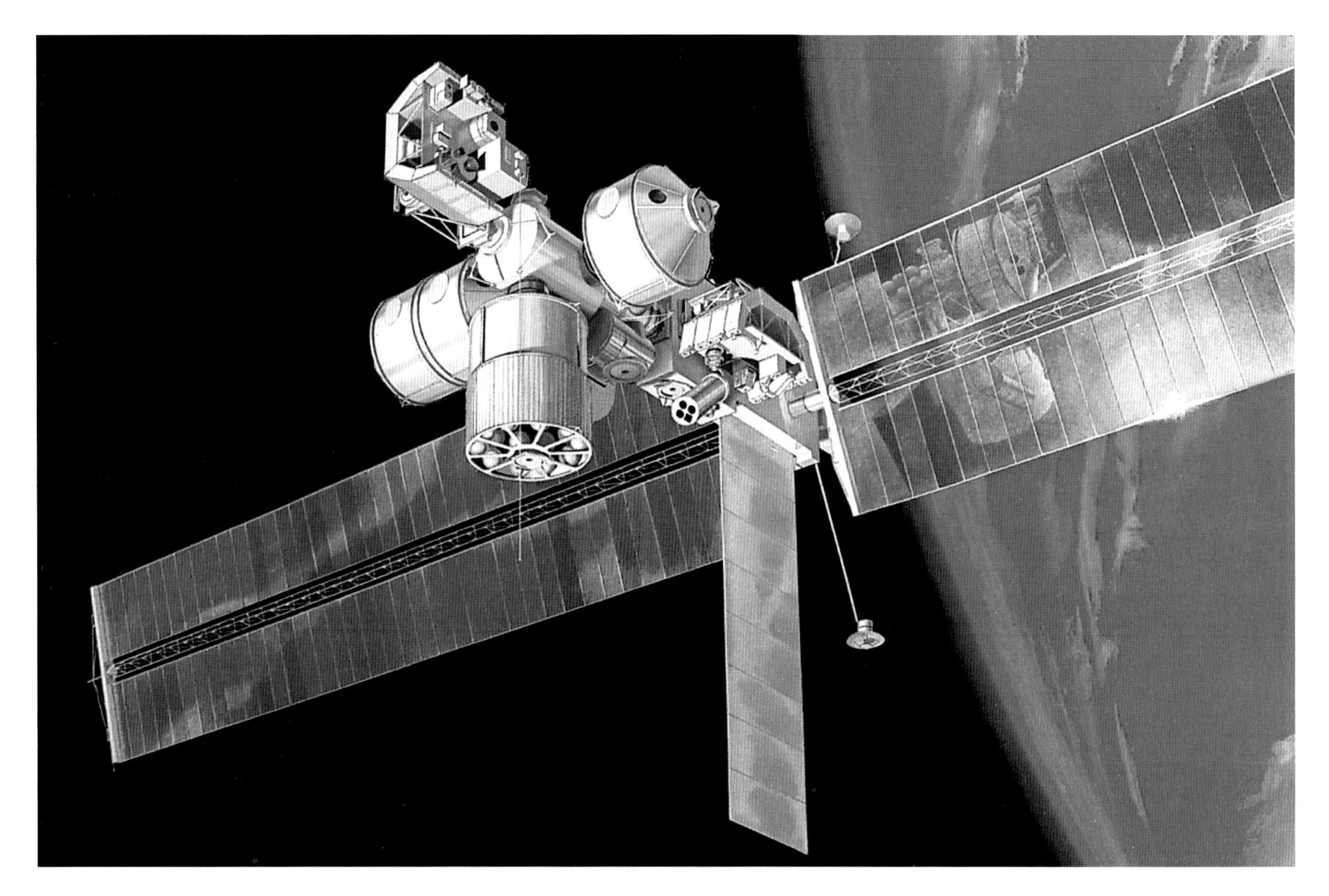

Eventually Spacelab-style modules and environmental systems could be added to allow three or four astronauts to stay on-board even when the shuttle departed.

Initially the MOF facility would be tended by visiting shuttles. The astronauts could maintain the systems in orbit and provide raw materials and supplies.

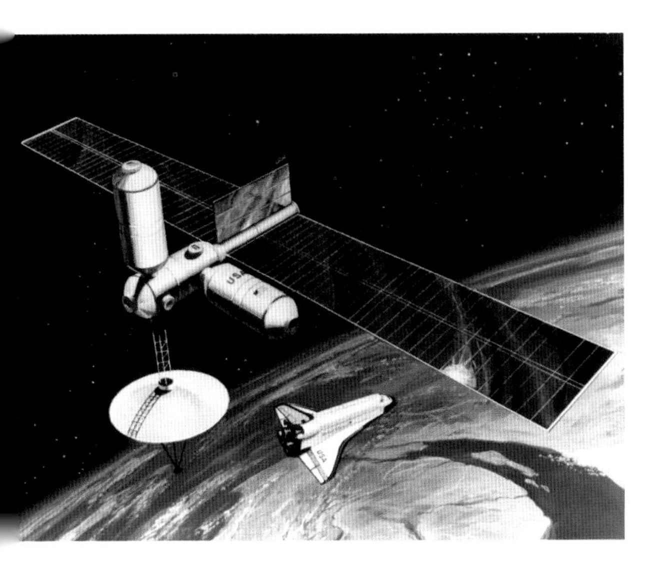

The shuttle would visit periodically and during these visits the onboard researchers would conduct experiments, plus gather the products of experiments for return and transfer new raw materials for additional experiments. The MOF would also augment shuttle capabilities and allow the shuttle to remain in orbit for 30-day missions. Eventually systems could be added so that a four-man crew was left on board between shuttle visits. Once operating with a permanent crew, the system had four units: a subsystem module, a habitability module, a logistics module, and a payload module. The modular design meant NASA could add additional modules to expand the station.

Once given a go-ahead, the MOF could have been been ready for launch within about ten years. Scientific investigations to be conducted from MOF would be an extension of research then being planned for shuttle–Spacelab missions: space manufacturing, advanced technology, life science/technology, Earth observations, and a host of astronomy and physics experiments: IR astronomy, UV astronomy, solar observations, magnetospheric and plasma physics, high-energy astronomy, and cosmic rays.

SPACE OPERATIONS CENTER 1979–82

In the late 1970s, there was considerable discussion about building large structures in space. Space colonies would permit hundreds to live beyond the bounds of our own planet. Large solar-powered satellites would beam, by microwave, energy to the Earth. Extensive operations with unmanned communications satellites had been underway for the last decade and so a facility for collecting and refurbishing satellites was considered as a possible requirement. There was also hope that once a space station was established, the Space Operations Center (SOC) would support the next program, which would be the assembly of a large spacecraft to carry people on a planetary mission.

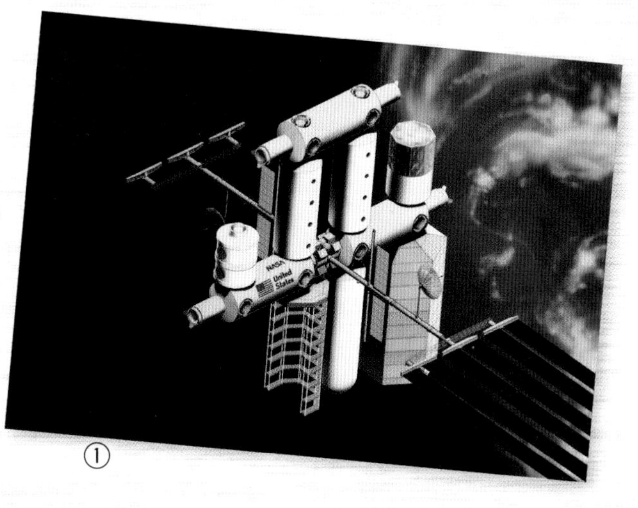

①

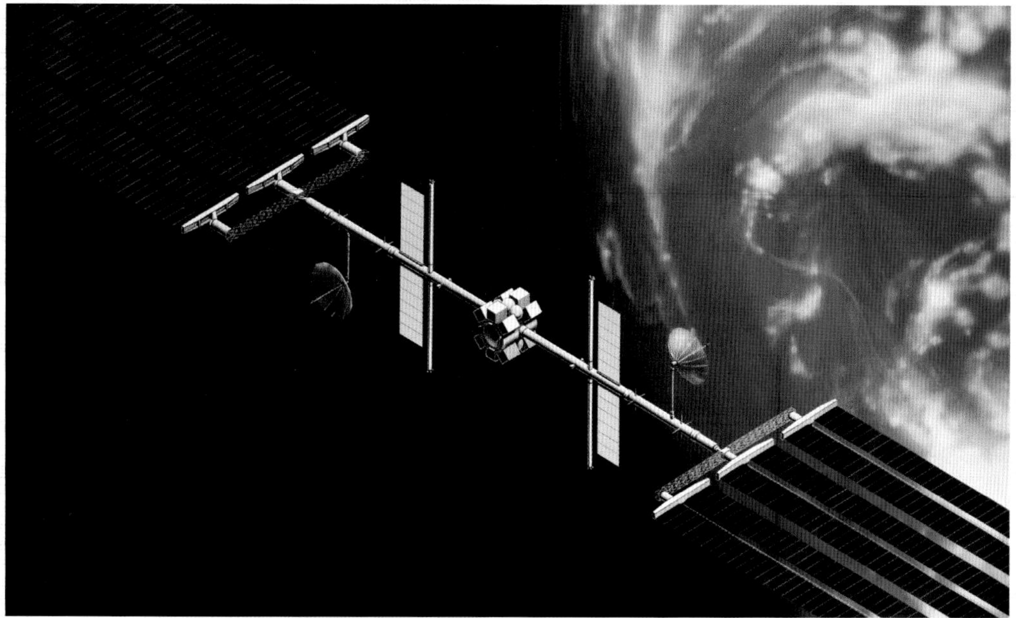

②

"I DON'T THINK THE HUMAN RACE WILL SURVIVE THE NEXT THOUSAND YEARS, UNLESS WE SPREAD INTO SPACE. THERE ARE TOO MANY ACCIDENTS THAT CAN BEFALL LIFE ON A SINGLE PLANET. BUT I'M AN OPTIMIST. WE WILL REACH OUT TO THE STARS."

STEPHEN HAWKING

The NASA Johnson Space Center (JSC) studied the SOC concept, beginning in 1979 and continuing through the early 1980s. The identified multiple applications for SOC: it was intended primarily as a facility for handling and maintaining satellites and assembling large, complex space vehicles. With suitable laboratories it could be used for scientific research. SOC was intended for a low-Earth orbit. Modules would be launched by the space shuttle, and shuttles would supply new systems and logistics. The mass of each payload brought to assemble or resupply the stations would be between 15,909 and 20,454 kg (35,000 and 45,000 lb).

EXPANDING CREW AND FACILITIES

SOC was intended to be evolutionary, growing in incremental stages. It would have an initial crew of two or four, but with more demand, power, life-support and supplies, and habitation module facilities, SOC could house up to 50. Each module would have a primary purpose, such as habitation, maintenance, research, or systems support. The first module launched would deploy booms, which would unfold to carry solar arrays for power, radiators to eliminate waste heat, an orientation system for attitude control, and a computer. Next would be a command-and-control module to complete a full initial set of systems: power, communications, command, control, stabilization, life support, and a data-management system for independent unmanned automated operations. When the shuttle was present, the crew would operate systems and conduct experiments and other functions from this module. Initially EVAs, or spacewalk, operations would be supported out of visiting shuttles, but an airlock with a hyperbaric chamber would be added during this first phase. Once completed, this initial phase would house a crew normally of two, but could accommodate four if required by the mission parameters.

During the second phase, a remote manipulator system, similar to that built by Canada for the shuttle, would be added for assembling platforms, and would later be used for maintaining and assembling satellites and large spacecraft. Another launch would bring up a second command-and-control module with duplicate computer systems. During a third phase, multiple habitation modules—connected by tunnels and with multiple docking modules to allow visiting shuttle and other spacecraft to attach—would provide completely redundant

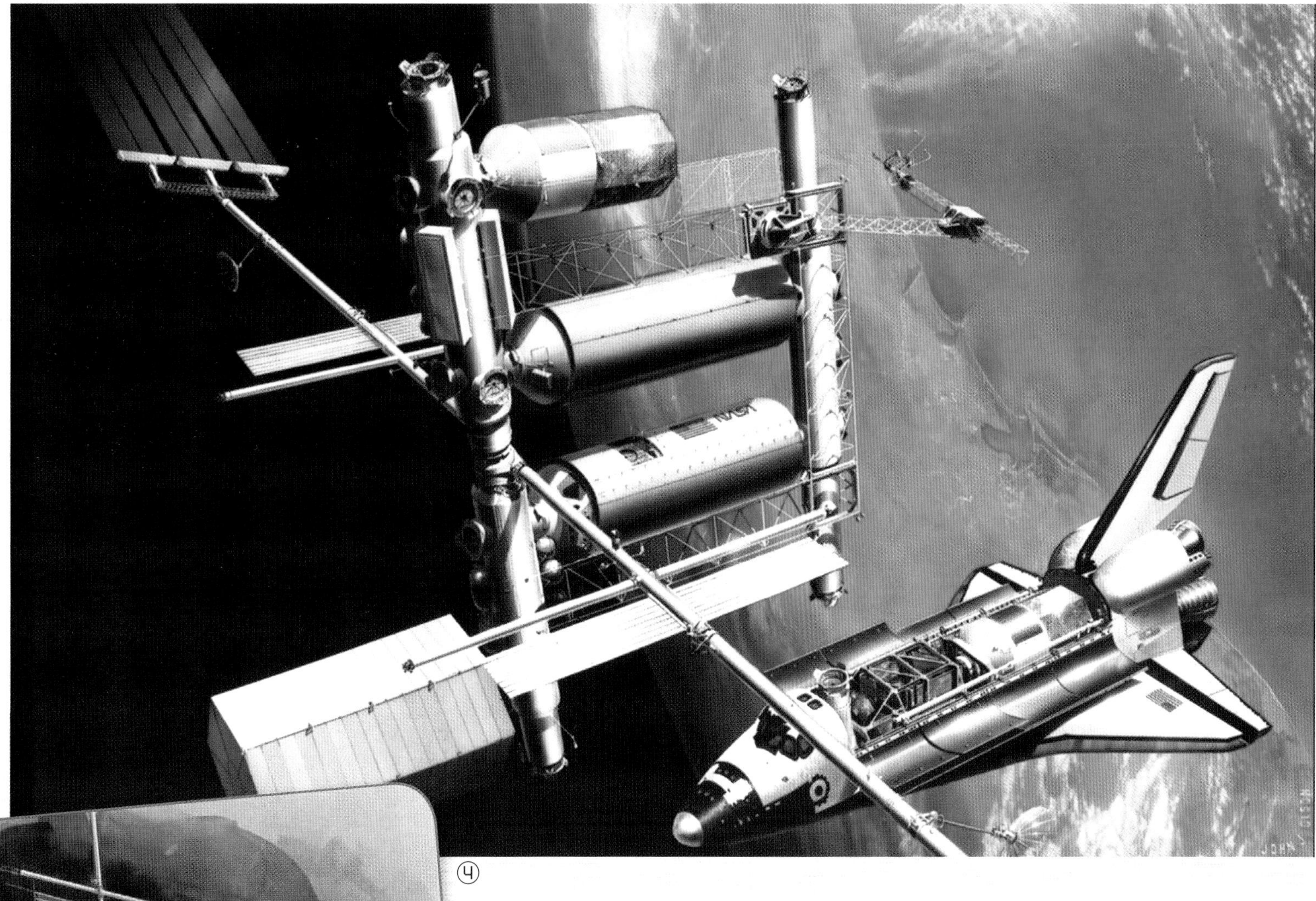

4

3

1. SOC's primary mission was the assembly and servicing of large spacecraft in orbit, rather than the conduct of scientific experiments.

2. SOC would be assembled in orbit with modules and other elements carried in the shuttle's payload bay. Astronauts would live on board for 90 days.

3. As additional modules, systems, and supplies were added, the crew size of the SOC could be expanded, eventually creating a significant off-planet community.

4. Over time, large hangers could be assembled to provide servicing bays for spacecraft, the bays protecting occupants from sunlight and providing handling mechanisms.

and separable living quarters in case they were needed. During phase three, crew size would grow to eight. Primary mission functions were satellite servicing and space-based construction.

CREW SYSTEMS

The crew would be involved in maintaining and servicing support vehicles. There would be two Orbital Maneuvering Vehicles (OMVs). For some missions, there could be Orbital Transfer Vehicles (OTVs)—these would be reusable and could carry spacecraft to geosynchronous orbit and other trajectories. Unpressurized hangers could be added to house and protect OMVs, OTVs, and satellites or spacecraft while they were undergoing maintenance or construction. Additional habitation modules and supporting systems would be added to allow crew growth as required.

The crew would be mixed male and female and would typically stay for 90 days. They would have private sleeping quarters, a galley with a dining table, medical facilities, and exercise equipment for health maintenance. For a crew of eight, six solar arrays would be mounted on booms that could be rotated to follow the Sun. Nickel hydrogen batteries would store energy for nighttime use.

SCIENCE AND APPLICATIONS

Science and applications could be conducted on vehicles independently, close to the SOC. The SOC crew would be involved in their maintenance, resupplying raw materials and harvesting product. Additional modules could be added to support biomedical studies and zero-g manufacturing, highly sophisticated communications, and detailed remote sensing of Earth's resources. With the addition of the SOC, the need for the number and length of shuttle flights that was then being planned could be minimized, reducing shuttle operations costs. On each visit, astronauts would scavenge for fuel, air, and water on the shuttle to help resupply the SOC, OMVs and OTVs.

CONFIGURATION DEVELOPMENT 1983–84

In 1983, the Configuration Development Group reviewed a variety of designs to define criteria that could be used for later evaluation. MSFC presented a configuration very similar to their MOF. It would start with a solar power platform that could be used to augment and extend space shuttle capabilities, and then add modules for work during man-tended and then free-flight phase. In time, additional systems would be added to permit a permanently crewed station. Johnson Space Center (JSC) presented several approaches. Their first was the Building Block, which resembled the SOC—a utility module would provide basic systems required first, like power, orientation, and propulsion systems, and over time more

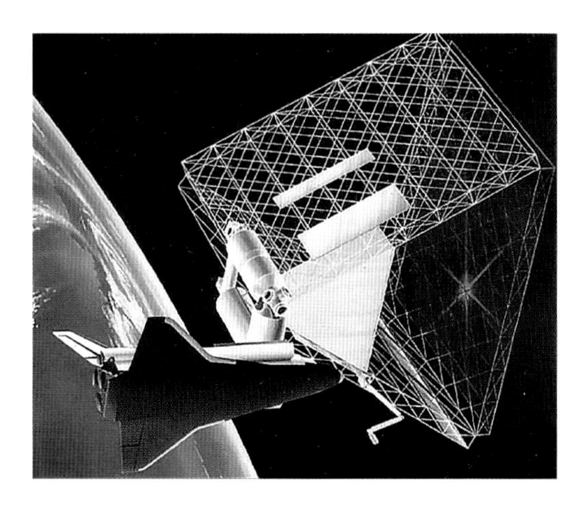

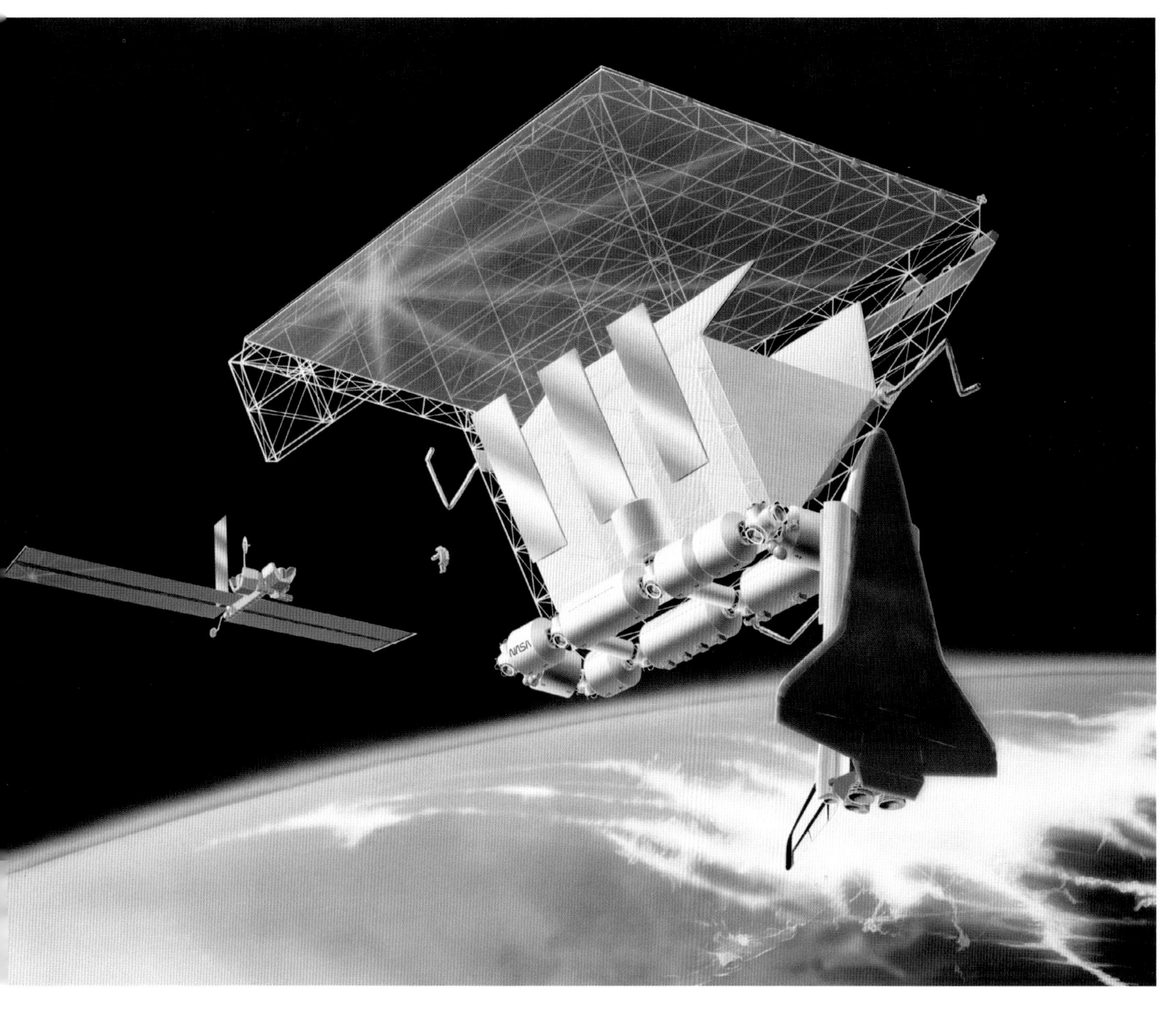

The "Delta" was a large triangular truss providing ample surface area for the mounting of modules, solar arrays, radiators, and various experiments.

The prime mission would be an operations base for lunar or Mars vehicle preparation. "Delta" could be expanded with servicing hangers, additional hab modules, and more power.

"Big T" had the advantage of using gravity gradient stabilization. Also shown in these views are the free-flying satellites that would accompany and be tended by astronauts from the station.

⭘ A General Dynamics/Convair proposal was intended for satellite servicing. Large hangers would provide servicing bays in which space-walking astronauts would maintain, refurbish, and refuel spacecraft brought from other orbits.

modules with additional functions would be attached. Hughes presented a unique design based on unmanned satellites. It was spin stabilized using empty shuttle external tanks. The tanks would be like four spokes of a wheel and a large circular solar array, 111 m (365 ft) in diameter, would lie flat on the tanks, the entire set of tanks and arrays spinning for stabilization. Habitation modules would be mounted by a bearing in the center and de-spun.

POWER TOWER 1984–85

Different NASA centers had different ideas of what kind of work would be important for a space station to perform. By 1983, all the NASA centers had identified their individual requirements, but no one had put together a concept for what the space station needed to look like. They hoped President Reagan would announce that the program was approved, but they needed a concept for people to see, so the Concept Development Group (CDG) was formed to establish a design reference based on costs.

The CDG reviewed several different configurations. A large triangular structure with modules along one side was called the Delta. Another design was the "Big T" with solar arrays flat on top and modules down near the bottom. Some had the parts held together by tethers and then the entire station spun for artificial gravity; gravity would vary at different places along the tether. In 1983, the Power Tower design was first studied by engineers and

Out of the 1983 Concept Development Group, the design selected for more in-depth study was the "Power Tower." It was gravity gradient stabilized. Mounting locations were provided for observation into space and modules would accommodate a crew of six astronauts.

NASA's Lewis Research Center in Ohio proposed solar thermodynamic generators that used parabolic mirrors to heat a fluid to drive electric generators.

Thermodynamic generators were more efficient, durable, and streamlined than solar cells, but required more extensive development and were more expensive.

The crew modules on the Power Tower, located on the right of this image, each had six berthing ports and were arranged in a square. Astronauts would be carried to and from the station by the shuttle, which is here seen docked to the end of the station. The large boxes in the center of the structure are servicing bays designed to house satellites. A free-flying satellite orbits nearby.

scientists. It was 122 m (400 ft) tall and had 75 kW of solar panels across a span of 82 m (269 ft). It used gravity gradient stabilization—differential gravity kept the modules toward Earth, reducing orientation fuel requirements. Viewing of the Earth was optimal and the location of the modules provided a clear path for shuttle to dock. There were attachment points for external payloads viewing Earth and space. There were two US Laboratory and two US Habitation modules, each 10.7 m (35 ft) long; all the modules were the same design which would reduce the cost of development. Construction would require 12 shuttle assembly fights, and six astronauts would work shifts of 12 hours each. The cost was based on development requirements for individual components and systems, which were in turn based on system complexity and mass. In 2017 dollars, cost was approximately $20 billion. Once the program officially started, the Power Tower became the point of departure for future developments.

DUAL KEEL 1985-86

In March 1986, the space station program held a systems requirements review (SRR). Engineers and scientists reviewed technical issues with the Power Tower design recommended changes in preparation for beginning Phase B, the preliminary design. The space shuttle had begun a series of materials-processing experiments and prospects were thought to be good for entering pharmaceutical production as early as the mid-1980s. The zero-gravity environment for space manufacturing would be a key to use of the station. On the Power Tower, the pressurized modules would contain the manufacturing experiments, with the modules at the end of a long lever arm, the central truss. This would magnify vibrations, however, destroying the microgravity environment and reducing the purity of experiment processing.

In response, a new design was established called the Dual Keel. It was 95.5 x 45.7 m (313 x 150 ft), rectangular, assembled from truss sections measuring 5 x 5 m (16.4 x 16.4 ft) section and 15 m (49.2 ft) long. The upper keel concept would support astronomy payloads. Earth observation payloads would be on the lower boom. A central 120 m (394 ft) truss ran between the two keels. The pressurized modules would be attached at the center, a location that would minimize vibration and acceleration forces induced by the structure

In order to reduce the vibrational disturbances for the crew, the modules were moved to the center of gravity. They would accommodate eight people.

The Dual Keel design initially would have had Solar Dynamic Turbine Generators for power generation. These were deleted in favor of solar arrays because of the costs associated with generators.

Habitable elements of the Dual Keel station. In this early view, six nodes and three connecting tunnels link two habitation modules and two laboratory modules in a figure "8" pattern.

The completed Dual Keel would offer a highly practical facility in space for engineering projects and logistics. It would provide ample "hanger space" for assembling spacecraft that would take humans back to the Moon and on to Mars, as well as serve as a satellite servicing facility.

As the Dual Keel was being discussed, the first agreements were reached to include laboratory modules of the ESA and Japanese.

and the crew. Solar arrays or power systems were placed to the sides and radiators near the center and hangers for servicing satellites were attached inside the truss. The Hubble Space Telescope would be brought to the station and serviced in the hanger. Two science satellites, and the OMV and OTV would be docked. The modules formed a figure "8." One US Laboratory module and one Habitation module were replaced by European and Japanese Labs as the countries joined in the ISS program.

UTILIZATION STUDIES

NASA, Europe, Japan, and Canada studied the multiple and varied uses of a space station and the implications of those uses on the design. A space station could be used as a base for launching lunar, asteroid, and interplanetary research spacecraft. There were advantages to having a human crew service instruments for Earth and planetary observation, plasma physics, and astrophysics. Human involvement in materials processing, life sciences, and solar physics research would also be valuable, and scientists with specialized research areas could fly as astronauts and conduct experiments. Nearby co-orbiting "free-flying" platforms could ensure isolation from disturbances or contamination in an inhabited location. Commercial opportunities for Earth observations lay in petroleum and mineral prospecting; agricultural forecasts; materials processing; and the launching and servicing of satellites being sent to higher orbits. Businesses might commercially exploit the processing of materials like pharmaceuticals, alloys, semi-conductors, and optical fibers in near-zero gravity, for the absence of gravitational forces in orbit permits the separation and purification of proteins and pharmaceuticals at rates 500 to 1,000 times faster than on Earth.

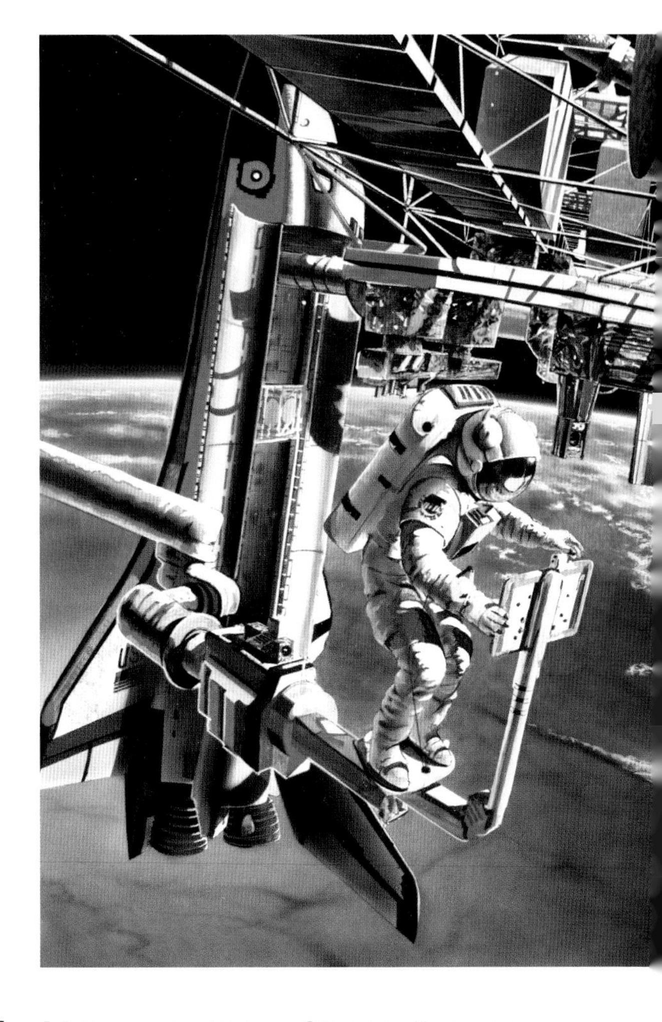

Concerns that the Dual Keel station would be too expensive, require more EVA assembly time, might have difficulty "closing" the keel structure, and would increase the number of shuttle assembly flights resulted in elimination of the two keels. The keels could be added back later if required.

The girder-like truss was going to be hand-assembled out of individual pieces by spacewalking astronauts working with robotic arms.

After the *Challenger* shuttle accident concerns arose that a new crew emergency return capability would be needed. To reduce costs the co-orbiting unmanned free-flyer was eliminated.

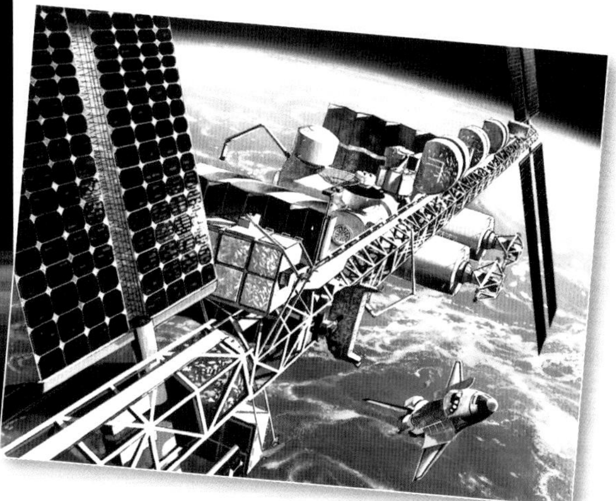

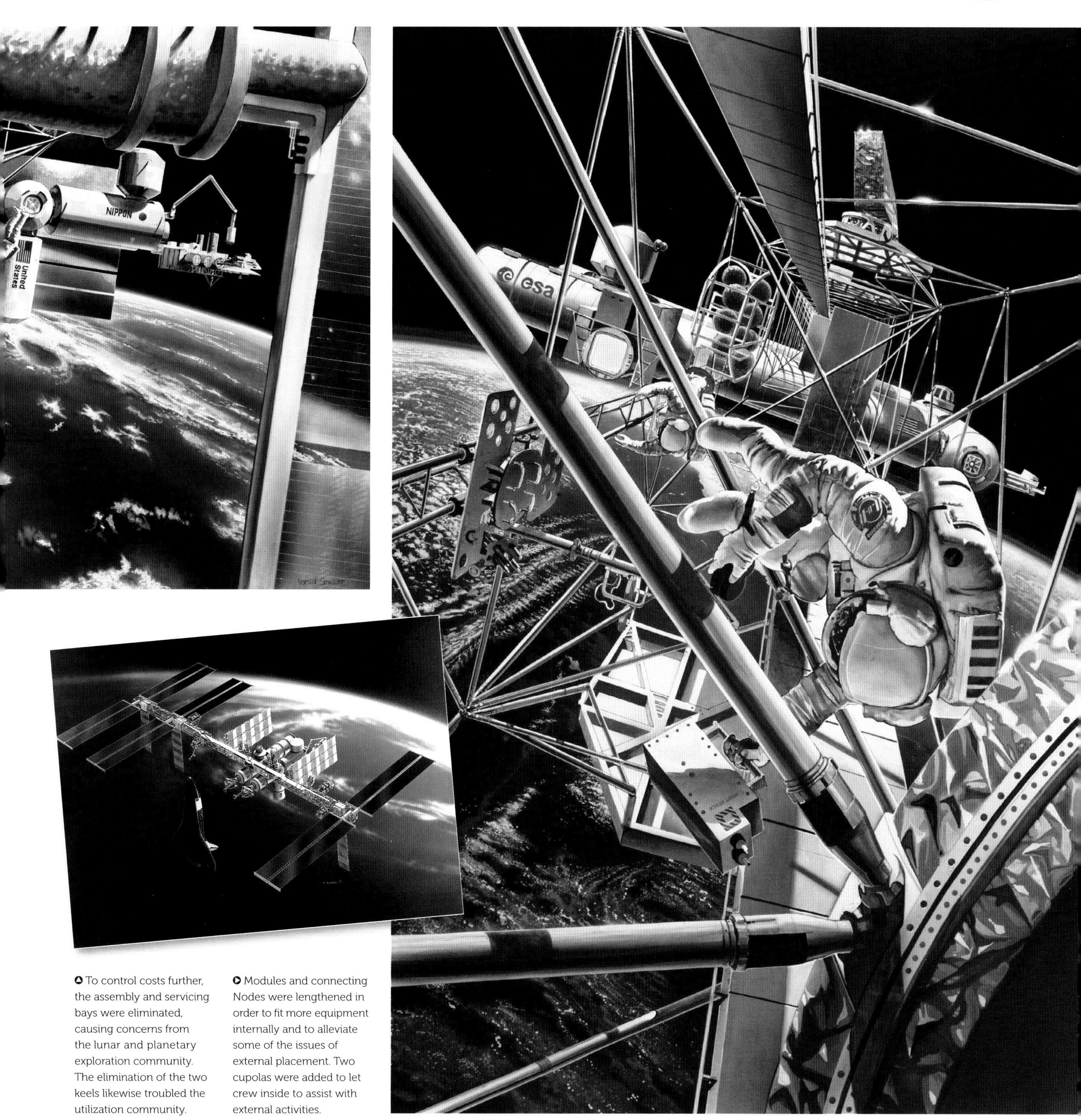

⮝ To control costs further, the assembly and servicing bays were eliminated, causing concerns from the lunar and planetary exploration community. The elimination of the two keels likewise troubled the utilization community.

⮞ Modules and connecting Nodes were lengthened in order to fit more equipment internally and to alleviate some of the issues of external placement. Two cupolas were added to let crew inside to assist with external activities.

REAGAN'S APPROVAL 1984

In 1980, Ronald Reagan became president of the United States. Reagan's NASA transition team lead, George Low, reported that the first Moon landing resulted in new knowledge, ideas, and technologies. US preeminence in space, however, had been eroded as Soviet space stations had realized economic and foreign policy advances. The first successful shuttle launch was three months after Reagan's inauguration. James Beggs now became NASA Administrator. At his confirmation, he said that the shuttle was successful, so the space station was the next step. It would open space to new long-duration experiments in a variety of scientific disciplines; a station would be required to test systems that would later be needed for outposts on the Moon or Mars, for example.

Reagan visited mission control, where he spoke with the astronauts, went to a shuttle landing, and voiced his interest in seeking returns on space investments. At the landing Reagan said he was charging NASA to establish a permanent human presence in space. Beggs established a space station task force, presenting the

During one Cabinet meeting, Reagan said that he expected people would remember his support of the space station when all his other accomplishments were forgotten.

Reagan speaks in California in front of the shuttle *Enterprise* while waiting for the landing of *Columbia* at the end of the fourth shuttle flight, July 4, 1982.

During the second shuttle flight, November 13, 1981, Reagan speaks with JSC Director Christopher Kraft, Jr. as Kraft describes in the Mission Control Center.

○ Reagan, with NASA Administrator James Beggs (standing second from right) and Christopher Kraft, Jr. (standing far right). Reagan speaks with the astronauts following the landing of the second shuttle flight.

○○ Reagan came into office just as space shuttle missions began to fly in 1981. The shuttle was a revolutionary step forward in space exploration, and for the possibilities of developing space stations.

potential of the space station to Reagan's Cabinet, yet only four out of 13 actually voted in favor of the station. Reagan got up and walked to the door; with his hand on the doorknob, turned around, and asked, "What do you guys think we're going to be remembered for 100 years from now?" The president intervened, using his authority, and personally approved the station. Beggs said that while the cabinet voted nay, Reagan voted aye. On January 25, 1984, in his State of the Union address, Reagan declared that "America has always been greatest when we dared to be great. We can follow our dreams to distant stars, living and working in space for peace, economic, and scientific gain. I am directing NASA to develop a permanently manned space station and to do it within a decade." Later, he named the space station "Freedom."

FREEDOM

President Reagan didn't announce his decision at the end of his Cabinet meeting. He asked if it were possible to do the space station internationally, to which James Beggs replied "of course." Reagan faced opposition from most of his Cabinet, his Science Advisor, and the Defense Department. Additional rationale was therefore gathered to gain more support. International participation looked good for purposes of peace, politics, and commerce, and to help potentially share costs. In his State of the Union address, Reagan asked NASA to "invite our friends and allies to strengthen peace, build posterity, and expand freedom."

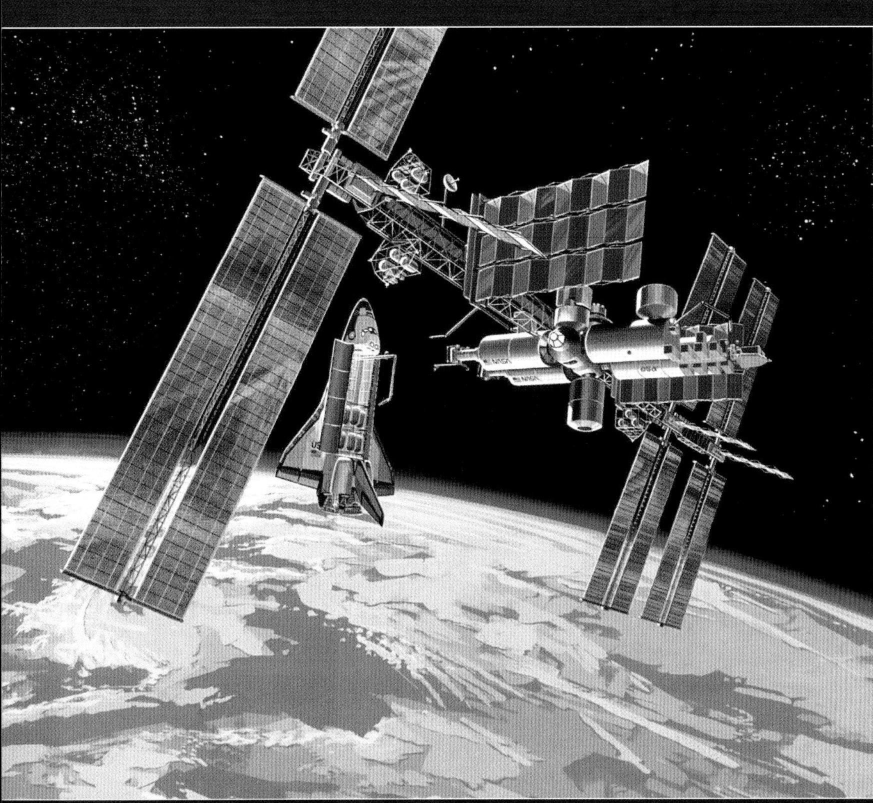

1. A vision of space station Freedom in 1991. Within two years, however, the focus of NASA's attention was directed toward the development of the ISS.
2. In 1991 the artist Alan Chinchar also provided a vision of the space station Freedom in orbit. He depicted the space station in its completed state, showing four pressurized modules (three laboratories and a habitat module) and six large solar arrays. The Moon and Mars are shown imaginatively in the background.

As the program formed, it committed to a two-year period to define requirements, cost, and schedule. Different NASA centers would have management roles, and each would write a request for proposal (RFP). Then companies would bid and identify their price and design. In August 1984, a reference station design was established. It was based on the Power Tower studied in 1983, and this was the basis for the initial cost estimate. The Dual Keel design emerged and was formally approved in March 1986, but it increased costs and delayed schedules. Another design review deferred the upper and lower keels. The revised design in mid-1987 left only the central truss with the modules in the center.

PEACEFUL PURPOSES

Negotiations over international contributions culminated in September 1988 with an Intergovernmental Agreement on the Space Station and resolved that it was for civilian, peaceful purposes. In a time of continuing global political tensions, this was an important step forward in cooperation. ESA and Japan would each provide a laboratory module, while Canada would provide a robotic arm based on the shuttle remote manipulator.

RUSSIA ENTERS THE PROGRAM

In the early 1990s, there was considerable interest in joint US–Russian space activities. Space station Freedom was facing several issues. Costs were growing and the space station was designed to provide a safe haven for astronauts in between shuttle flights, but after the *Challenger* disaster in 1986 the shuttle did not fly for 30 months. After the accident, the program decided it needed a rescue vehicle, so astronauts would always have a way to return, but there was no budget for a new vehicle. Simultaneously, as the Soviet Union fell in 1991, Russian government support of space companies, like Design Bureau 1 (OKB-1), dropped precipitously. The Russians did not want to lose their space program, but the government was no longer funding it.

"WE WENT TO THE RUSSIAN ENERGIA FACTORY. THEY SHOWED US THE BURAN. IT WAS EXACTLY LIKE THE US SHUTTLE. ANY PAYLOAD THAT COULD FLY ON THE BURAN COULD FLY ON THE SHUTTLE."

ARNOLD "ARNY" ALDRICH
NASA PROGRAM MANAGER

1. NASA began discussions with the Russians over the potential use of Soyuz spacecraft to serve as an emergency rescue vehicle for the Freedom space station.

2. The Soviet Buran was modeled on the US shuttle, but with differences, such as liquid-fueled booster rockets and placing the main engines at the base of the core stage.

In 1991, the US Senate asked NASA officials whether the space station could use the Russian Soyuz as a rescue spacecraft, then in 1992, President George H. Bush sought to rejuvenate US–Russian space cooperation. The two countries agreed to launch a Russian cosmonaut on the US shuttle and an American astronaut to Russia's space station Mir, with a docking between the US shuttle and Mir.

Beginning in early 1992, meetings started between NASA and Russian personnel to identify more possible areas of cooperation. During a visit to the Design Bureau 1 (OKB-1), where Sputnik, Vostok, and Mir had all originated, US officials were shown test and operations facilities for Soyuz. They were also shown a Russian space shuttle. The Buran shuttle had flown a single unmanned space mission in 1988. A docking module was mounted inside the OKB-1 shuttle, intended for use with the Mir Orbital Station. The Americans examined the Buran and found it was built to the same specifications as the US shuttle: the payload bay was the same size, the rails to which the docking module was attached were the same dimensions and used the same kind of latches. The Russian docking module was therefore a perfect fit for the US shuttle.

OPTIONS AND DANGERS

This correspondence opened options for a series of joint missions. First a Russian cosmonaut would fly on a US shuttle. Then a US shuttle would rendezvous with the Russian Mir Orbital Station. Then a US astronaut would live on Mir. A US shuttle would use the docking module to dock with Mir and take the astronaut back to Earth.

The Russian technical personnel were skeptical at first, but the Russian political leadership agreed to the plan. In 1993, Bill Clinton became the US president just as space station Freedom's costs rose and schedules slipped. The program was in political danger. A congressional vote missed killing the program by just one vote. Dan Goldin became the NASA Administrator and ordered a redesign to reduce program cost. Three options were proposed:

- Option A used many of the components of Freedom. A significant change was that the solar arrays did not track the Sun, and the entire vehicle therefore had to point toward the Sun.
- Option B was a refinement of the Freedom design.
- Option C was a large cylindrical module that

3. The original plan had been for the US shuttle to return to pick up astronauts on Freedom, but after the *Challenger* accident the shuttle did not fly for two-and-a-half years.

4. President Clinton asked NASA to redesign the station in 1993, but the focus turned to adding two Russian modules to form the keystone elements for rest of the station.

would be launched in a single mission. The large station module would replace a shuttle orbiter; it would use the shuttle's solid rocket boosters, external tank, and main engines. Some versions were based on a shuttle orbiter modified by replacing its wings with solar arrays.

By the summer of 1993, the White House expressed support for an emerging design that simplified the station design and eliminated some components. This was followed with a dramatic announcement on September 2, 1993 that Russia would join the space station program as a partner. Clinton had vastly expanded the nature of the cooperation by merging the US and Russian space station programs. The announcement may have saved the program because the new Russian partnership attracted votes in Congress. NASA had already contemplated Russian participation as a paid hardware supplier during the redesign process.

CREATING THE MODULES

Russia agreed to build several station modules. Two would serve as the basis of the station, providing critical propulsion, orientation, and habitation functions. The Russians would also launch two Soyuz spacecraft a year to serve as "lifeboats" and several Progress spacecraft each year to "reboost" and keep the station in the correct orbit. These elements put the Russians in the critical path and required their continuing commitment. Russian involvement helped the technical aspects of Alpha, which otherwise would have required further redesign.

On September 7, 1993, the new design—Alpha, with Russian support—was announced.

There were technical issues, however. Freedom was going to orbit at a 28.5° inclination. The ISS had to fly at 51.6° because the Russians could not launch to 28.5°. Increasing the shuttle's inclination dramatically reduced the amount of mass the shuttle can place in orbit by about 6,818 kg (15,000 lb) each flight. The shuttle could not support station assembly at that inclination, so the shuttle capability had to be upgraded. They were able to increase the shuttle's capacity by about 5,454 kg (12,000 lb), which meant the capability had been reduced by only 1,363 kg (3,000 lb). Assembly could be supported, but over the duration of 35 missions about three additional shuttle flights were required. There were other issues such as the lighting and temperature regime at the 51.6° inclination, which provided good rationale for reassessing the solar array rotation requirement.

MIR ORBITAL STATION

Although Mir was based on the earlier Salyut DOS design, and its specialized modules were based on the Almaz/TKS module design, it came to establish a fundamentally different principle. As the Mir concept was being established in the early 1970s, it was at first called the Permanent Orbital Station. Later, plans were scaled back and it was redesignated as DOS-7.

The significant change in the structure of the Mir from its Salyut predecessors was the addition of four lateral docking ports. The earlier Salyuts had two docking ports along the long axis, permitting the docking of Soyuz spacecraft carrying crew, Progress spacecraft carrying supplies, and Almaz/TKS-type spacecraft, which could bring large-scale systems. Any two of these three vehicles could be docked at one time on Salyut. Mir would incorporate six docking ports; the old-style Salyut axial ports plus the four new lateral ports, permitting six different vehicles or modules to be docked at once.

Significant changes were also made on the interior of Mir. Although the core module of Mir appeared from the outside to be similar to previous Salyuts, the earlier Salyut modules had all of the systems required to support a crew throughout their stays of several months in orbit. The Mir core module was configured principally as the cosmonaut's habitation module, with private crew quarters for sleeping, a galley for food preparation, a private waste management compartment for hygiene, and exercise equipment through most of the nodule's center.

A photograph of the completed Mir Orbital Station in 1998 taken through the window of a visiting space shuttle.

The Mir in its final configuration after the addition of the Spektr and Priroda modules during the NASA-Mir program.

Photograph taken on the floor of the Krunechev factory in Moscow in 1994, a dozen Proton rockets await use to launch large payloads such as modules of the Mir and the ISS.

In the Krunechev factory in 1994, shown here are several modules that will become the last modules of the Mir Orbital Station and the first modules of the ISS.

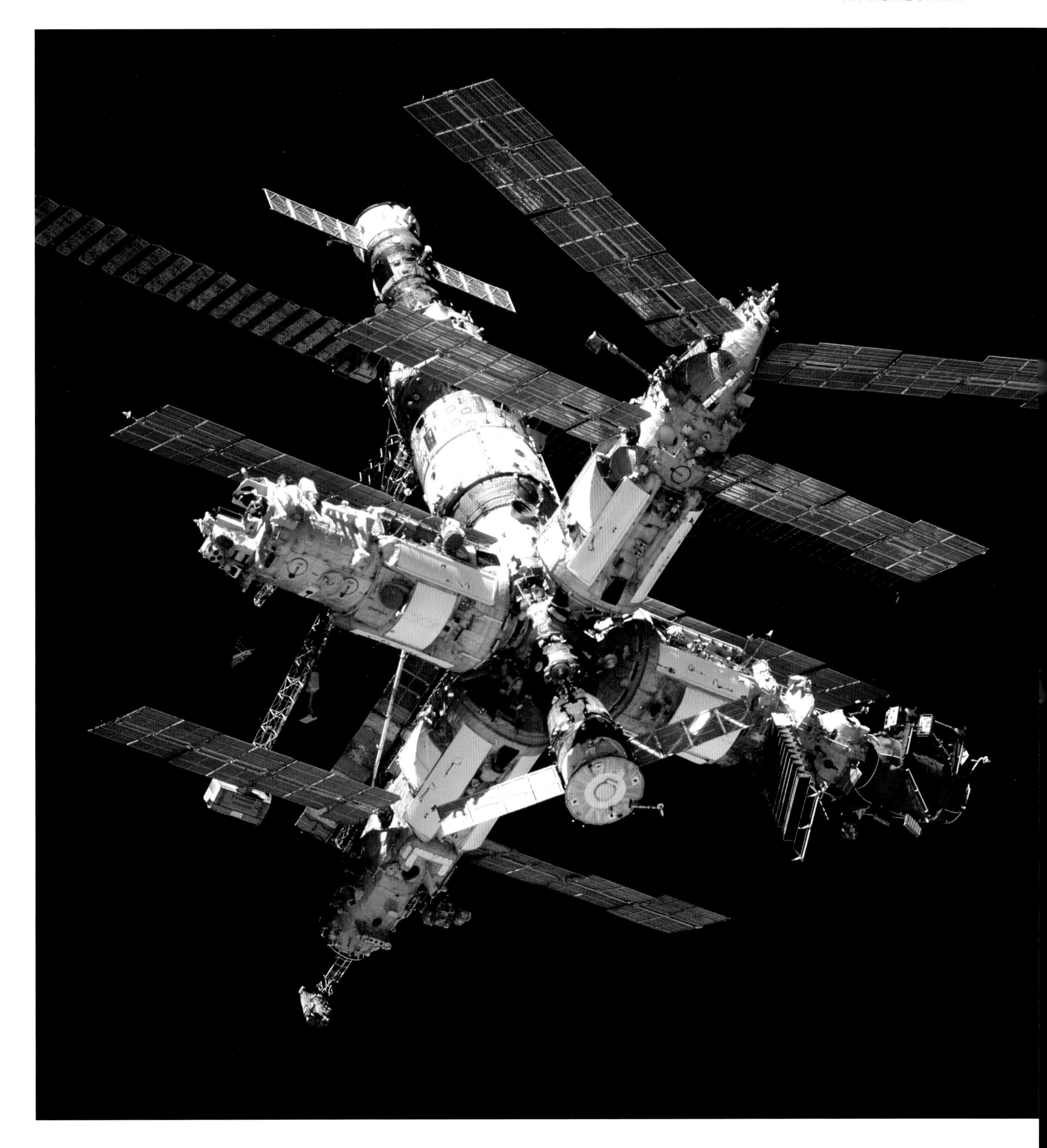

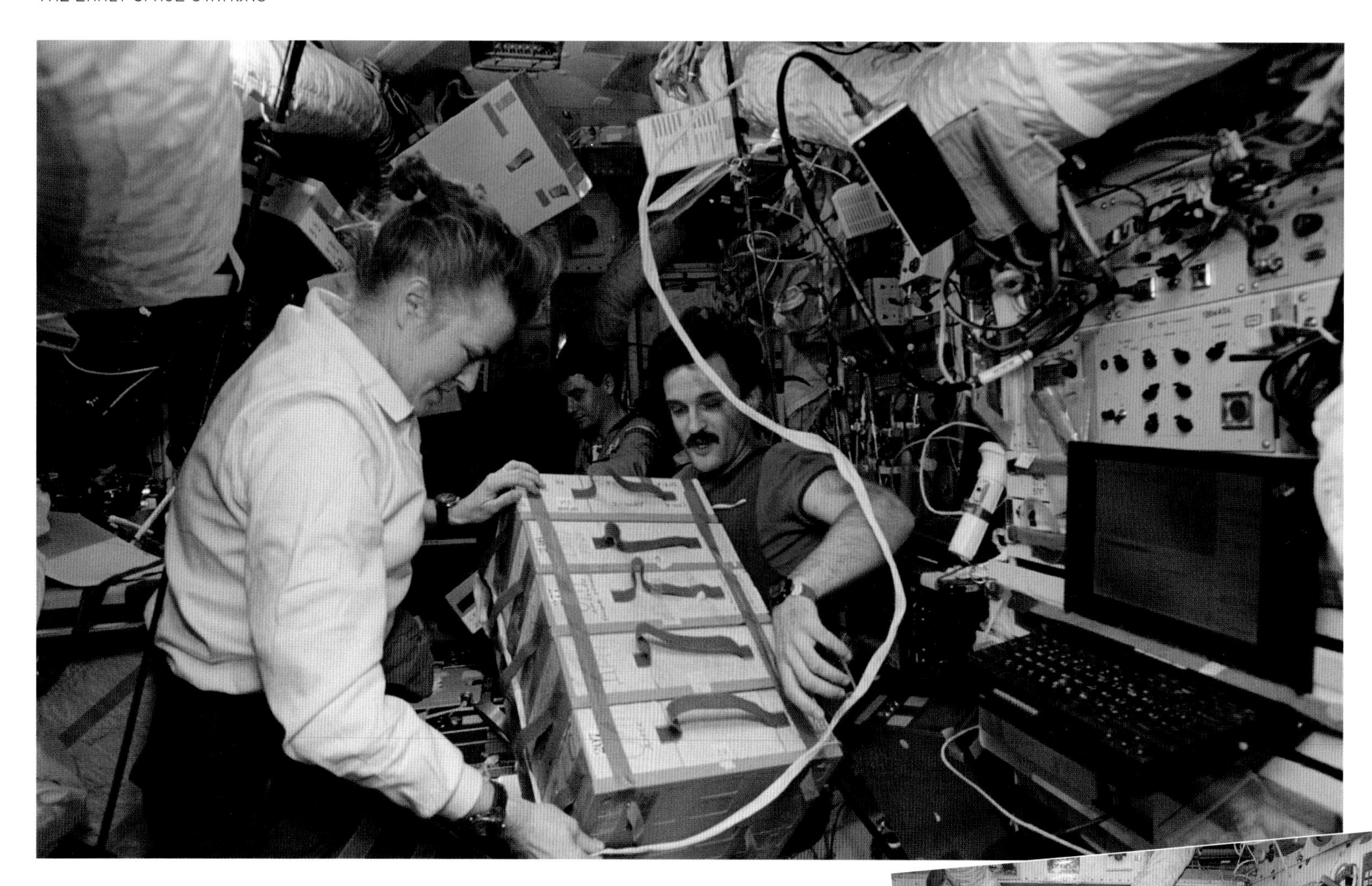

The core module was launched by a Proton rocket from Baikonur on February 20, 1986. The first crew arrived less than a month later. Leonid Kizim and Vladimir Solovyev would stay for two months. Solovyev returned to Earth to become one of the primary Russian flight directors in Russia's mission control throughout the Mir program. In 1987, a year later, the first specialized module, called Quant, was added. Quant and a second module, Quant II, added in 1989, were both used for astrophysics and environmental monitoring experiments. A third specialized module, Kristal, was added in 1990 and would be used for materials processing, Earth observation, and biological experiments.

The Russians had hoped to provide logistical support to the Mir using the Buran space shuttle. Its first test flight was in 1988—it flew a one-orbit unmanned flight—but as the Soviet Union collapsed and funding to the Russian space program slowed, the Russians found themselves unable to afford the Buran and it never flew again.

Mir continued to operate, relying on the logistical support of Soyuz and Progress spacecraft, remaining almost continuously occupied for nine years. Throughout Mir's 15-year lifespan, there would be only five brief periods during which it was unmanned.

OO Shannon Lucid with Alexander Kaleri and Valeriy Korzun on board Mir. Lucid prepares empty food trays to be returned on the US shuttle. During her fifth spaceflight, Lucid spent a total of 188 days in space, including 179 days aboard Mir. At the time, Lucid held the record for the longest duration stay in space by an American.

O Astronaut Shannon Lucid exercises on the treadmill in the Mir Base Block or core module. Her replacement, John Blaha, is to the right, studying the daily mission plan. Former USAF colonel Blaha went on to become a NASA astronaut, flying numerous space shuttle missions and spending four months aboard Mir.

Mir 22 and STS-81 crews. Front, left to right: Mike Baker, John Grunsfeld, Aleksander Kaleri, Middle row: Valeri Korzun, Marsha Ivins and John Blaha. Back row: Jerry Linenger, Peter Wisoff and Brent Jett.

Yuri Onufrienko, Mir expedition 21 commander, in the Base Block (core module) of Mir in March 1996. This view looks forward into the Mir's node to which most of its other modules are berthed.

On June 25, 1997, an unmanned Progress resupply collided with the Spektr, causing damage to the solar array shown here and depressurizing the module. The air leak was never repaired, and the module's use never recovered.

SHUTTLE-MIR

In 1991 the Americans asked the leaders of the Russian shuttle program to consider whether a Russian crew member could fly on the shuttle and conduct a joint spacewalk. Early the following year, the US program assessed joint shuttle and Mir missions. Several meetings between Americans and Russians resolved technical issues, and the use of the Russian docking system on the US shuttle was agreed upon. The program would consist of the flights of astronauts and cosmonauts on each other's spacecraft and culminate with a single shuttle and Mir docking mission.

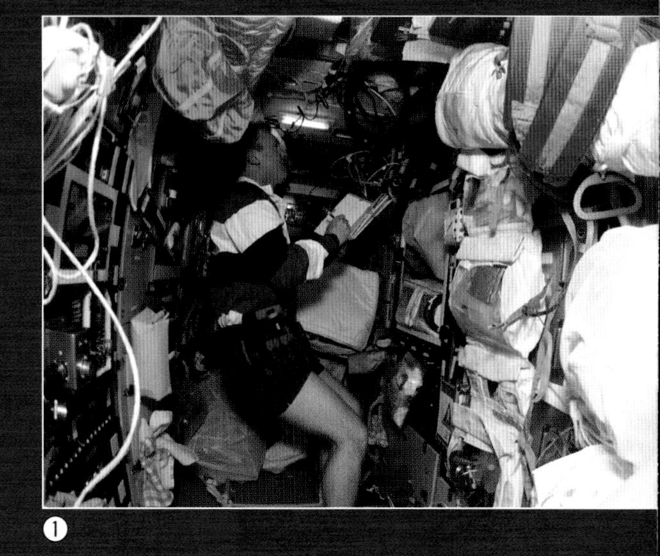

1

2

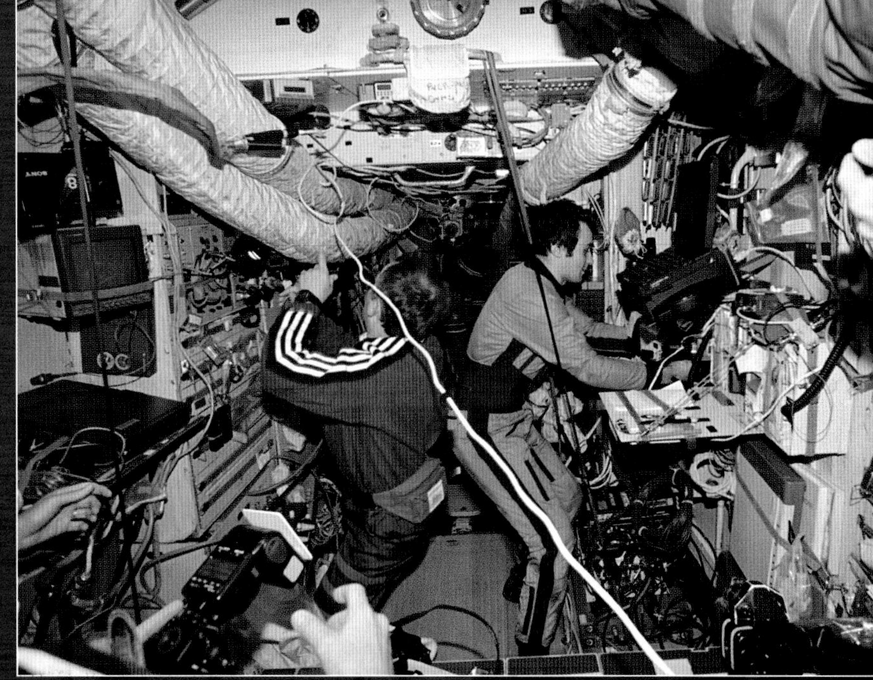

3

When the US discovered that the Russians, desperate for cash, were selling their technology, the new US president, Bill Clinton, offered to pay for a series of shuttle flights to Mir. The flights would also help to integrate the Russians into the ongoing space station program. This was called Phase I of the ISS.

CULTURAL ISSUES

There were delicate cultural issues. Europeans, Japanese, and Canadians had been part of space station since 1988; they did not understand how the new series of flights became Phase I of an ongoing program. The Russian philosophy of the purpose of a space station was different; Mir was thought of as a frontier outpost for humans. The US philosophy was that a space station was a research facility for science, technology, and commerce. The US shuttle short-duration missions were jam-packed, focused missions on which performance was measured by success carrying out scheduled activities. Russian long-duration missions operated with a long-term philosophy—if activities did not get done today, then the crew could always work on them tomorrow.

Phase I would provide practice for the cultures to work together to develop mutual working relationships, agreements, contracts, and the approval of hardware safety on each other's vehicles. The initial agreement was for a single shuttle–Mir docking flight. The NASA Administrator said to add two more missions. Then contracts were negotiated for the conversion of Russian Mir modules to US research modules and for ten shuttle missions to Mir.

4

5

1. Andy Thomas, the last long-duration US crewman on Mir, moves into his quarters in the Priroda module, January 26, 1998. The white cargo transfer bags to the right hold his belongings and experiment content. As they were progressively emptied, they could be collapsed down for more convenient storage.
2. Shannon Lucid and John Blaha hard at work inside the Priroda module of the Mir station during handover activities. The two astronauts are inspecting the US microgravity glovebox, a contained environment for research with hazardous materials. Priroda was modified to include a large number of sophisticated US systems to support research on the Mir.
3. Robert "Hoot" Gibson (left) and Vladimir Dezhurov, commanders of shuttle STS-71 and Mir Expedition 18, work in the Base Block (core module) of Mir in July 1995.
4. Joe Edwards, pilot of shuttle STS-89, and Salizan Sharipov, inside the Spacehab module prepare to carry a gyrodyne to Mir for replacement. The gyrodyne contains a massive spinning flywheel and uses electrical power rather than rocket fuel to orient the space station.
5. The US space shuttle *Atlantis* and the Mir Orbital Station linked together for the first time, during the mission of STS-71, on July 4, 1995.
6. Shannon Lucid watches growing wheat inside the *Svet* or greenhouse, located inside of the Kristal module aboard Mir.

6

Chapter Five

THE INTERNATIONAL SPACE STATION

The largest spacecraft ever built, the International Space Station (ISS) is humanity's foothold in low-Earth orbit. The outpost spans the length of an American football field and has an inhabitable volume larger than the inside of a Boeing 747. Since 1998, the space station has hosted a continuous human occupancy, supporting vast numbers of scientific investigations and providing the baseline for future missions into deep space.

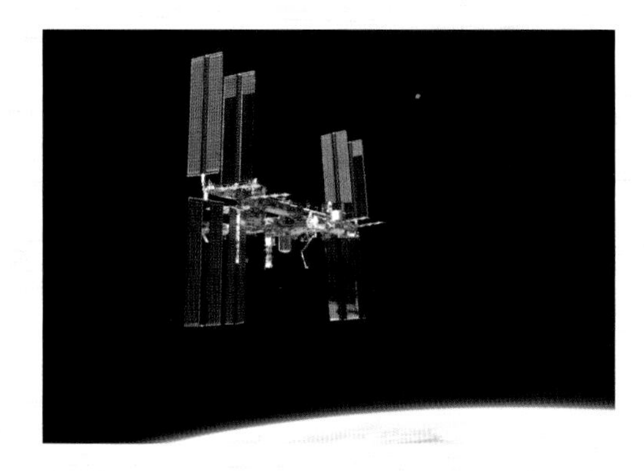

(Left) Backdropped by Earth's horizon and the blackness of space, the ISS is photographed by an STS-130 crew member aboard the space shuttle *Endeavour* on February 19, 2010.

(Above) The ISS is seen from an oblique angle with the Earth below and Moon above, as photographed on July 19, 2011 during the final mission of NASA's space shuttle program.

The International Space Station

THE INTERNATIONAL SPACE STATION WAS HUMANITY'S FIRST SUCCESS AT ESTABLISHING A PERMANENT OUTPOST OFF THE PLANET. ULTIMATELY MADE POSSIBLE BY INTERNATIONAL COOPERATION, THE APTLY TITLED ISS WAS MORE THAN JUST A PLATFORM FOR SCIENCE EXPERIMENTS; THE ASSEMBLY OF THE ORBITING LABORATORY REQUIRED THE EVOLUTION AND MATURITY OF IN-SPACE OPERATIONS, AS WELL AS NEW FOUNDATIONS FOR DIPLOMACY.

In November 2015, cosmonaut Oleg Kononenko, on board the ISS, observed that "The main achievement, in my opinion, is that people on the ground fail to hear each other, see each other. Here in space, this is impossible." His profound comment marked 15 years of the ISS being continuously crewed. "Everyone is important, and the success of the program—and sometimes even life—depends on what each and every one of us does. This is the perfect example of how cooperation can be achieved," he radioed to Earth.

PIECED TOGETHER

The construction of the ISS began simply enough. A propulsion module originally intended for Russia's space station Mir was repurposed and connected in Earth orbit to a US-built, multiport node in December 1998. From there, additional Russian and American "rooms" were added, as were components from Japan, Canada, and the multination European Space Agency (ESA). These elements were connected to each other and mounted to a multisegment truss—a backbone spanning the length of an American football field.

At each end of the truss, four sprawling solar array wings were deployed to generate the power needed to operate the life-support and science facilities aboard the space station.

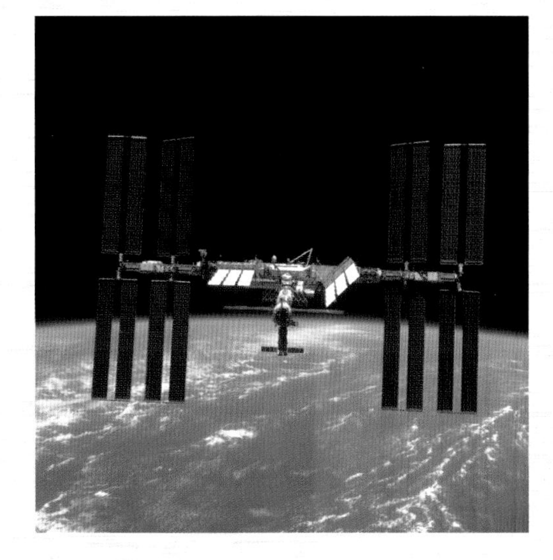

The ISS as seen in dramatic counterpoint against Earth by the crew of STS-119 aboard space shuttle *Discovery* on March 25, 2009, after installing the ISS's fourth and final set of solar array wings.

"WE DO A LOT OF EXPERIMENTS UP HERE, BUT I THINK THE MOST IMPORTANT EXPERIMENT IS THE STATION [ITSELF] AS AN ORBITING VEHICLE THAT KEEPS HUMANS ALIVE IN SPACE FOR LONG PERIODS OF TIME."

SCOTT KELLY, ISS EXPEDITION 45 COMMANDER

In total, it took more than a dozen years, more than 100 rocket launches, and more than 200 astronauts and cosmonauts to build the ISS. "Space station assembly is complete," declared NASA astronaut Mark Kelly in May 2011, though more components were added later, and even more are still planned. Perhaps most remarkably, most of the major components that compose the station could not be tested with their other connecting parts until they were in space. It was like assembling a giant jigsaw puzzle, where each of the pieces were individually designed by a different puzzle maker and the first time they were fitted together was after the puzzle was purchased.

OFF THE EARTH, FOR THE EARTH

Building the ISS provided new lessons for how countries could work together. It also advanced techniques in extravehicular activities (EVAs, or spacewalks), maintaining critical life-support systems in orbit, and troubleshooting problems when things went wrong. After its major assembly was complete and the focus changed to utilization, the ISS began to fulfill its stated purpose—providing a laboratory unlike any other available.

In its first two decades, the ISS supported more than 2,000 investigations from more than 100 countries in disciplines as varied as

The American, Russian, and Italian crew members of the ISS's Expedition 45 pose together with the flags of the 15 nations that are partners in the international outpost.

The ISS is seen as it grew large in the window from on board Russia's Soyuz TMA-12M spacecraft approaching a docking at the outpost on March 27, 2014.

Bathed in green LED light, the inside of the ISS glows at night in this moody photograph taken by ESA astronaut Alexander Gerst in March 2014.

biology and biotechnology, to technology development and demonstration. It has helped produce a vaccine for salmonella; advanced our understanding of how flames and fluids behave in microgravity; and served as a platform for the detection of dark matter, the "missing" mass of the universe.

Commercial research aboard the ISS has pioneered new and improved material manufacturing techniques, proven designs for future spacecraft, and contributed to better and more capable robotic systems. Above all else, it has taught us more about our own bodies. The astronauts and cosmonauts, serving as study subjects, have exposed new risks from long-duration exposure to microgravity—in addition to bone loss and muscle atrophy, internal pressure has led to impaired eyesight—and revealed effective ways to combat some of those hurdles to future space exploration. Research on the genetic level has advanced our understanding of how gravity affects human development, as well.

EXTENDING HUMANITY BEYOND EARTH

An engineering marvel and a world-class laboratory, the ISS also, at its most basic level, serves as a home away from Earth for the crew members who spend months at a time on board. After more than 50 expeditions, the ISS has incorporated some of the best of the astronauts' and cosmonauts' traditions from their respective home countries and begun to develop a new culture that is uniquely space-based.

The Sunday night "come to supper" still exists, but has introduced a new element of barter, as Russians trade canned meats for American thermostablilized desserts. Movies and television serve as more than just a source of entertainment, but also a mechanism by which to monitor communication links with Earth. Pastimes like playing musical instruments and personal photography have taken on new meaning as well, beyond a means of artistic expression to a conduit for sharing the experience of spaceflight with everyone back on Earth.

Social media and live video have provided a continuous pipeline for "Earthlings" to get a taste of life in space, while the view of our home planet below continues to redefine our sense of self and place within the universe.

TESTBED FOR THE FUTURE

In its later years, the ISS has served as a testing ground for the future of humanity in Earth orbit, and beyond. Self-funded missions to the orbiting outpost have demonstrated a market for space tourism, while the expansion of commercial payloads—in particular through the establishment of a National Laboratory within the US operating segment of the space station—has provided the early test cases for establishing independently owned, commercially run space stations.

In addition to ongoing studies on the effects of microgravity on the human body, the ISS has provided a platform for testing the self-contained life-support systems that will be needed for future missions back to the Moon and onwards to Mars and the asteroids. Water recycling, oxygen generation, and carbon dioxide scrubbing systems will be critical for keeping crews alive far away from Earth. On the space station, the trials and tribulations of keeping such hardware running have refined the ongoing development of advanced equipment on Earth.

The ISS's open docking ports have also contributed to the future, providing berthing locations for two countries' and two companies' first automated cargo spacecraft, and the development of the first privately built and run crewed vehicles. The ISS has even possibly changed the shape of space stations to come, with the addition of its first expandable, or inflatable, module.

Whereas once space stations were precarious perches on edge of humanity's reach, the ISS has provided the solid grounding for our ultimate push outward.

DESIGNING AN INTERNATIONAL SPACE STATION

As was demonstrated by the outposts that preceded it, designing a space station to be built and run by one country was a challenge unto itself. The ISS needed not only to address the engineering and logistical needs common to any such orbital platform, but also to integrate the ideas and tenets of multiple partners that entered into the project with distinct design philosophies and practices.

Some aspects of the ISS's design were dictated by how the pieces would reach orbit. The size of the US operating segment modules, whether developed by the United States, Japan, or Europe, needed to fit within the payload bay of the US space shuttle for launch, whereas all but one of the Russian segment components needed to be compatible for flight on the nation's rockets.

Common and complementary engineering systems were required so that the modules could work as a single space station. Inside each of the rooms on the US side, a modular rack system was designed so that science and evolving life support equipment could be more easily interchanged throughout the life of the outpost. There was also a psychological aspect to the design. As a living space, the modules' layout needed to take into consideration such details as color schemes and hardware orientation. There may be no "up" in space, but the ISS largely had a common orientation with floor, ceiling, and walls, and with the floor toward Earth.

The size of the ISS's rooms was limited by the dimensions of the space shuttle's payload bay. Here *Discovery* delivers Japan's Kibo laboratory, the largest of the station's modules, in June 2008.

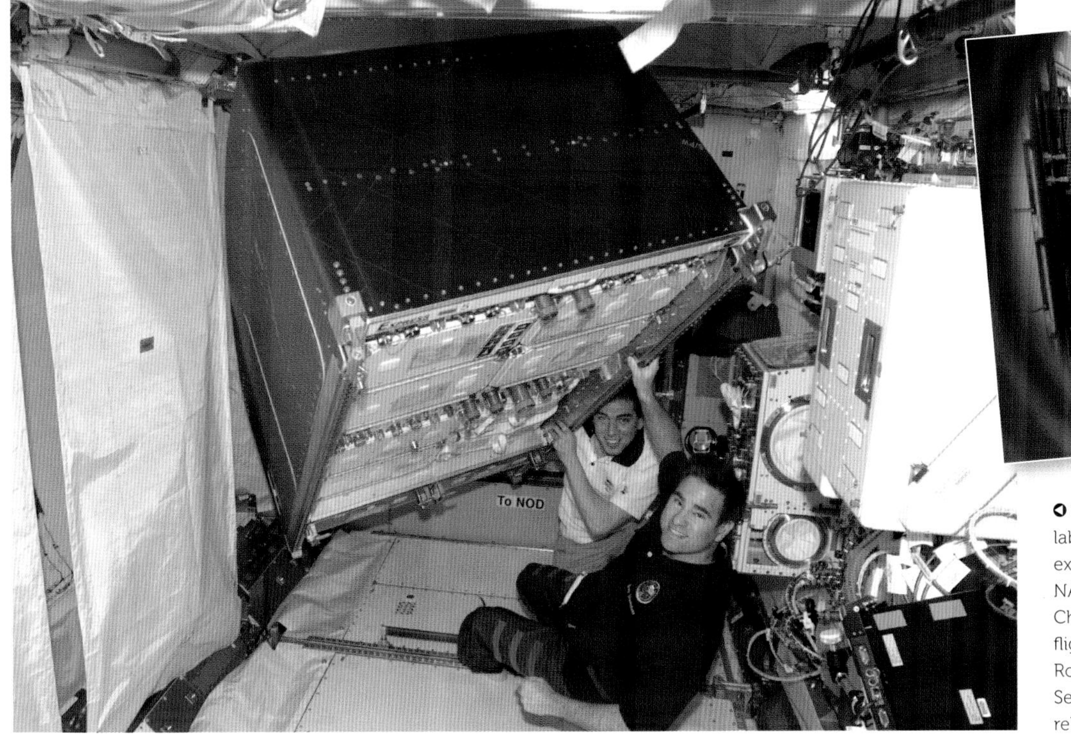

The walls of the ISS's laboratories are lined with experiment racks. Pictured, NASA astronaut Greg Chamitoff, Expedition 17 flight engineer, and Roscosmos cosmonaut Sergei Volkov, commander, relocate a rack in 2008.

The interior of Node 2 "Harmony" was photographed after it was attached to the ISS during space shuttle mission STS-120 in 2007. The nodes have multiple ports to connect modules to the outpost.

Expedition 32 flight engineer Aki Hoshide is seen behind a common berthing mechanism (CBM) hatch, the largest "door" to be launched into space. The CBM connects all of the modules on the US segment of the ISS.

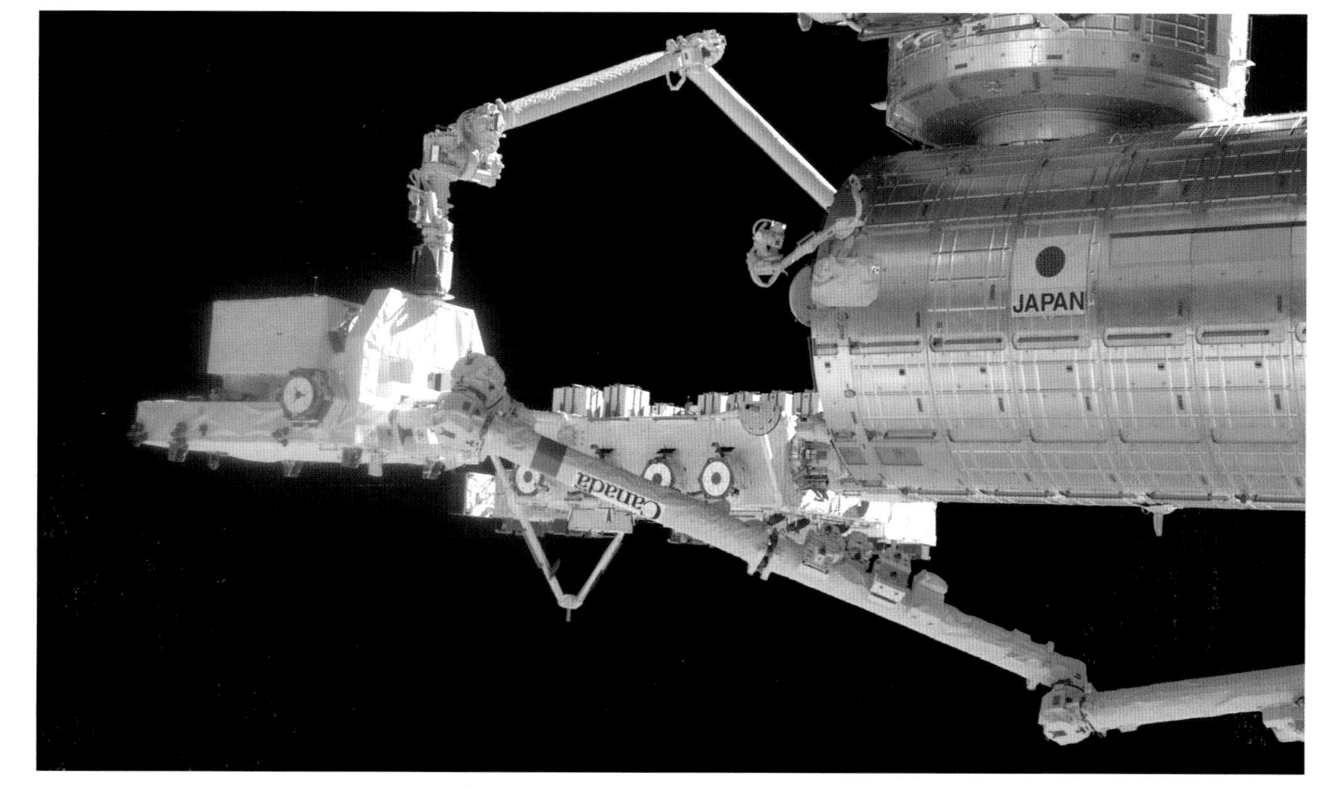

Robotic arms were used on the exterior of the ISS to assemble and maintain the outpost. Here, Japan's JEM-Robotic Manipulator System (JEM-RMS) outside of the Kibo laboratory and the space station's Canadarm2 are seen together in 2009.

TESTING AND DEVELOPMENT

Assembling the ISS required the preparation and launch of more than 150 separate components that all had to fit together, many for the first time only after reaching orbit. The sheer size and international nature of the ISS, plus the time required for construction, precluded building the complete station on Earth.

For the modules, truss segments, and other hardware to be launched by the US space shuttle, NASA constructed a 42,500 m² (457,000 square ft) Space Station Processing Facility (SSPF) at the Kennedy Space Center in Florida. The building provided a clean-room class facility for the preparation and emulated integration testing of key components prior to launch. Three major tests were performed over the course of the construction of the ISS. The US

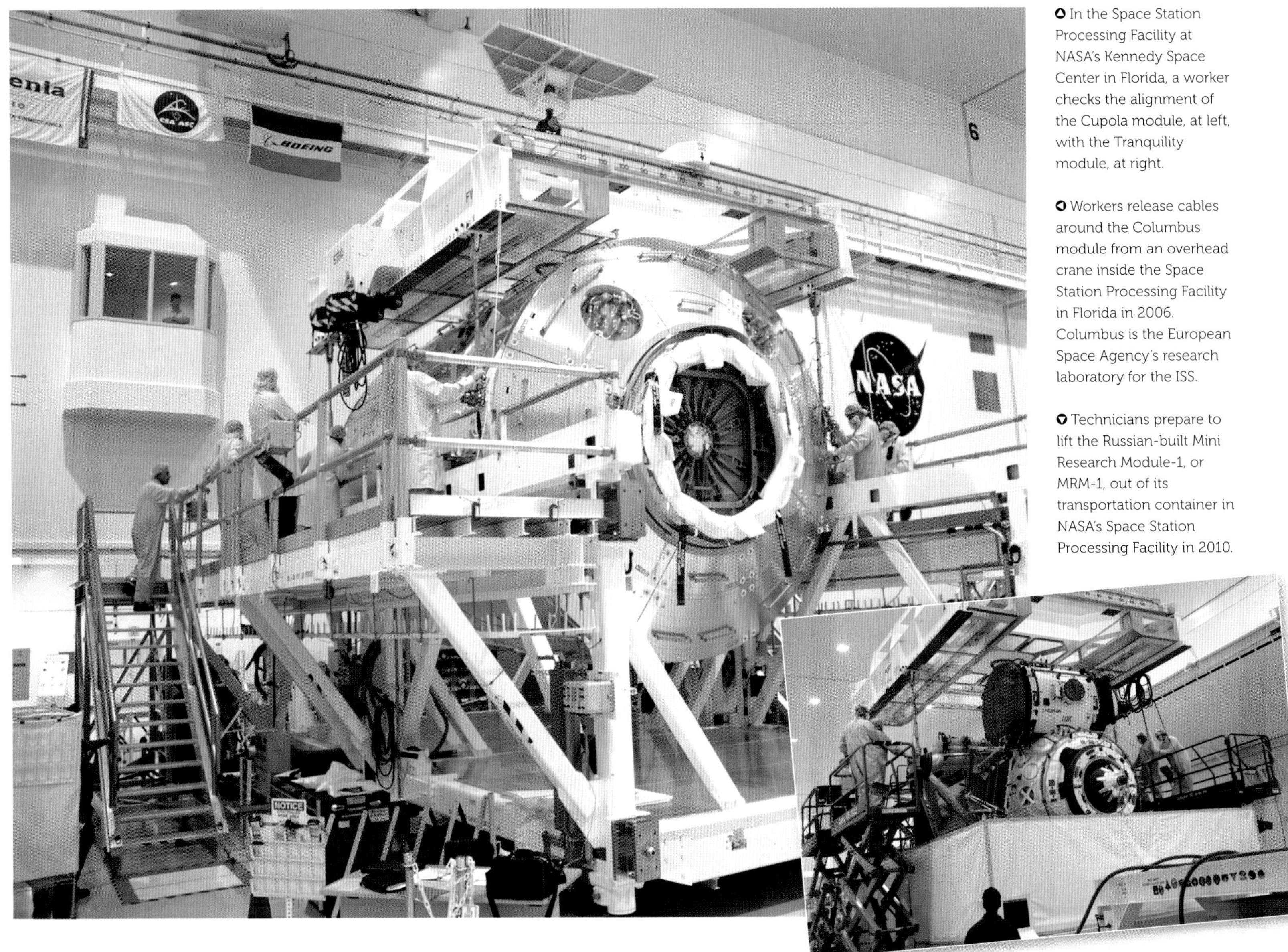

In the Space Station Processing Facility at NASA's Kennedy Space Center in Florida, a worker checks the alignment of the Cupola module, at left, with the Tranquility module, at right.

Workers release cables around the Columbus module from an overhead crane inside the Space Station Processing Facility in Florida in 2006. Columbus is the European Space Agency's research laboratory for the ISS.

Technicians prepare to lift the Russian-built Mini Research Module-1, or MRM-1, out of its transportation container in NASA's Space Station Processing Facility in 2010.

The US Node 2 Harmony, at lower left, and the first pressurized module of the Japanese experimental module Kibo are prepared for launch in the Space Station Processing Facility in Florida in 2003.

An overhead crane in the Space Station Processing Facility lifts a segment of the ISS's integrated truss from its workstand in preparation for its launch in 2007.

laboratory "Destiny" was integrated with two backbone truss segments, as well as an emulator for the already launched Node 1 "Unity." Similar Multi-Element Integration Testing (MEIT) was conducted with other parts of the truss and the mobile transporter that would serve as a railcar for the space station's Canadarm2 robotic arm and with Japan's experiment module and Node 2 "Harmony." These tests and subsequent emulated trials helped to identify issues before the modules actually reached orbit, and it verified that critical power and cooling systems were functioning as they had been designed.

ZARYA ENTERS ORBIT

The first piece of the ISS to reach Earth orbit lifted off on board a Proton-K rocket from the Baikonur Cosmodrome in Kazakhstan on November 20, 1998. The Functional Cargo Block (FGB), aptly christened "Zarya" in Russian or "Dawn" in English, was built by the Khrunichev State Research and Production Space Center in Moscow under a contract with The Boeing Company. Although it was assembled in and launched by Russia, it was financed and owned by the United States. The module was 13 m long by 4 m wide (41 × 13 ft 6 in) and was equipped with three docking ports and provided guidance, propulsion, power (through two solar arrays), and storage for the early phases of ISS assembly. The FGB was originally developed for use with the former Mir space station but was repurposed for the ISS. Zarya was equipped with engines and thrusters that were used to maintain the space station's orbit until the addition of later modules. Its propellant storage tanks were subsequently used to hold fuel reserves for the station's other Russian segment engines. As the ISS's oldest component, Zarya was originally anticipated to have an on-orbit lifetime of 15 years. With the ISS expected to continue operation through 2024, it will have exceeded its design by more than a decade.

A Russian Proton-K rocket launches from the Baikonur Cosmodrome in Kazakhstan carrying the Zarya FGB, the first component of the ISS, on November 20, 1998.

NASA astronaut Nancy Currie and cosmonaut Sergei Krikalev, STS-88 crewmates, use power tools to tighten and loosen fixtures inside the Zarya module during its initial setup in 1998.

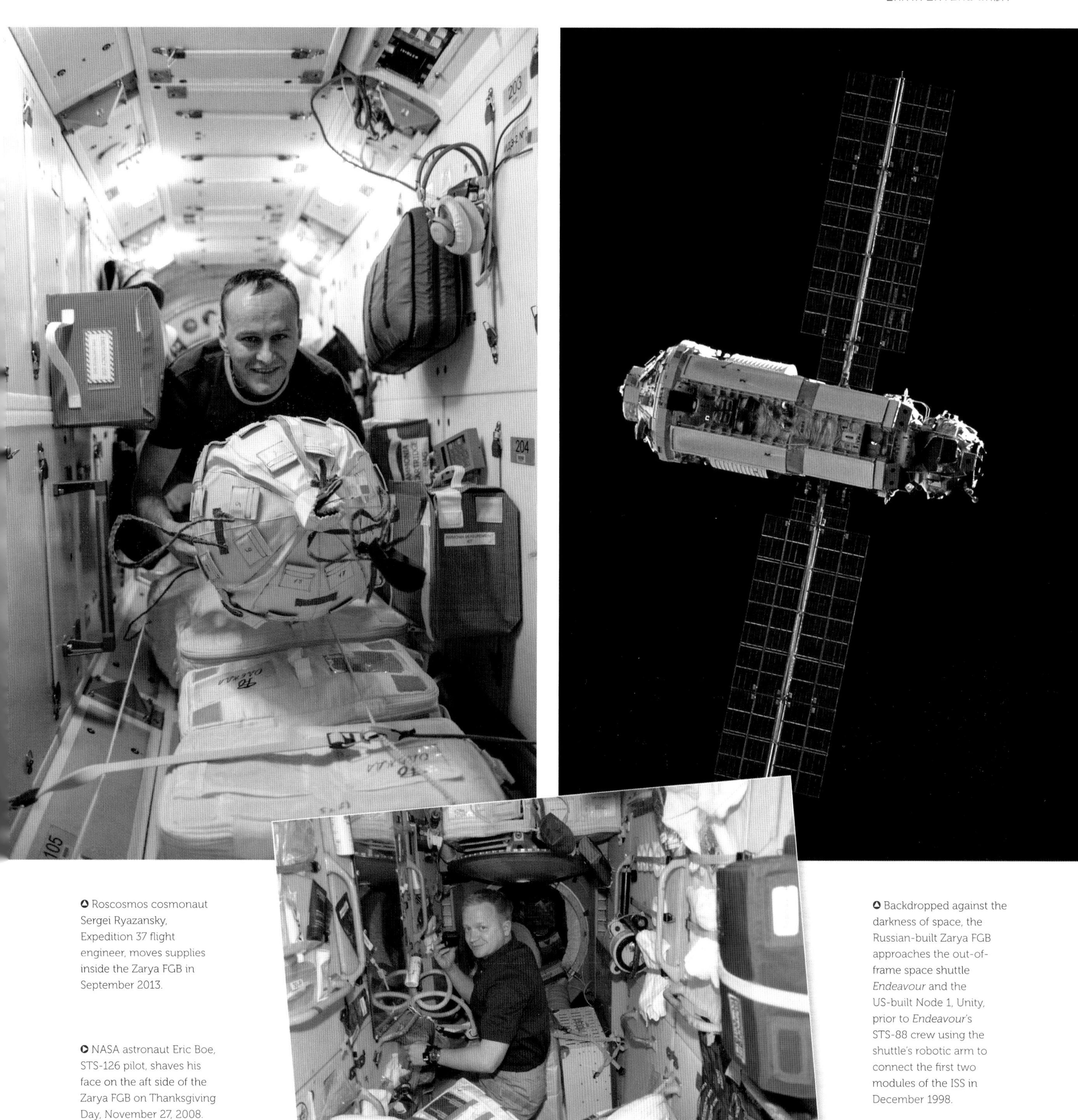

Roscosmos cosmonaut Sergei Ryazansky, Expedition 37 flight engineer, moves supplies inside the Zarya FGB in September 2013.

NASA astronaut Eric Boe, STS-126 pilot, shaves his face on the aft side of the Zarya FGB on Thanksgiving Day, November 27, 2008.

Backdropped against the darkness of space, the Russian-built Zarya FGB approaches the out-of-frame space shuttle *Endeavour* and the US-built Node 1, Unity, prior to *Endeavour's* STS-88 crew using the shuttle's robotic arm to connect the first two modules of the ISS in December 1998.

ASSEMBLY

The assembly of the ISS has been called an engineering marvel. It all began with the mating of the Russian "Zarya" FCB with the US Node 1 "Unity" on December 6, 1998. NASA astronaut Nancy Currie-Gregg used the space shuttle *Endeavour*'s robotic arm to grab hold of Zarya and aligned it with the docking mechanism on Unity. STS-88 mission commander Bob Cabana then fired the orbiter's thrusters to marry the two modules. "We have capture of Zarya," radioed Cabana as the newly born ISS and space shuttle flew over the South Pacific."Congratulations to the crew of the ship *Endeavour*," replied Mission Control in Houston. "That's terrific."

It would take another dozen years for the space station's additional nine modules to be added, as well as the 11-piece backbone truss, four primary power-providing solar array wings, and support components, including Canada's Canadarm2 robotic arm and Japan's exposed facility, the latter a porch for experiments exposed to the vacuum of space. Components continued to be added to the space station even after it was declared "complete" in 2011. An experimental commercial expandable (or inflatable) module was added in 2016 and a Russian multipurpose laboratory module, named "Nauka," was slated to in the 2018–19 timeframe after a two-year delay.

Astronauts aboard the ISS use the Canadarm2 robotic arm to install the S0 (S-Zero) segment of the space station's backbone truss after its delivery by the space shuttle in 2002.

In December 1998, the STS-88 crew aboard space shuttle *Endeavour* began construction of the ISS, joining the US-built Unity Node 2 to Russia's Zarya functional cargo block.

On February 10, 2001, the crews of the space shuttle *Atlantis* and the ISS successfully installed the US Destiny Laboratory onto the space station. The lab added 108 m^3 (3,800 cubic ft) of volume to the station, increasing the on board living space by 41 percent.

The Bigelow Expandable Activity Module (BEAM) was added to the ISS on April 16, 2016. The inflatable room was attached to the aft port of Tranquility Node 3.

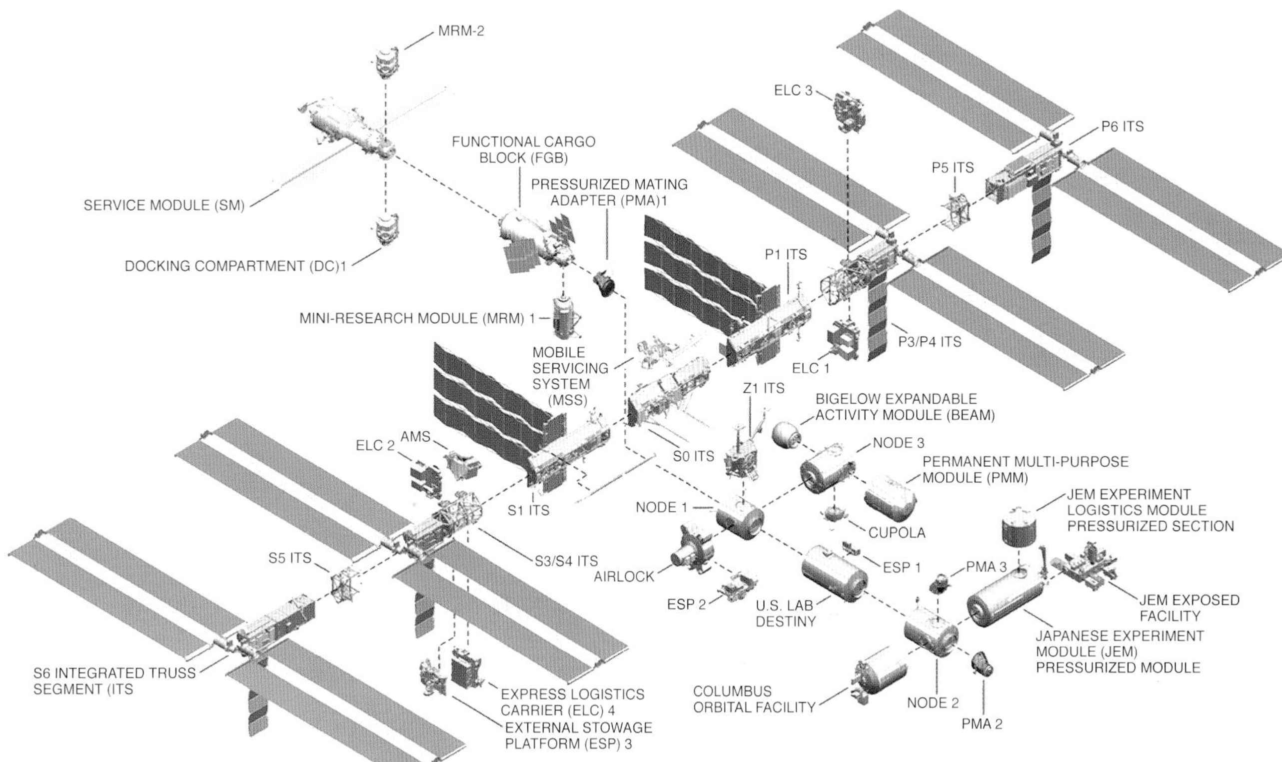

Exploded-view diagram of the ISS, showing how the orbiting outpost came together between 1998 and 2016. The Russian segment can be seen above the backbone truss; the US Operating Segment in front.

EXTRAVEHICULAR ACTIVITIES

"I remember, pre-ISS, talking about the hundreds, or more than one hundred, EVAs that are going to be required for assembly and thinking that was a huge mountain to climb," Expedition 16 astronaut Dan Tani said in December 2007 after performing the 100th spacewalk in support of ISS assembly and maintenance. EVAs, or extravehicular activities—commonly known as spacewalks—were crucial to assembling the ISS and keeping it running. Robotic arms could be used to mate modules and components together, but it took astronauts' and cosmonauts' hands to connect the many power and cooling system cables that allow the space station to operate a single spacecraft. Spacewalks also installed exterior cameras, mounted experiment packages and, when needed—sometimes urgently—repaired and replaced components that malfunctioned.

Two types of spacesuits are used for spacewalks outside the ISS. From the Russian segment, cosmonauts wear Orlan spacesuits, similar to the ones used on Salyut and the Mir space station. From the US airlock "Quest," NASA astronauts and their international partners wear modular Extravehicular Mobility Units (EMUs), a spacesuit that was first used during the space shuttle program. On May 12, 2017, NASA astronauts Jack Fischer and Peggy Whitson completed the 200th spacewalk in support of the ISS. By coincidence, Whitson had also been on the 100th spacewalk with Daniel Tani ten years earlier.

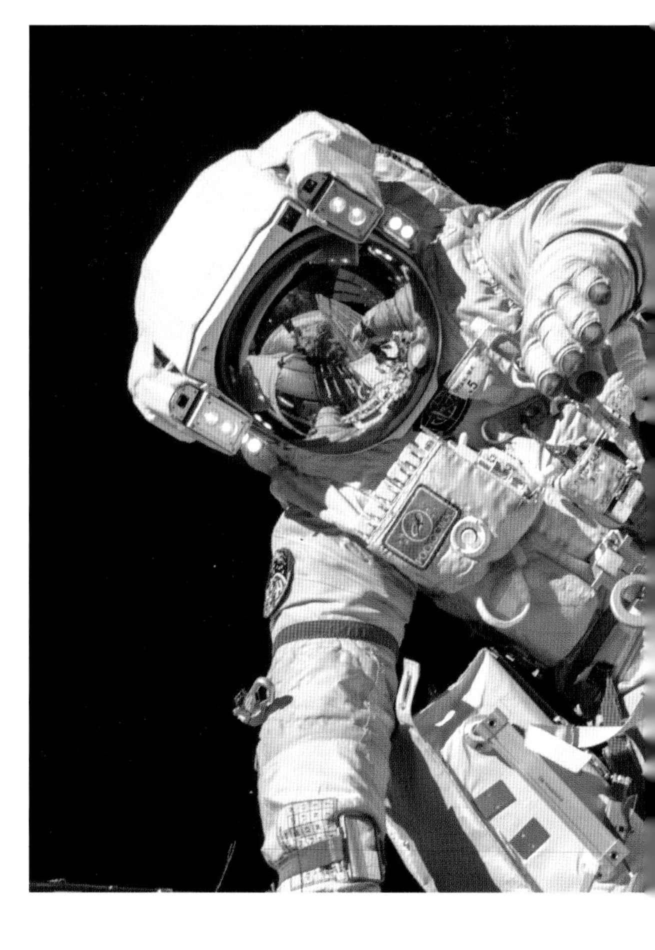

Oleg Kotov, Expedition 37 flight engineer, conducts a spacewalk to set up a combination EVA workstation and biaxial pointing platform, November 2013.

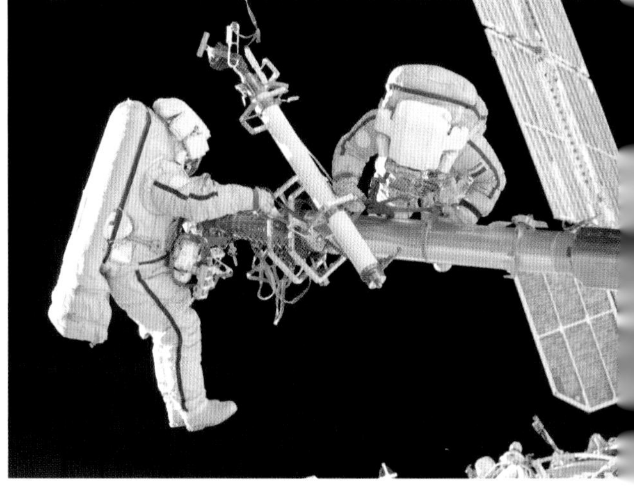

Cosmonauts Oleg Kononenko (left) and Anton Shkaplerov, Expedition 30 flight engineers, move the Strela-1 crane from the Pirs docking compartment using the Strela-2 boom.

NASA astronaut Jack "2Fish" Fischer works outside the Japanese Kibo laboratory module during the 200th spacewalk in support of ISS assembly and maintenance on May 12, 2017.

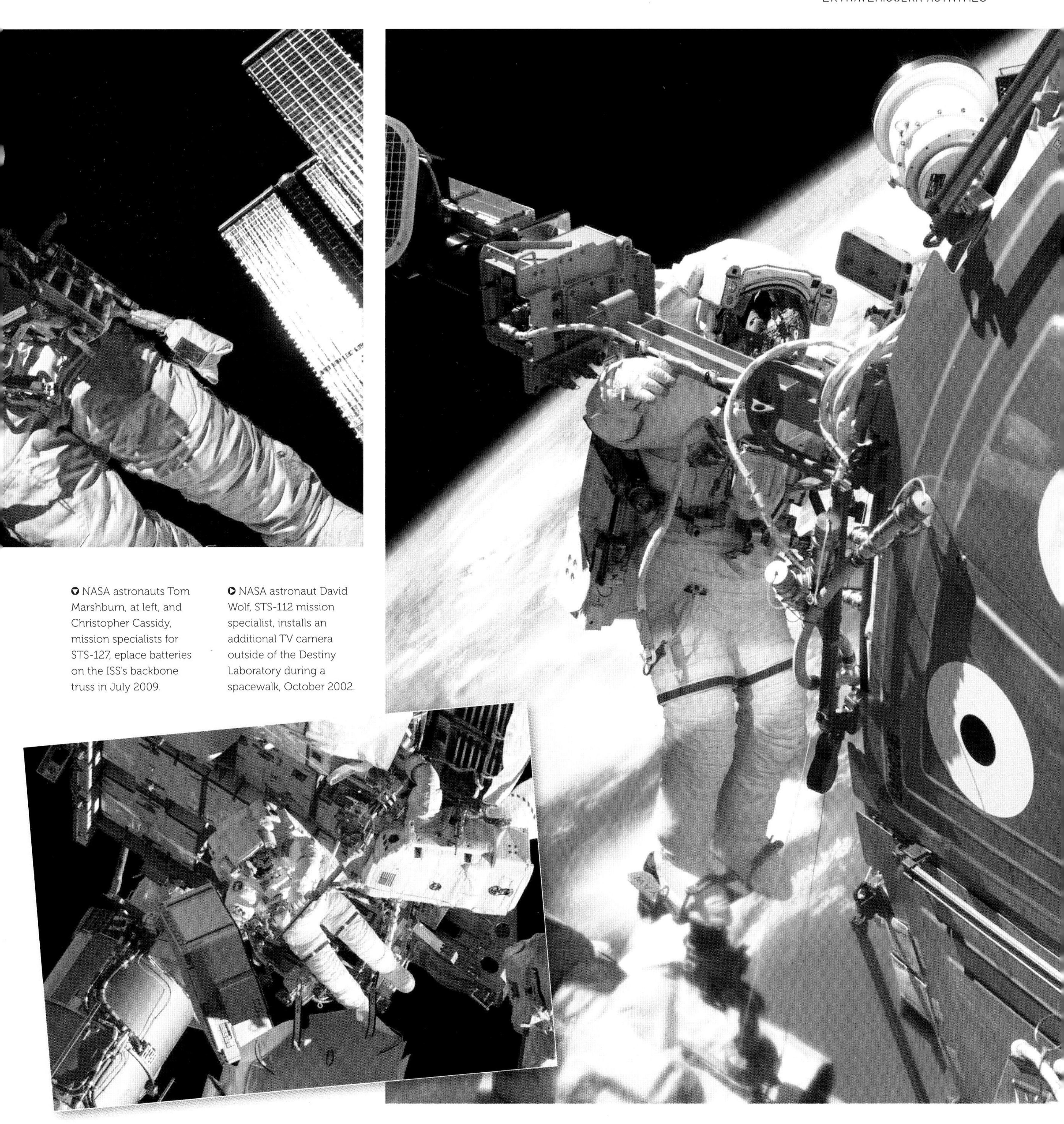

⭘ NASA astronauts Tom Marshburn, at left, and Christopher Cassidy, mission specialists for STS-127, eplace batteries on the ISS's backbone truss in July 2009.

⭘ NASA astronaut David Wolf, STS-112 mission specialist, installs an additional TV camera outside of the Destiny Laboratory during a spacewalk, October 2002.

INSIDE THE ISS

The ISS serves as a house for its crews, a microgravity laboratory for science investigations, a platform for observatories pointed at Earth and the universe, and a testbed for future exploration technologies. Thus, looking inside any of its modules can leave a mixed impression as to its overall purpose. "It's definitely our home at this point in time," observed NASA astronaut Sunita "Suni" Williams, as she surpassed her first 100 days aboard the ISS in October 2012.

1

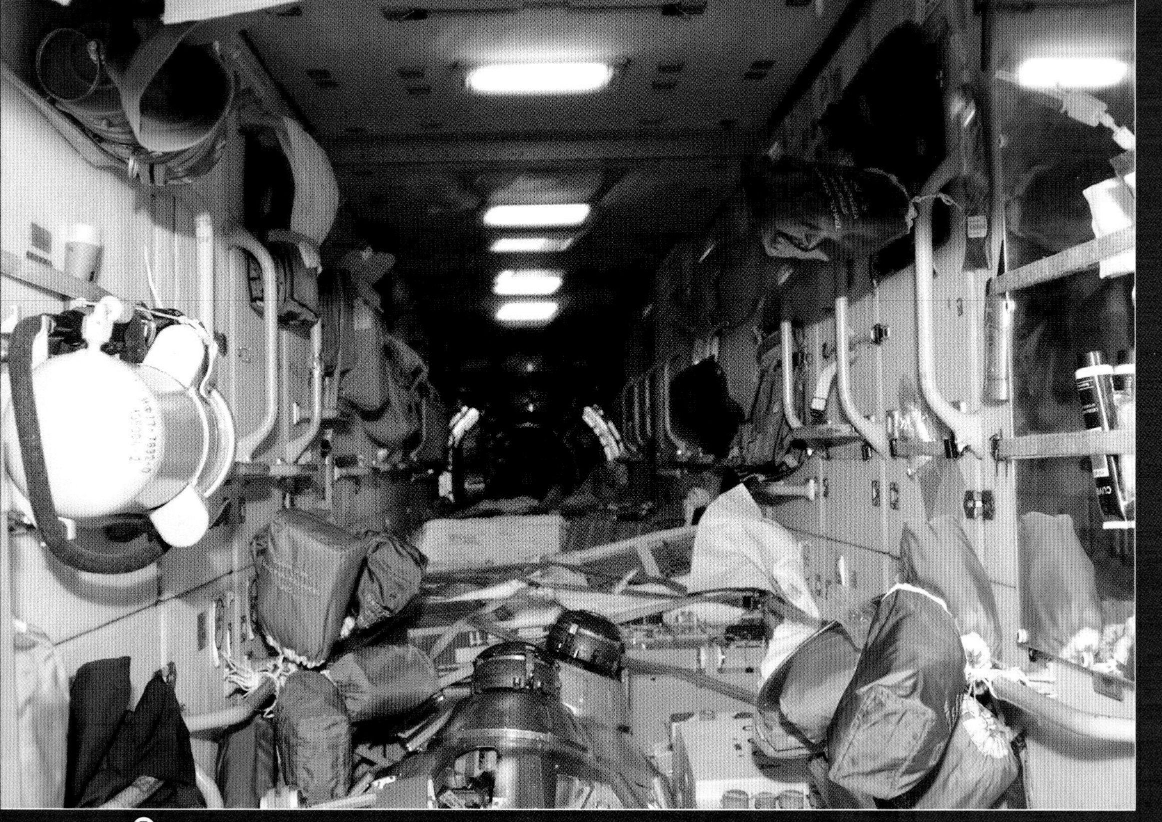
2

3

1. An overall interior view of Russia's Zvezda service module photographed by an Expedition 17 crew-member on the ISS in August 2008.
2. This view shows supplies and equipment stored inside Russia's Zarya functional cargo block as photographed by an ISS Expedition 10 crew member in April 2005.
3. Stowage bags are packed into the Pressurized Mating Adapter 1 (PMA-1) and food canisters lined up against the wall in Node 1 Unity of the ISS, as photographed in 2005.
4. A view inside the Destiny laboratory taken by ESA astronaut Alexander Gerst in 2014. NASA astronaut Reid Wiseman can be seen cleaning the space station, a chore all crew members do every Saturday.

Science equipment and life support hardware are mounted in refrigerator-size modular racks that line the walls of much of the US operating segment's rooms. These removable fixtures can be brought forth from the hull to allow for maintenance and for the contingency that the outer wall of the module is damaged by a micrometeorite impact. Most of the time though, they form the inner boundaries of the modules' workspace.

These racks support combustion chambers, plant growth facilities, animal research habitats, and microgravity science gloveboxes, the latter a windowed container with integrated gloves for handling hazardous materials while isolating them from the atmosphere aboard the space station. The racks also serve as mounting points for other science equipment, as well as the computer laptops used to operate and record data from the experiments, control the space station's exterior robotic arms, and monitor the outpost's systems. Throughout the decades of crews living and working on board the ISS, the racks have also gained some character with the addition of decals and small knickknacks brought and left behind by their users. We have our little things scattered here and there and we know where they are," said Williams.

CREW QUARTERS

To provide crew quarters for the astronauts, four small "cabins" are installed on the US segment with another two similar booths on the Russian side. The quarters offer a storage space for the crew members' personal items, a laptop and tablet for communication with the ground (including their family members), a sleep restraint (similar to a sleeping bag), lighting, and sound isolation.

Two waste and hygiene compartments provide the bathroom facilities for the crew. The toilets rely on airflow, rather than gravity, to collect waste, which is then directed into storage containers for subsequent disposal (in the case of solid waste) or recycling (for liquid waste). "It occurs to me that our regenerative life support equipment is really just a fancy coffee machine," wrote NASA astronaut Don Pettit in

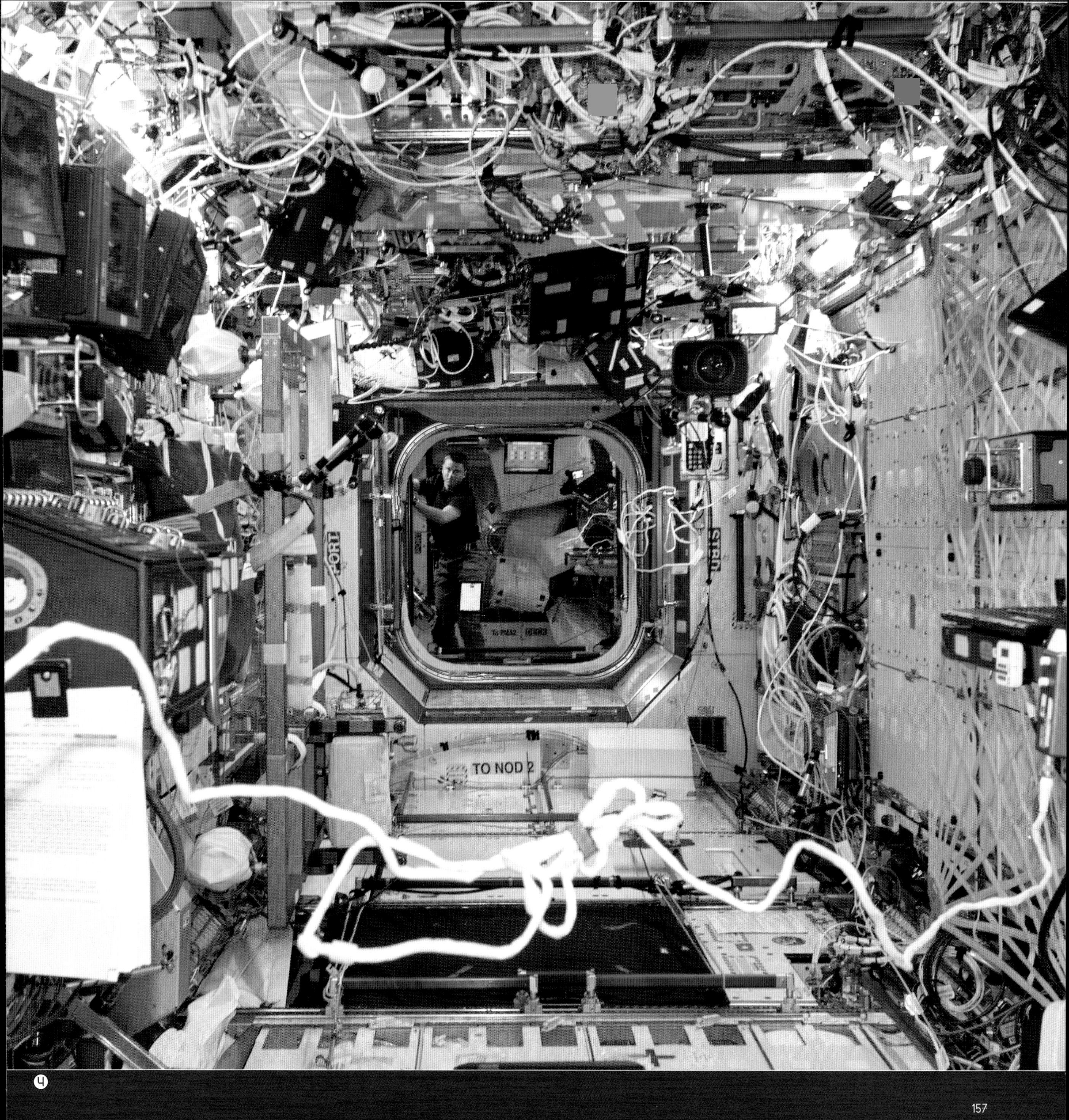

4

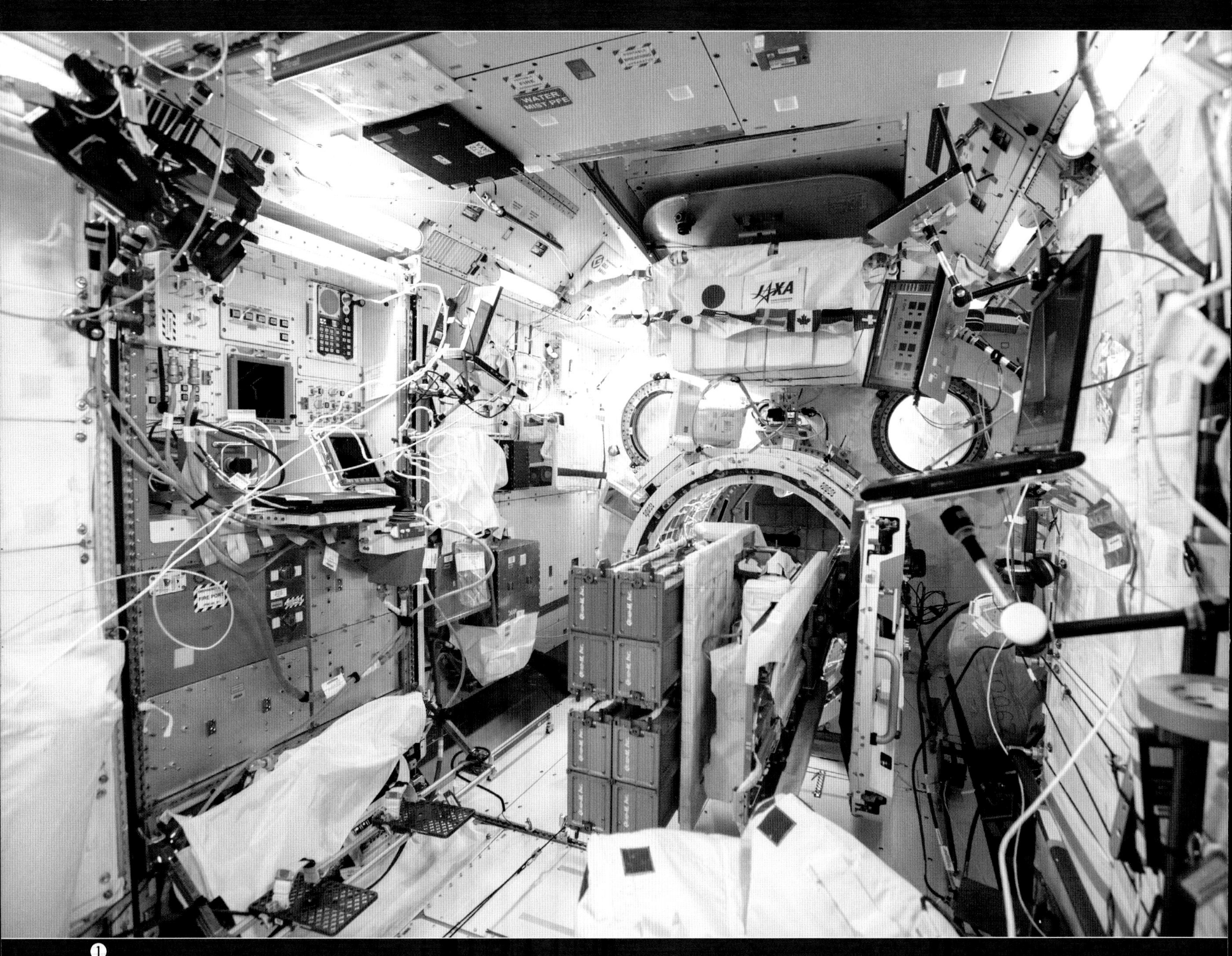

1

his journal from on board the space station in February 2012. "It makes yesterday's coffee into today's coffee."

FIXTURES AND FITTINGS

Other major fixtures aboard the space station include the crew's exercise devices, necessary for maintaining their muscle and bone mass in the microgravity environment of space. The ISS has been equipped with US-designed treadmills and ergometers (similar to a stationary bicycle), as well as a NASA-developed Advanced Resistive Exercise Device (ARED), which simulates weight lifting without the need for gravity.

Much of the other wall (ceiling and floor) space is taken up by storage lockers and bags. Spare equipment, water reserves, and tools are strapped to the sides of the modules, especially within the connecting passageways between laboratories. Additional stowage is provided inside two designated logistics modules, as well as within a prototype expandable room following its first two years being verified safe for crew use. Several windows allow crew observations of Earth and the exterior of the space station. In addition to the seven windows in the ISS's Cupola, the Russian service module ("Zvezda") has 14 windows, including one in each of the crew quarters. A science-dedicated window is in the US laboratory ("Destiny").

The ISS is also equipped with four airlocks—three for supporting spacewalks and one for deploying payloads robotically (a fifth, commercial payload airlock was slated to be added to the space station in 2018). Russian spacewalks begin from the Pirs docking compartment. US excursions have been made from the Quest joint airlock, but could faciltate Russian spacewalks. Science payloads and small satellites, called nanosatellites, are deployed using the Kibo airlock in the Japanese experiment module.

As the space station has grown and evolved, more "creature comforts" have been introduced to make the crew members more comfortable away from Earth. Two projectors and projection screens provide the astronauts and cosmonauts with access to television and movies, as well as

2

3

shared training experiences. An experimental coffee machine and a prototype bread maker have sought to expand the crew's menus, while the introduction of personal tablet computers has provided the residents with greater access to multimedia and the internet.

4

1. An interior view of the Japan Aerospace Exploration Agency's Kibo experimental module, including its airlock used to expose various science payloads to the vacuum of space.
2. NASA astronaut Scott Kelly described his crew quarters, pictured here, in 2015: "My bedroom aboard ISS. All the comforts of home. Well, most of them…"
3. Interior of the European Space Agency's Columbus laboratory as photographed in 2014 by an Expedition 40 crew member on the ISS.
4. NASA astronauts Ron Garan (bottom) and Cady Coleman, ESA astronaut Paolo Nespoli (left), and Russian cosmonaut Alexander Samokutyaev, all serving as Expedition 27 flight engineers, pose for a photo by popping symmetrically out of their crew quarters in the Harmony node of the ISS.

CUPOLA

Ask any astronauts who visited the ISS after February 2010 what was their favorite room aboard the outpost, they are almost certain to reply, "the Cupola." Designed by NASA, largely built by ESA, and launched by NASA's space shuttle *Endeavour* together with the Node 3 "Tranquility" to which it is attached, the Cupola (Italian for "dome") is part panoramic control room and part observation deck for the space station. The 3 m (10 ft) diameter Cupola supports seven windows in a hexagonal layout, providing a sweeping view of the Earth rotating below. (The top window is the largest window ever put on a manned spacecraft.) "The astronauts,

Tranquility Node 3 and its Cupola, photographed during the STS-130 mission that installed both of the components on the ISS.

This July 12, 2011, view shows the Cupola, back-dropped against some parts of solar array panels, on the ISS.

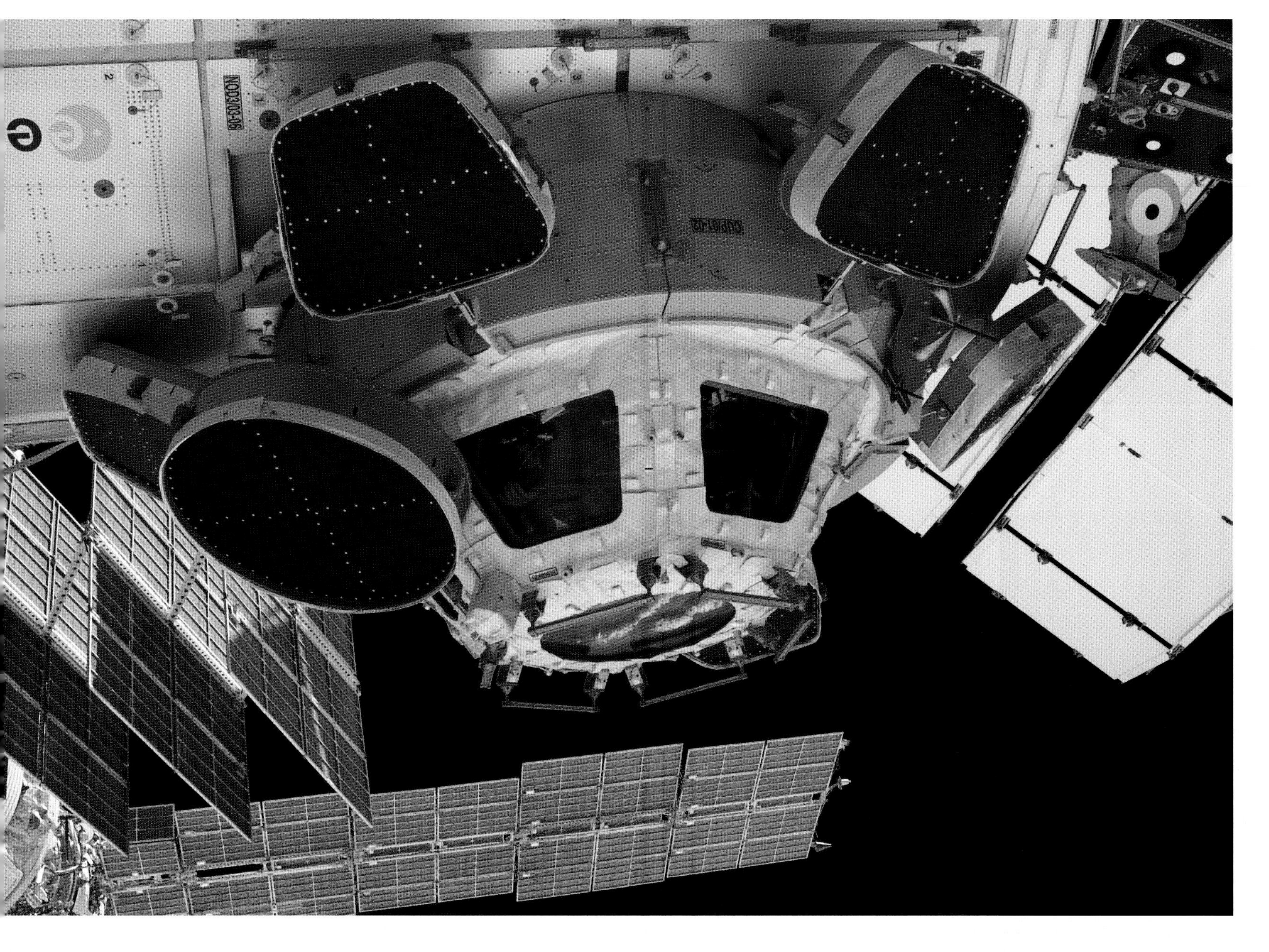

who are accustomed to views you and I can't really describe, were moved to tears when they looked outside the windows of the Cupola for the first time," said flight director Bob Dempsey. In addition to being a preferred location for Earth photography and reflection, the astronauts use the computer stations mounted in the Cupola to control the Canadarm2 robotic arm to capture visiting vehicles bringing cargo and supplies to the US side of the ISS. The Cupola is outfitted with large shutters to protect the windows from damage when not in use. Still, that has not prevented the multilayer panes from being chipped by micrometeorites over the course of the Cupola's time on the ISS.

European Space Agency astronaut Alexander Gerst, Expedition 40 flight engineer, enjoys the view of Earth from the windows in the Cupola in June 2014.

Interior view from the Cupola, which houses one of the ISS's two robotic work stations used by astronauts to manipulate the large robotic arm, seen through the right window.

CREWS

In its first 15 years of human habitation, the ISS hosted 223 individuals from 17 nations for periods ranging from a week to six months and longer, an incredible achievement in human and mechanical engineering The space station is capable of supporting a crew of nine for extended stays. Typical crew complements during the ISS's first 50 expeditions included a minimum of two and a maximum of six astronauts and cosmonauts. The crews were generally evenly divided between Russian and US contingents, the latter including NASA, ESA, Canadian Space Agency and Japan Aerospace Exploration Agency (JAXA) astronauts. With the introduction of US commercial crew vehicles servicing the space station in 2018–19,

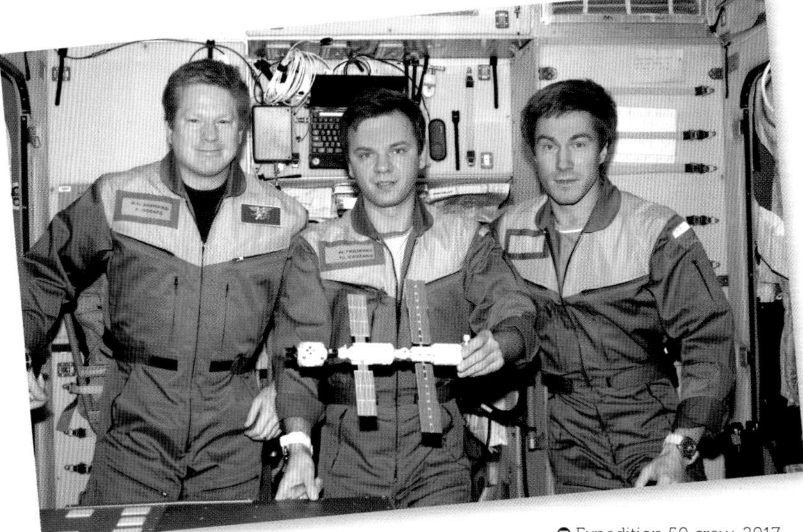

Expedition 1 crew members (left to right) Bill Shepherd, Yuri Gidzenko, and Sergei Krikalev pose with a model of the ISS.

Expedition 50 crew, 2017. Front row (left to right): Andrei Borisenko, Shane Kimbrough, and Sergei Ryzhikov. Back row: Thomas Pesquet, Peggy Whitson, and Oleg Novitsky.

Nine Expedition 37 crew members gather for a group portrait in 2013. Pictured clockwise from bottom right are Luca Parmitano, Karen Nyberg, Fyodor Yurchikhin, Mikhail Tyurin, Sergei Ryazansky, Rick Mastracchio, Oleg Kotov, Koichi Wakata, and Michael Hopkins.

Expedition 7 crewmates Ed Lu, at the musical keyboard, and Yuri Malenchenko share a light moment during off-shift time in 2003.

the orbiting laboratory's crew was expected to increase by one to seven, with the additional person being a USOS crew member.

In July 2009, the ISS set yet another a record for the most people aboard the same spacecraft at the same time, including the six members of the Expedition 20 resident crew and the seven people on space shuttle *Endeavour*'s STS-127 crew. The 13 people represented five countries: the United States, Russia, Canada, Belgium, and Japan. The feat was repeated a year later with the arrival of space shuttle *Discovery*'s STS-131 crew during the space station's Expedition 23.

TRAINING

To orbit the Earth as crew members aboard the ISS, astronauts and cosmonauts have to first circle the world—on the ground—to prepare for their expeditions. Astronauts and cosmonauts typically spend two to three years training to spend five to six months aboard the ISS (though that training has been reduced to as little as six months in at least one case of a late-assigned NASA astronaut). Given the international nature of the space station, that preparation includes traveling to training centers in the United States, Russia, Germany, Canada, and Japan. Crew members are instructed in operating the space station's systems, controlling the robotic arms, flying the spacecraft that will take them to and from the orbiting laboratory, operating the science experiments they will conduct on orbit, and language lessons in English or Russian (or both, as needed.) "It's like getting a full four-year college degree compressed into two years," Don Pettit, a NASA astronaut who served on two expeditions totaling more than 300 days on the ISS, told CNN in a 2017 interview. "There are no summer breaks."

Astronauts and cosmonauts practice for spacewalks underwater in neutral buoyancy labs. They also undergo survival training in the case of off-course landings, and train in mockups of the space station's modules to become familiar with the layout of the sprawling orbital complex.

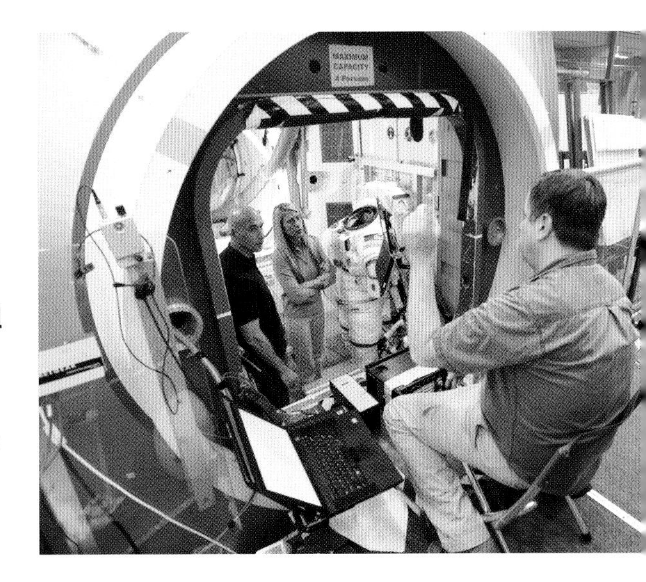

Instructor Regan Cheney (right) assists Expedition 36/37 crew members Luca Parmitano and Karen Nyberg in the Space Vehicle Mock-up Facility at the Johnson Space Center.

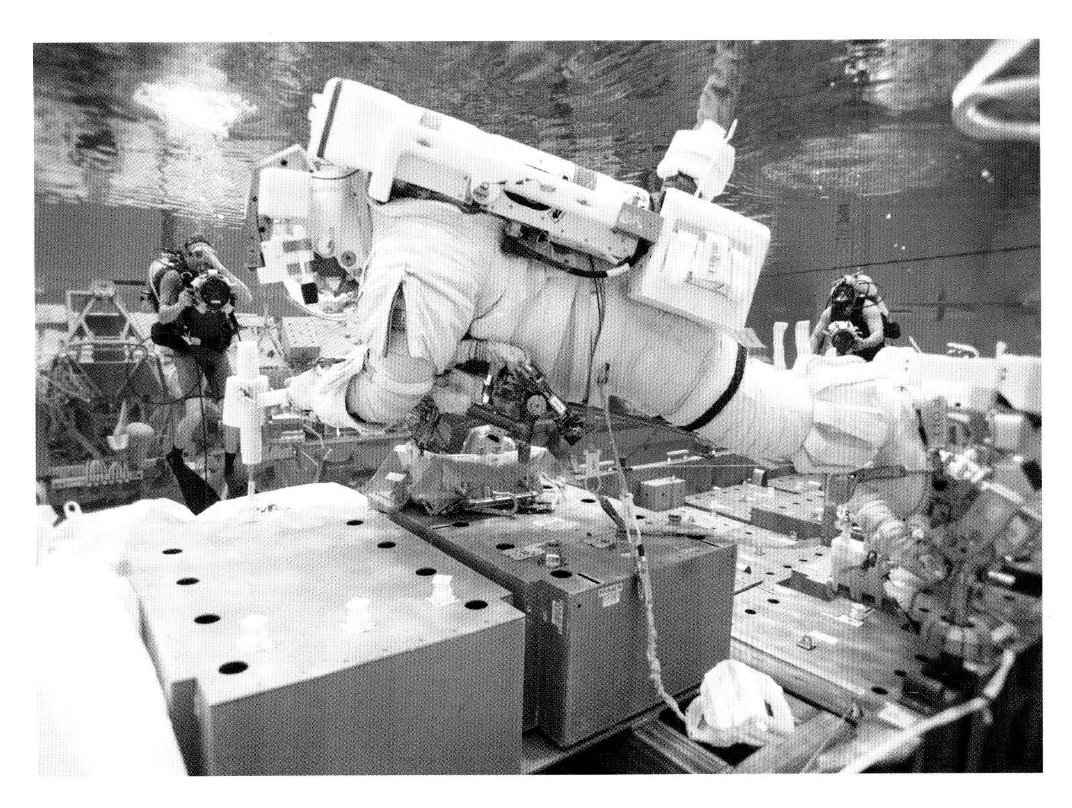

Training underwater, astronaut Terry Virts simulates an EVA, or a spacewalk, in Johnson Space Center's Neutral Buoyancy Laboratory in Houston, Texas.

Expedition 46 crew members Tim Kopra (left), and Tim Peake engage in emergency scenario training in the Space Vehicle Mockup Facility at Johnson Space Center.

Expedition 31 crew members Joe Acaba (right) Sergei Revin and Gennady Padalka enter a mockup Soyuz spacecraft to train in Star City, Russia.

Expedition 52 flight engineer Randy Bresnik of NASA is seen inside the Soyuz simulator at the Gagarin Cosmonaut Training Center in Star City, Russia.

PARTNERS

As its name implies, the ISS differs from prior orbiting outposts in its multicountry approach. A treaty signed in January 1998 established the framework for the space station as a partnership between 15 nations. Canada, Japan, Russia, the United States, and 11 member states of ESA—Belgium, Denmark, France, Germany, Italy, The Netherlands, Norway, Spain, Sweden, Switzerland, and the United Kingdom—agreed to establishing and operating a "permanently inhabited civil space station for peaceful purposes." Subsequent agreements between the United States and each of the four other partners (Canada, Japan, Russia, and ESA) established NASA as the manager of the ISS.

The signatories of the 1998 Intergovernmental Agreement on Space Station Cooperation pose in front of ISS Unity Node 1, which is being prepared for launch.

POCKOCMOC

Agencies partnered in the ISS. From top: NASA, Roscosmos, Canadian Space Agency, Japan Aerospace Exploration Agency (JAXA), European Space Agency (ESA).

⊙ NASA astronaut Scott Kelly, Expedition 26 commander, works with the flags of the international partners inside the Destiny laboratory on the ISS in March 2011.

⊙ A commemorative certificate presented at the official signing of the ISS agreements in January 1998, featuring the flags of the nations participating in the program.

In 1997, NASA signed an additional agreement with Brazil to bring the South American country in as a space station partner. After ten years, however, Brazil was unable to afford its hardware commitments and left the program. (A Brazilian astronaut, Marcos Pontes, later visited the ISS on a ten-day Russian Soyuz taxi mission.) In 2007, China expressed interest in participating in the space station, but due to political concerns, primarily from the United States, did not become a partner and instead pursued the construction of its own orbiting laboratory.

PARTNER FACILITIES

The ISS is operated as one spacecraft with two distinct physical segments. The Russian side comprises the first element of the ISS, the Zarya FGB, and the service module Zvezda, as well as the Rassvet and Poisk mini research modules and Pirs docking compartment. A Russian multipurpose research laboratory, named Nauka, is planned for addition to the ISS in the 2018 to 2019 timeframe.

The US Operating Segment (USOS) includes NASA's Destiny Laboratory and three connecting nodes, Unity, Harmony, and

Japanese astronaut Koichi Wakata, Expedition 18/19 flight engineer, works inside the Kibo laboratory airlock in April 2009.

The Japanese Experiment Module Kibo laboratory and Exposed Facility, viewed through a window on the ISS, 2009.

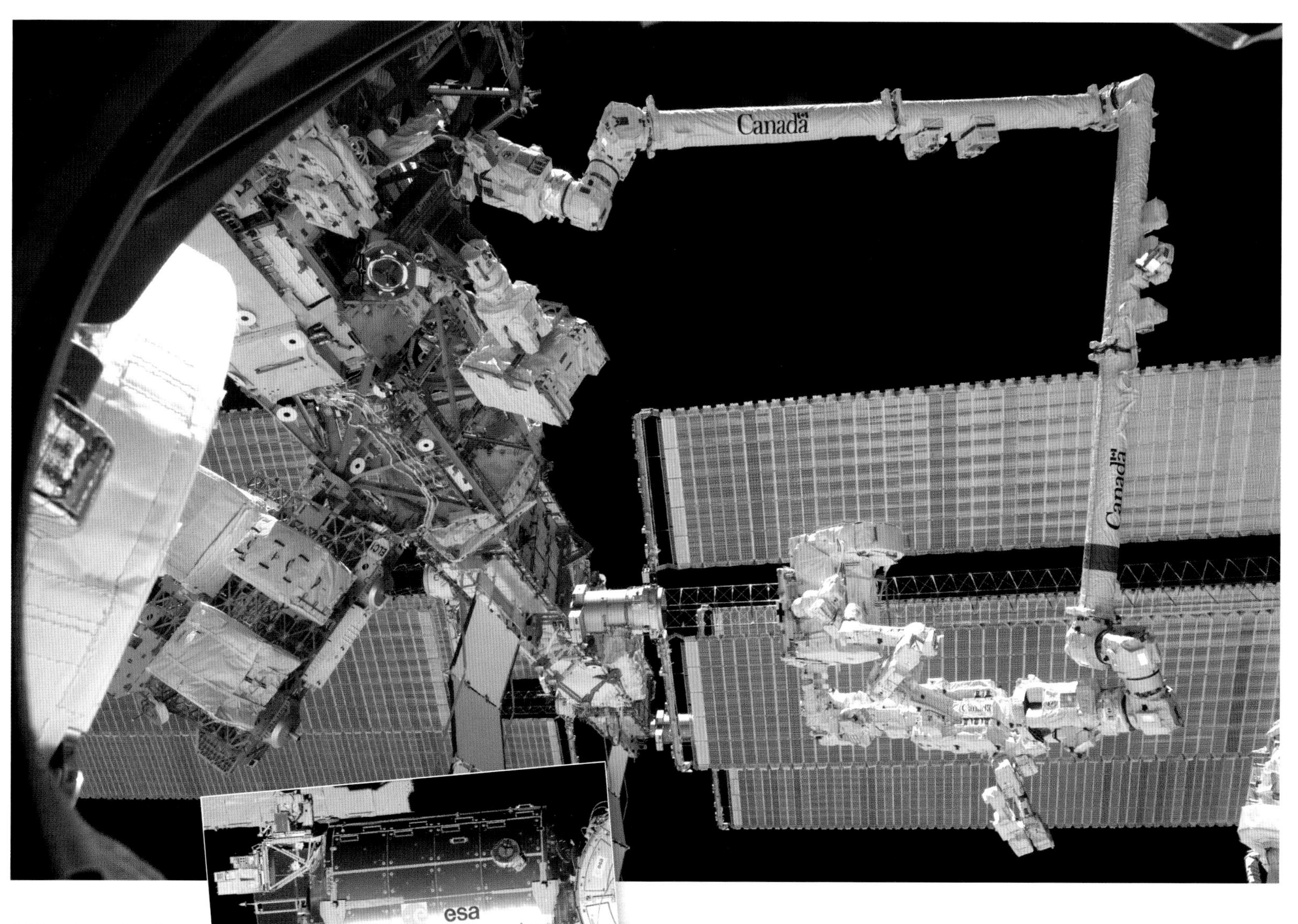

⭘ While attached on the end of the Canadarm2 robotic arm, Dextre, the CSA's robotic "handyman," is photographed by an Expedition 26 crew member in 2011.

⭘ An exterior view of ESA's Columbus laboratory, as seen in 2009. The lab is the largest ESA contribution to the ISS.

Tranquility. The USOS also encompasses the Columbus laboratory contributed by the European Space Agency (ESA) and the Kibo experiment module, including a logistics module and exposed facility, from JAXA. Canada's participation in the space station includes the Canadarm2 robotic arm, the primary manipulator mounted on the ISS's backbone truss, and the Dextre special-purpose dexterous manipulator, a two-armed robot that is used to move spare equipment and install hardware in place of spacewalking astronauts.

In addition to their in-space facilities, the partners also support the space station's crew from a number of ground-control buildings. NASA's Mission Control manages USOS operations from Houston, Texas, while a payload operations center oversees science experiments from Huntsville, Alabama. The TsUP located in Korolyov, outside of Moscow, controls the Russian segment. Additional control centers are operated by ESA in Cologne, Germany, and by JAXA in Tsukuba, Japan.

⭘ The Japanese Experiment Module Kibo includes an external platform for payloads, an airlock, and a robotic arm.

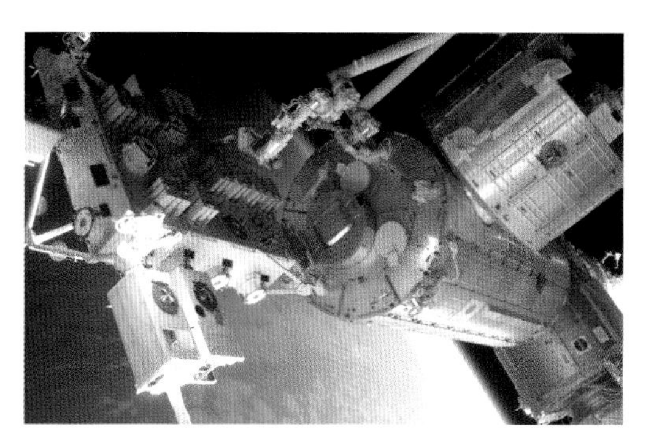

ZVEZDA SERVICE MODULE

The third component to be added to the ISS was the Russian-built service module, "Zvezda," or "Star" in English. Launched on a Proton-K rocket (a version of the Proton), Zvezda became the primary module on the space station's Russian segment after docking to the Zarya FGB on July 26, 2000. Due to financial constraints, the launch of Zvezda had been delayed by a year and a half, presenting an early hurdle for the developing space station program. Originally designed and partially built to serve as the core of a second Mir space station that was never realized, the service module is about the size of a city bus, measuring 13 m (43 ft) long and 4 m (13 ft 6 in) in diameter. It is equipped with two solar arrays for power and four docking ports (three as part of a transfer compartment at the module's front where it is connected to Zarya and one at its aft). Zvezda is also outfitted with engines used to reboot the space station in orbit and the antennas used by the Russian segment to communicate with visiting vehicles and the TsUP mission control center outside of Moscow. The habitable volume of the service module includes the crew quarters for two cosmonauts, a waste and hygiene compartment (one of the two bathrooms on the space station), life support systems for the Russian segment, and a total of 14 windows.

A Russian Proton-K rocket lifts off on July 12, 2000, with the Zvezda service module, the third major component to be added to the ISS.

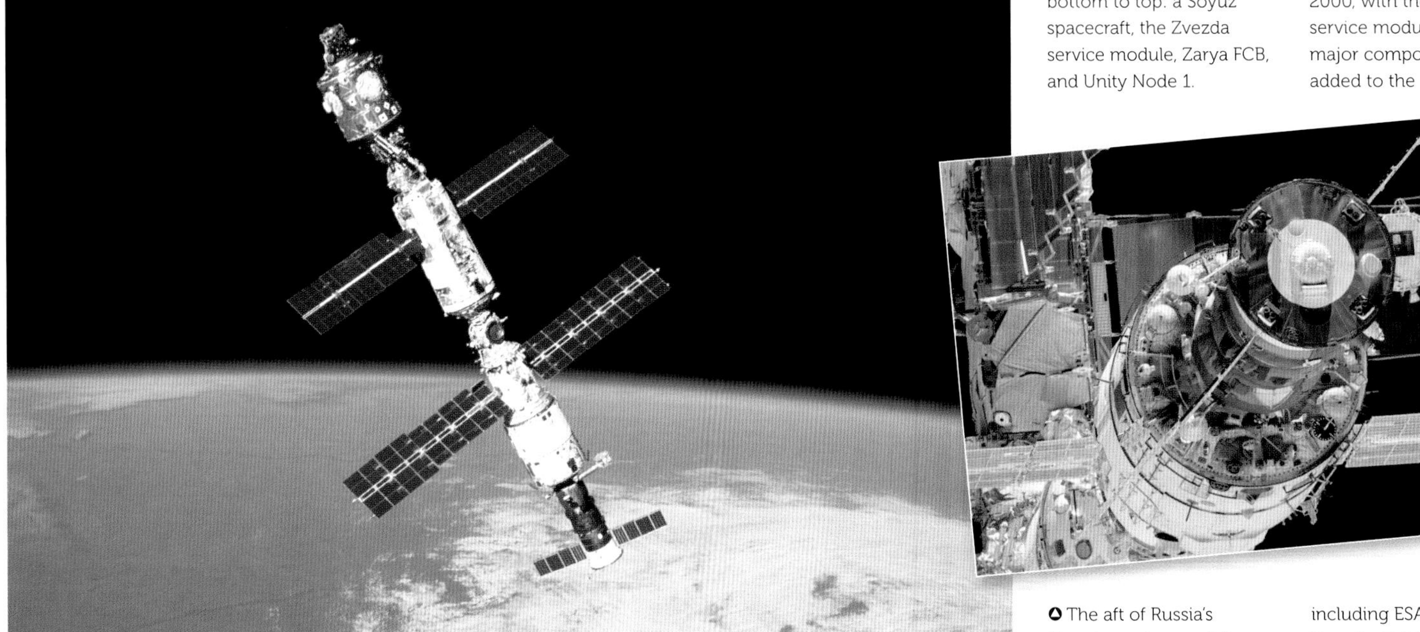

The ISS as seen in September 2000. From bottom to top: a Soyuz spacecraft, the Zvezda service module, Zarya FCB, and Unity Node 1.

The aft of Russia's Zvezda service module provided a docking port for visiting vehicles to the ISS, including ESA's Automated Transfer Vehicle (ATV) and Russia's Progress (pictured) and Soyuz spacecraft.

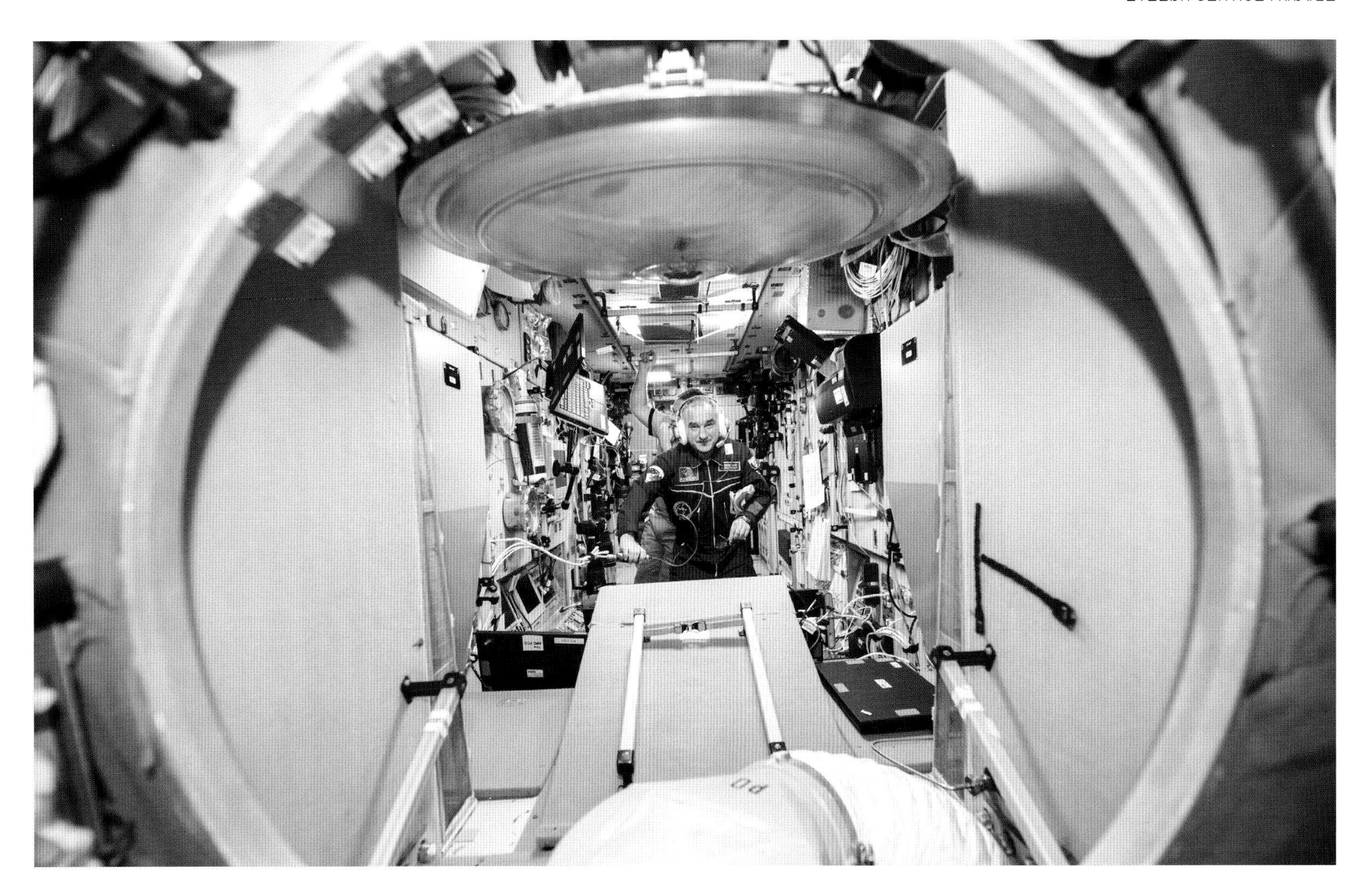

Russian cosmonauts Alexander Skvortsov and cosmonaut Maxim Suraev work in the Zvezda service module during Expedition 40 in September 2014.

Expedition 52 crew members Fyodor Yurchikhin, middle foreground, and Jack Fischer in the Zvezda service module monitoring the docking of a Russian Progress cargo ship in June 2017.

SUPPLYING THE ISS WITH CREW AND CARGO

Unlike the Salyut, Skylab, and Mir space stations that preceded it, the ISS was not designed to be crew-tended. From the moment it was capable of doing so, the ISS began hosting its first residents and since then it has never been vacant. ISS crews are organized into expeditions. For its first 12 crews, Russian cosmonauts and American astronauts served on a single expedition and then returned to Earth.

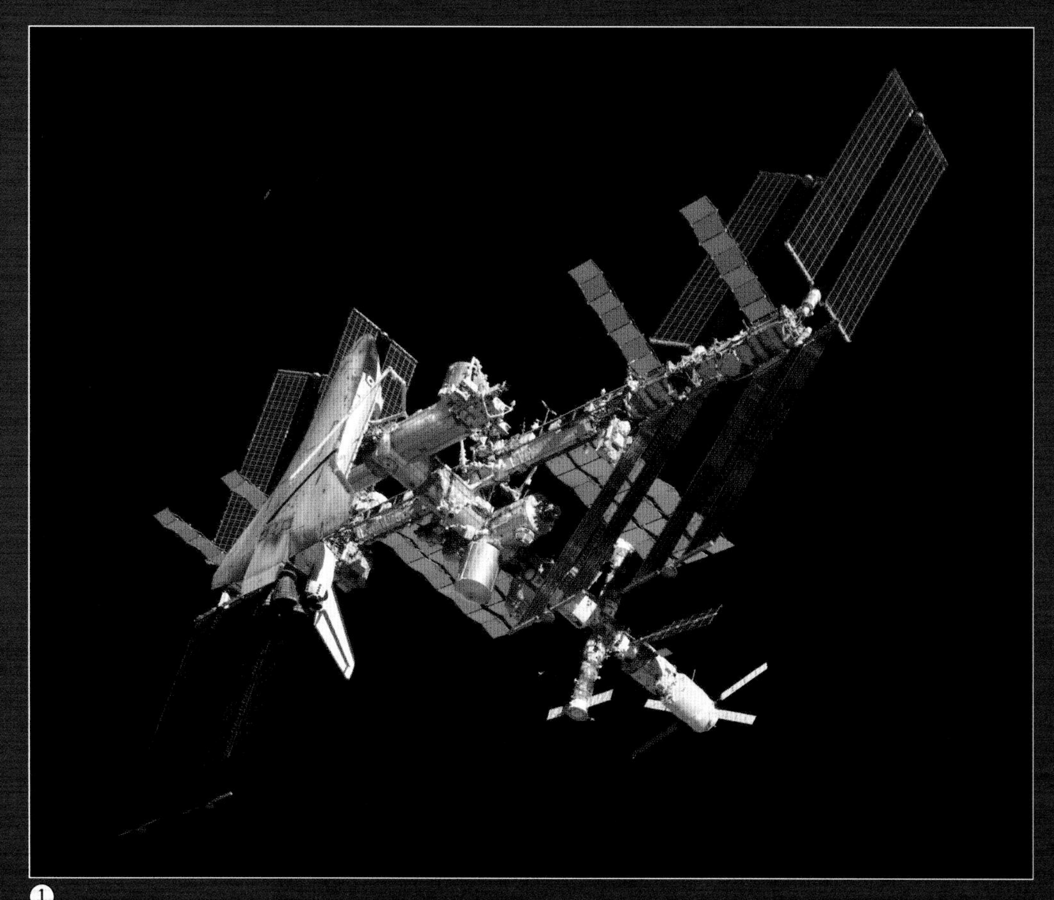

❶

❷

For example, NASA astronaut William Shepherd and Roscosmos (the Russian space agency) cosmonauts Sergei Krikalev and Yuri Gidzenko comprised the Expedition 1 crew from October 31, 2000, through March 21, 2001. On March 8, 2001, Yuri Usachev of Roscosmos and Jim Voss and Susan Helms of NASA arrived for a brief handover period and then took over as the Expedition 2 crew after the Expedition 1 crew departed.

That rotation continued until July 4, 2006, when German astronaut Thomas Reiter of ESA arrived at the space station aboard the US space shuttle *Discovery* and transferred to the Expedition 13 crew, already three months into their stay. Two months later, Reiter joined the Expedition 14 crew, beginning a staggered, overlapping pattern that eventually expanded to include three crew members.

DELIVERING AND ROTATING CREW

By 2011, when assembly of the space station was declared complete, the norm was for crews of three members to serve on two consecutive expeditions, arriving at the outpost about half way through the first expedition and departing about half way through the next. That staffing plan was only disrupted by the decision to fly two crew members for a yearlong stay, and for a brief period when Roscosmos decided to decrease its cosmonaut contingent to two as a cost-cutting maneuver in 2017.

The primary means of delivering crews to and from the ISS fell to Russia's three-seat Soyuz spacecraft. Launched from the Baikonur Cosmodrome in Kazakhstan—most of the time from the same pad used by the first human to fly to space, Yuri Gagarin—the Soyuz also returned under parachute to the steppe of Kazakhstan at the end of its mission. From November 2000 through December 2012, the Soyuz flew a two-day, 34-orbit journey to the space station. Beginning with the Expedition 35/36 crew in March 2013—when possible based on orbital mechanics—the rendezvous was reduced to six hours and four orbits.

The US space shuttle orbiters were also used to rotate expedition crew members, beginning

1. The space shuttle *Endeavour* is seen docked to the ISS in a photo taken by Expedition 27 crew member Paolo Nespoli from Soyuz TMA-20 after its undocking on May 23, 2011.
2. A Russian Soyuz-U rocket launches from the Baikonur Cosmodrome in Kazakhstan carrying the Soyuz TM-31 spacecraft with the first crew for the ISS on October 31, 2000.
3. An artist's rendering of both the crew version, at left, and cargo version of SpaceX's Dragon spacecraft docking to the ISS.
4. View of Russia's Soyuz crew spacecraft (in the foreground) and Progress cargo ship docked to the ISS while orbiting over Canada and the Great Lakes in 2015.

3

4

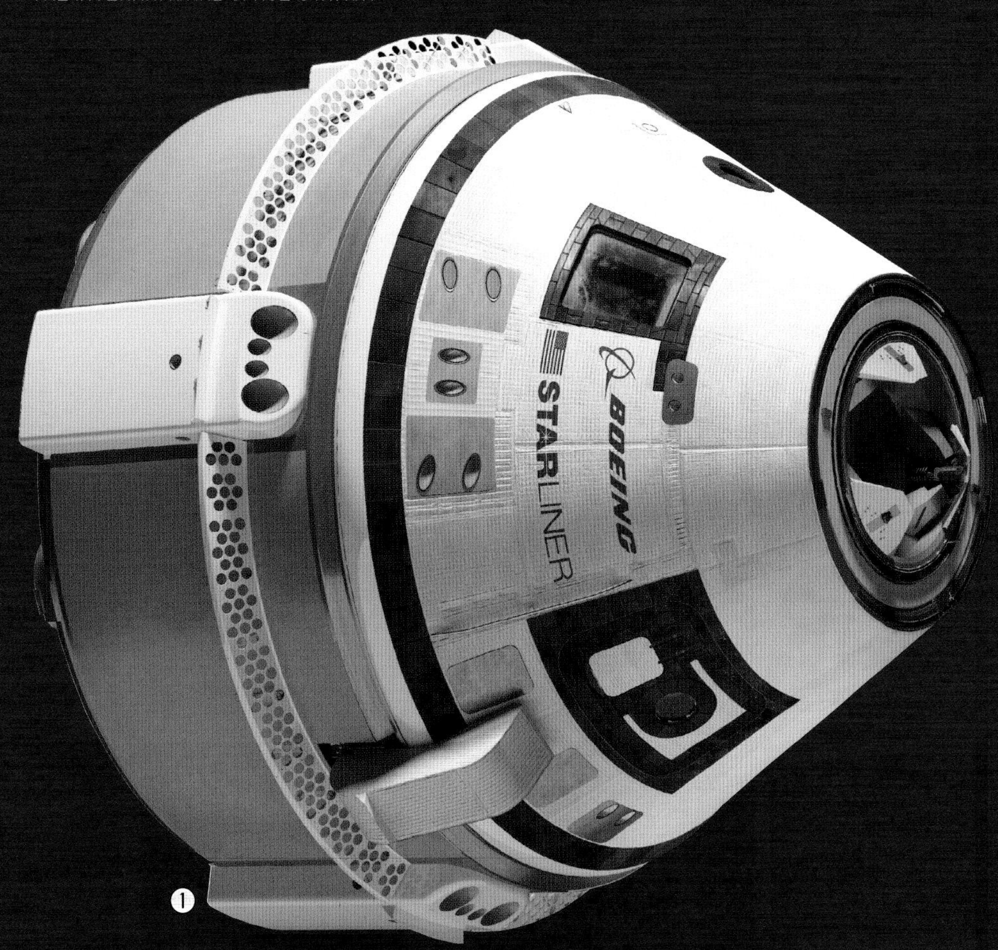

1. An artist's rendition of The Boeing Company's CST-100 Starliner spacecraft, which NASA contracted to deliver astronauts to and from the ISS, the transits beginning in 2019.
2. Packed with crew supplies and science experiments, Orbital ATK's Cygnus cargo spacecraft (pictured here in 2016) is captured by the Canadarm2 robotic arm to be berthed on the US segment of the ISS.
3. The JAXA H-II Transfer Vehicle, or HTV, is seen here in 2013 delivering both pressurized and unpressurized cargo to the ISS.
4. Artist's rendition of Sierra Nevada Corporation's Dream Chaser winged cargo spacecraft. NASA contracted for Dream Chaser resupply missions to the ISS, with flights beginning in 2020.
5. ESA launched five Automated Transfer Vehicle (ATV) resupply ships to the ISS between March 2008 and July 2014. Pictured here is ATV-4, the "Albert Einstein," in June 2013.

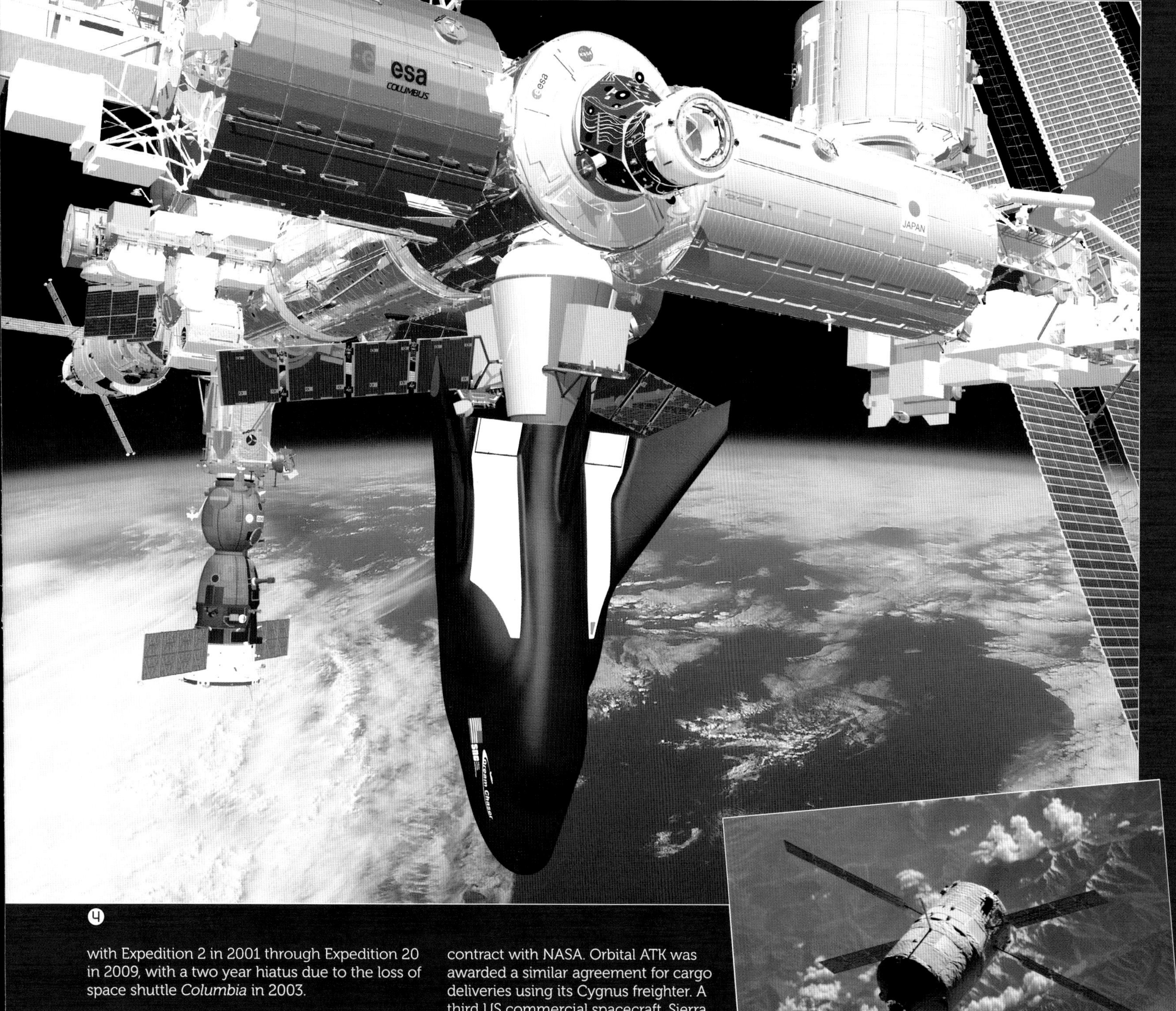

4

with Expedition 2 in 2001 through Expedition 20 in 2009, with a two year hiatus due to the loss of space shuttle *Columbia* in 2003.

COMMERCIAL SPACECRAFT

With the retirement of its winged orbiters in 2011, NASA contracted with The Boeing Company and SpaceX to design, build, and operate commercial spacecraft to resume flights to the space station from the United States. After several delays, Boeing's CST-100 Starliner and SpaceX's Dragon (V2) were expected to enter service in 2018.

Since May 2012, SpaceX has also flown Dragon spacecraft without crews to the space station under a commercial resupply services contract with NASA. Orbital ATK was awarded a similar agreement for cargo deliveries using its Cygnus freighter. A third US commercial spacecraft, Sierra Nevada Corporation's winged *Dream Chaser* space plane, is also contracted by NASA to convey supplies and science to and from the ISS starting in 2020.

US resupply missions were complemented by five flights of ESA's Automated Transfer Vehicle (ATV) between 2008 and 2014, and by JAXA's H-II Transfer Vehicle (HTV), beginning in 2009. Russia's Progress spacecraft, first used to resupply the Salyut 6 space station in 1978, continued service to the ISS, with more than 70 launches through 2018.

5

With the exception of the Dragon, which splashes down in the ocean, and the Dream Chaser, which lands like a glider on a runway, the other cargo-carrying visiting vehicles are purposely destroyed during their reentry into Earth's atmosphere.

TECHNOLOGICAL ADVANCEMENTS

As a one-of-a-kind laboratory operating in a unique environment, the ISS has been a testbed for the development of new technologies for use in orbit and on the ground. If humans are ever to venture beyond Earth orbit on long-duration missions to Mars or beyond, they will need spacecraft systems that are self-contained and reliable. The ISS offers a platform by which to test such systems while still being close enough to home to not put the crew members lives at greater risk. A good example of these are the Environmental Control and Life-Support Systems (ECLSS) on the ISS. Launched in 2008, NASA's Water Recovery System recycles wastewater, including the crew's urine, such that the space station's crew has a continuous source of potable water. "Recycling will be an essential part of daily life for future astronauts, whether on board the space station or living on the Moon," said Mike Suffredini, NASA's ISS program manager, in 2008. Some of the same technologies have been put into use in providing clean water for Third World developing nations. Other new technologies developed for the space station include advanced robotics, additive manufacturing facilities (also known as 3D printers) capable of operating in microgravity, and the use of ultrasound hardware as a diagnostic tool for astronauts' health.

NASA astronaut Leland Melvin is seen reflected in a water bubble during the STS-129 mission to the ISS in 2009. ISS experiments have revealed much about the way that fluids behave in zero-gravity conditions.

Expedition 19 crewmates Koichi Wakata (left) Gennady Padalka and Mike Barratt toast the welcome functioning of the ISS's water recovery system with drink bags filled with recycled water.

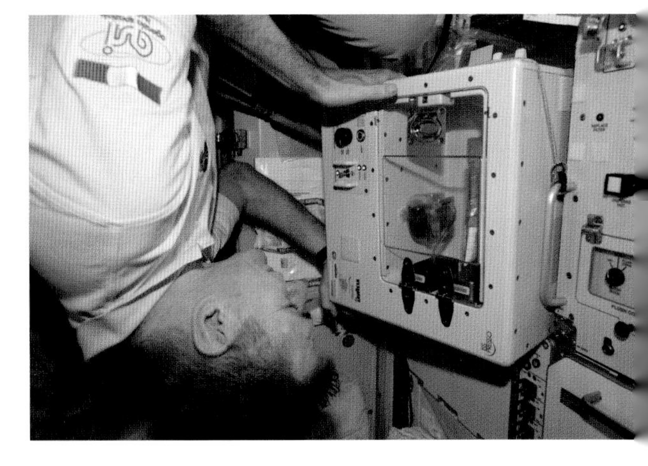

Taking the comforts of home into space, ESA astronaut Paolo Nespoli fills a 3D printed "zero-g" cup with freshly brewed Italian coffee using Lavazza's and Argotec's ISSpresso machine.

▲ Astronaut Tom Marshburn (background), Expedition 35 flight engineer, tele-operates Robonaut 2, or R2, at a taskboard on the ISS in April 2013.

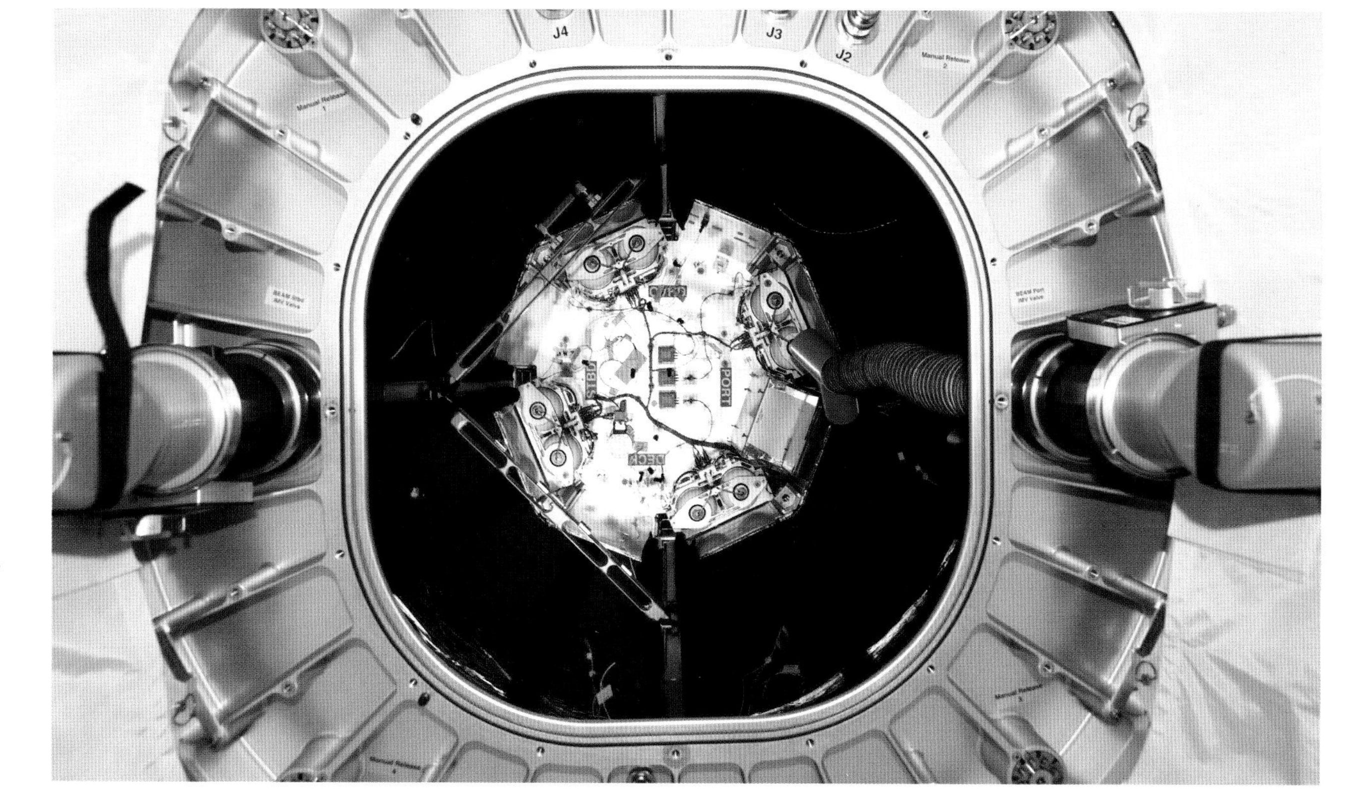

▶ The view looking into the Bigelow Expandable Activity Module, or BEAM, from inside Tranquility Node 3 on the ISS.

MISSION CONTROLS

Just as the ISS has been continuously crewed since November 2000, an even larger contingent has held constant watch over the orbiting outpost from facilities on the ground. While the astronauts and cosmonauts conduct most of the hands-on work to run and maintain the space station, flight controllers at five mission control centers distributed around the globe are responsible for monitoring the status of the ISS's on-board systems and planning the crew members' daily schedules.

Primary management of the space station resides within the Christopher C. Kraft, Jr. Mission Control Center at NASA's Johnson Space Center in Houston, Texas, for the US operating segment and

ESA's Columbus Control Center in the village of Oberpfaffenhofen, Germany. The center, call sign "Munich," is responsible for European science activities on the ISS.

Moscow Mission Control, (TsUP). Pictured on the large central screen is the Expedition 48 crew soon after the docking of Russia's Soyuz MS-01 spacecraft in July 2016.

at the TsUp mission control center in Korolyov, located outside of Moscow, for the Russian side of the ISS. NASA flight controllers rotate on nine-hour shifts. The "front room" teams can number between six or more than a dozen, depending on the activities that are planned in space. Russian flight controllers work 24-hour shifts, followed by three days off. Up to 50 people can staff the TsUP on any given shift. In addition to the MCC in Houston and TsUp near Moscow, additional mission control centers are located in Huntsville, Alabama, for US science operations; in Cologne, Germany, for ESA science operations; and in Tsukuba, Japan, for science operations in JAXA's Kibo laboratory.

NASA's Mission Control Center in Houston, Texas is the main control center for the ISS. Flight controllers manage daily operations, including vehicle arrivals and departures.

Japan's Mission Control Room at the Tsukuba Space Center manages science and logistics aboard the Kibo Japanese Experimental Module (JEM) on the ISS.

ASTRONAUT HEALTH AND LIFE SCIENCES

A primary mission for the ISS has been to study and learn how to counteract the effects of astronauts' long-term exposure to the microgravity environment. As prior space stations and spacecraft programs have revealed, extended weightlessness has a negative impact on the human body, resulting in decreased muscle mass and bone loss, among other medical concerns. Not only do these detriments pose a risk to astronaut health on their return to Earth, but they also present a hurdle to sending crew members on long-duration missions to Mars or elsewhere in the solar system and beyond. Experience on the ISS has shown that an exercise regimen of at least two hours daily can effectively combat most skeletal and muscular long-term impacts. Other medical concerns, though, such as diminished eyesight as a result of intercranial pressure, are still in need of a countermeasure to be developed.

In addition to studying the macro effects of microgravity on astronaut health, the ISS is also being used to further genetic research into how the human body reacts to changing environments. One such study, conducted between twin astronauts—one on the space station and one on the ground—revealed that spaceflight results in an increase in the process of turning genes on and off. "Some of the most exciting things that

Astronaut Karen Nyberg exercises on the Advanced Resistive Exercise Device (ARED) in Tranquility Node 3 during Expedition 37. The ARED helps astronauts prevent a loss of bone mass in space.

NASA astronaut Randy Bresnik explains: "Aboard the ISS in 0 g . . . Sasha [A. Misurkin] runs on the forward wall while Sergei Ryazansky lifts 'weights' on the back wall."

Tim Peake uses ESA's Mares machine on the ISS: "No, not testing a new rollercoaster ride—research into muscle atrophy and how this may help patient rehabilitation on Earth."

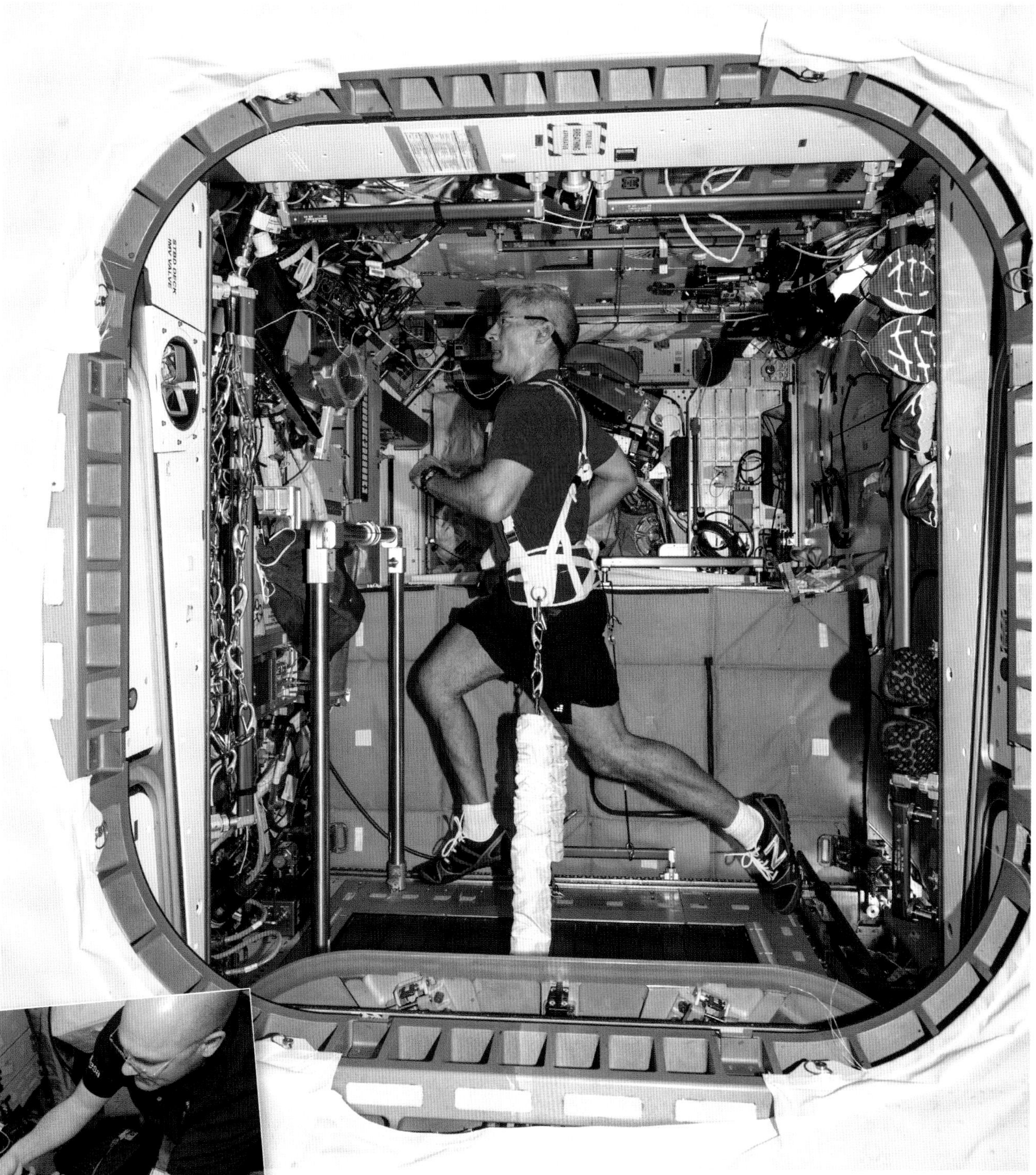

NASA astronaut Mark Vande Hei jogs on the Combined Operational Load Bearing External Resistance Treadmill (COLBERT), named after the American comedian and late night TV host.

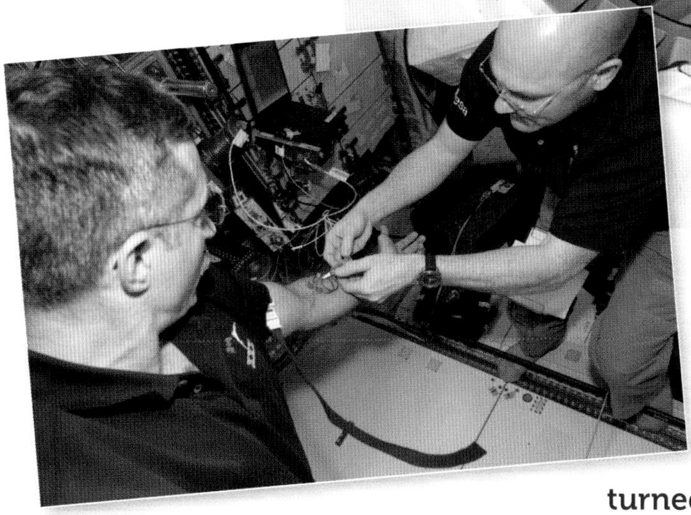

Expedition 30 crew members Dan Burbank (left) and Andre Kuiper perform blood draws for an integrated immune study aboard the ISS.

we've seen from looking at gene expression in space is that we really see an explosion, like fireworks taking off, as soon as the human body gets into space," said Chris Mason, Ph.D., of Weill Cornell Medicine. "With this study, we've seen thousands and thousands of genes change how they are turned on and turned off. This happens as soon as an astronaut gets into space, and some of the activity persists temporarily upon return to Earth." The health effects are still being analyzed.

LENGTHS OF STAY

Not all visits to the ISS are equal. The typical stay for an expedition crew member has been about five-and-a-half months. The Russian Soyuz spacecraft, which for the space station's first two decades has served as the primary conveyance and emergency lifeboat for the space station's crews, has an on-orbit lifetime of six months. Hence, the vehicles need to be swapped out every half year, leading to a rotation of crew members.

Before the space station expanded to a six-person expedition crew contingent, the Soyuz rotation schedule allowed for additional astronauts and cosmonauts to visit the space station on short, two-week "taxi" flights, flying to the ISS with new expedition crew members on a new Soyuz and then returning to Earth with outgoing crewmates on the older vehicle. This rotation also opened up seats to self-funded and non-partner spaceflight participants, leading to privately-financed "space tourist" visits and countries like South Korea and Malaysia sending astronauts for short stays. Astronauts have also resided on the space station for longer than six months. NASA and Roscosmos undertook the first year-long mission aboard the ISS between March 2015 and March 2016 in an effort to learn more about the human effects of long-duration microgravity exposure.

Sunrise as viewed from the ISS on November 17, 2015. The crews aboard the space station are treated to 16 sunrises and sunsets every 24 hours, with truly spectacular visual effects.

Astronauts Kimiya Yui (left) and Kjell Lindgren celebrate their first 100 days in space together. The pair received special commemorative patches for the milestone.

One-year-mission crew members Scott Kelly (left) and Mikhail Kornienko celebrated their 300th day in space on January 21, 2016. The pair spent a total of 340 days on the ISS.

◆ Zero g gives objects a life of their own. Creating a strange, almost sentient appearance, a pair of trousers floats above the crew quarters for one member of the Expedition 50 crew aboard the ISS.

◆ Terry Virts (left) Samantha Cristoforetti, and Anton Shkaplerov marked their 100th day aboard the ISS on March 2, 2015, with this photo and the comment: "Celebrating flight day 100 of our mission."

WOMEN ON THE ISS

On July 6, 2016, NASA astronaut Kathleen "Kate" Rubins became the sixtieth woman to launch into space, heading to join Expedition 48 aboard the ISS. Of those 60 women, more than half flew to the ISS; 28 were American and one was from Russia. The others were from Canada, Japan, Italy, France, and South Korea. In the same period, the 200 or so other visitors to the ISS were men. "I do not keep track if a female is doing this or a male is doing that, and my colleagues really don't [either]. I think that's really a testament to where we are

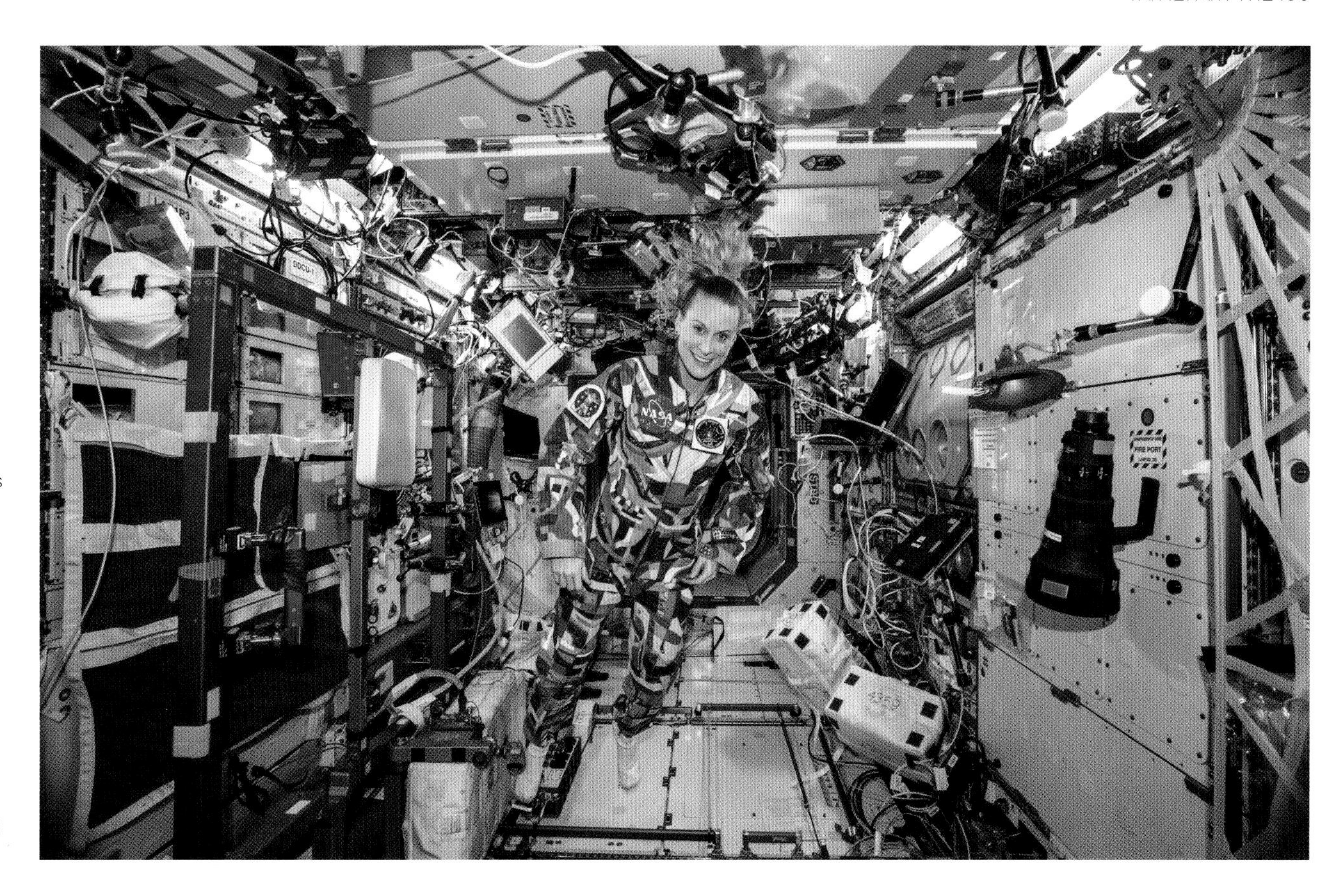

Astronaut Nancy Currie-Gregg became the first woman to enter the ISS as a mission specialist aboard space shuttle *Endeavour's* STS-88 crew in December 1998.

"Astronaut Kate Rubins became the sixtieth woman to fly into space on her first ISS mission in 2016. Here, Rubins wears a flight suit hand painted by pediatric cancer patients as part of the Spacesuit Art Project..

Astronaut Peggy Whitson, seen here during a 2017 spacewalk, was not only the first woman to command the ISS (and the first to do so twice), but also set records for the most time in space by a US astronaut, at 665 days.

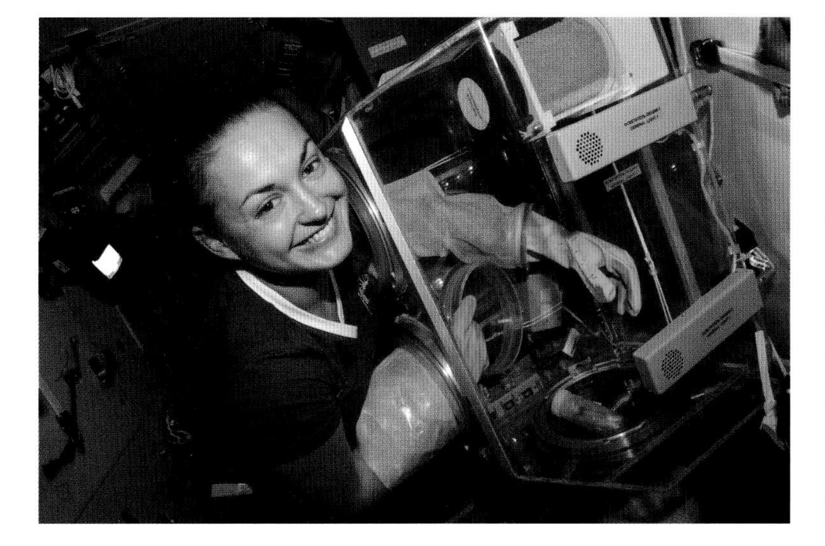

Cosmonaut Yelena Serova became the first Russian woman to enter the ISS in 2014 and only the fourth Russian woman to fly into space.

The most women in space at one time, as of 2018, has been four. Here Dottie Metcalf-Lindenburger (left), Naoko Yamazaki, Stephanie Wilson, and Tracy Caldwell Dyson (top) pose for a photo on the ISS in April 2010.

right now," said Rubins in 2016. The first woman to come aboard the ISS was Nancy Currie-Gregg, a NASA astronaut and member of space shuttle *Endeavour*'s STS-88 crew, which docked the first two components of the space station on December 6, 1998. The first woman to serve on an ISS expedition crew was NASA astronaut Susan Helms, who, with Roscosmos cosmonaut Yuri Usachev and NASA astronaut James Voss, comprised the second contingent to live aboard the space station. The woman with the most time aboard the space station as of 2018 was NASA astronaut Peggy Whitson with 665 days spread over three expeditions stays between 2002 and 2017.

PHOTOGRAPHY OF THE EARTH

Orbiting between 355 and 560 km (220 and 250 mi) above Earth, the ISS has provided some spectacular views of our home planet. "Recording what we see as astronauts from space of Earth is one way of sharing the experience with folks who do not get a chance to go into space," said NASA astronaut Don Pettit, who pioneered several new photography techniques aboard the space station. The space station crew has access to an armada of cameras and lenses to capture Earth. In the first 15 years of the ISS being on orbit, astronauts and cosmonauts took more than three million photographs. One astronaut, Terry Virts of NASA, set a record in 2014, snapping hundreds of thousands of photographs—319,275, to

Oleg Kononenko poses with photography equipment aboard the ISS in 2015. Crew members took more than 2.5 million images over the first 15 years of station occupancy.

This photo, taken by an Expedition 30 crew member in 2012 above the southeastern Tasman Sea, is believed to be the one millionth still image captured by ISS crew.

Astronaut Don Pettit has described the photographic process behind creating such incredible images. "My star trail images are made by taking a time exposure of about 10 to 15 minutes.... I take multiple 30-second exposures, then 'stack' them using imaging software, thus producing the longer exposure."

Alexander Gerst, flight engineer, uses a still camera at a window in the Cupola as a SpaceX Dragon cargo craft approaches the ISS in September 2014.

Astronauts have a wide selection of cameras and lenses to use when photographing their view from the Cupola and other windows on the ISS.

be exact—with the majority focused on Earth. "My most important work in space, in my opinion, was photography and the impressions of space that I can share with others," Virts later wrote in his book, *View from Above* (2017).

Astronaut Earth photography has provided new insight into life-threatening storms, the effects of climate change, and the impact of humans on their surroundings. Contrary to popular belief, astronauts on the ISS cannot see the Great Wall of China or other manmade structures unaided, but due to the presence of electrical lighting and, conversely, the lack thereof, they have been able to photograph the outline of political borders between countries.

SCIENCE AND UTILIZATION

Although science operations on the ISS began as soon as it was first crewed, it was not until after the orbiting laboratory was declared "complete" in 2011 that the astronauts and cosmonauts were able to turn their primary focus to utilization. As of 2017, expedition crew members on average spend about 40 hours a week split between conducting science, performing maintenance on the space station, and exercising for 2.5 hours a day. A record was set though during the week of February 5, 2018, when the six members of the Expedition 54 crew put in a combined 100 hours of science work. "This new record for science shows the crew is spending more time using the station for its intended purpose as a weightless space laboratory," stated the European Space Agency.

The science conducted aboard the ISS is divided into several disciplines, including biology and biotechnology, Earth and space science, educational activities and outreach, human research, physical science, and technology development and demonstration. A typical expedition crew will work on two- to three-hundred different investigations during their stay, with as many as one hundred of the science experiments being new to that increment.

The "Veggie" plant growth facility on the ISS is used to cultivate edible greens and flowers to better understand how plants respond to microgravity, important work for the future of space travel.

Astronaut Randy Bresnik works on lung tissue studies inside one of the ISS's microgravity science gloveboxes. "Amazing research developing what [astronaut] Jack Fischer dubbed 'cancer seeking missiles.'"

The Microgravity Science Glovebox (MSG) is one of the dedicated science facilities inside the US Destiny laboratory. It has a large front window and built-in gloves to provide a sealed environment for conducting science and technology experiments and is particularly suited for handling hazardous materials.

◀ Astronaut Kate Rubins poses for a picture by the side of the minION device that she used to perform the first DNA sequencing in space aboard the ISS in 2016.

▲▲ Astronaut Chris Cassidy replaces a fuel reservoir in an ISS combustion chamber. The reservoirs contain the liquid fuel used during droplet combustion experiments.

▲ A close-up image of a flame during a combustion experiment studying how to extinguish fires in microgravity. Flames in space burn at slower rates, with lower temperatures, and with less oxygen than in normal gravity.

CULTURE ON THE ISS

From May 2008 through September 2009, the ISS gained a crew member unlike any other—a Star Command Space Ranger that was no stranger to going "to infinity and beyond!" Through an educational partnership with the Walt Disney Company, NASA launched a 30.5 cm (12 in) Buzz Lightyear action figure to the space station. The toy was used to tell stories about real spaceflight in videos distributed to classrooms and students around the globe.

Buzz Lightyear, though, is not the only artifact (since donated for display at the Smithsonian National Air and Space Museum) from pop culture to make it to the space station. In its first 15 years, the ISS played host to LEGO building kits, including one that formed a model of the space station itself; the jerseys from all of the NFL professional football teams; a plush red "Angry Bird" from the popular mobile game; and Luke Skywalker's lightsaber from *Star Wars, Episode IV: Return of the Jedi.* And even as Earth's pop

An action figure of Disney's *Toy Story* character Buzz Lightyear "plays" with liquid aboard the ISS as part of an educational outreach project.

Japanese astronaut Satoshi Furukawa, flight engineer on Expedition 28/29, poses with a large-scale LEGO model of the ISS that he assembled while on the space station.

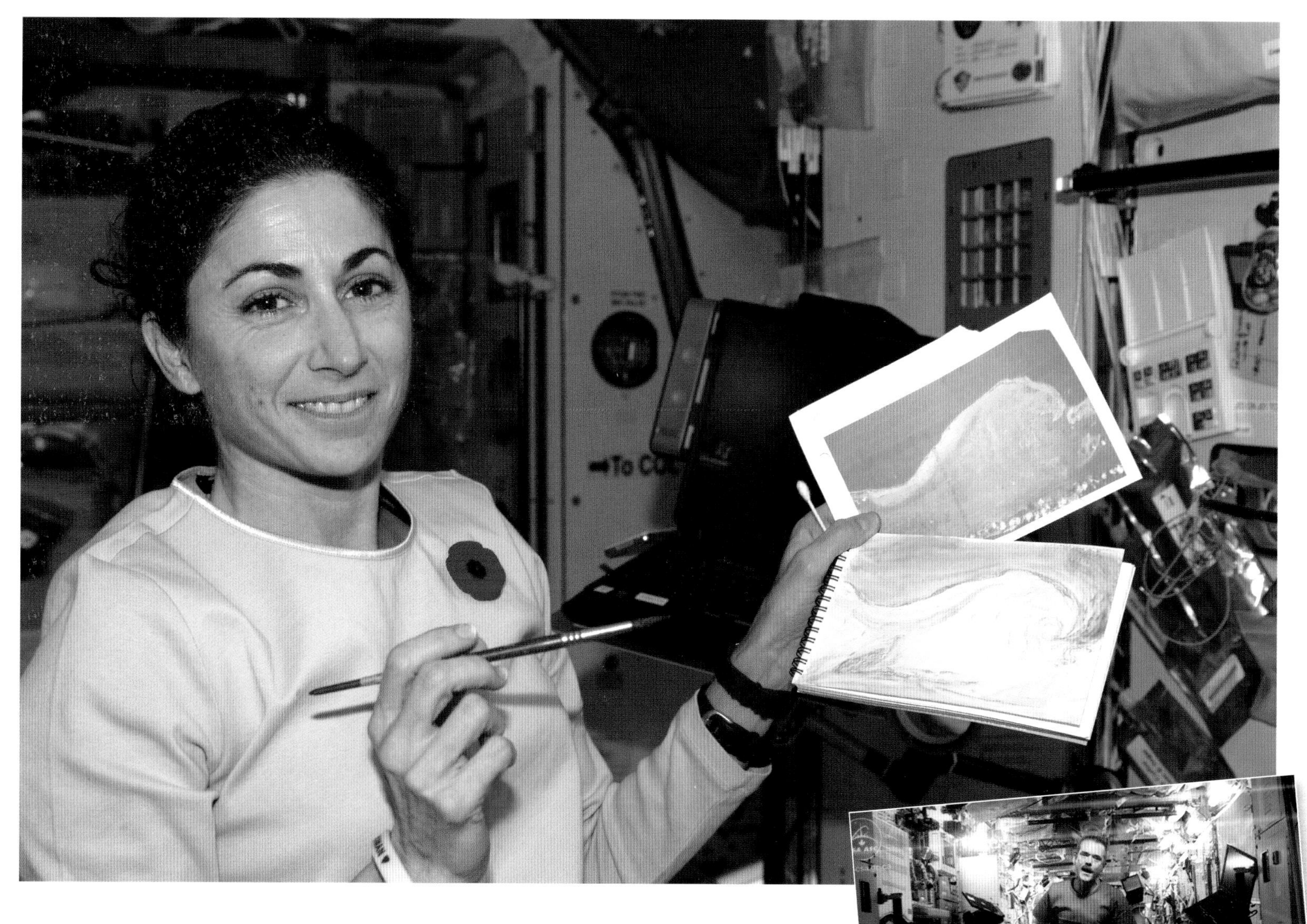

During her time as an ISS Expedition 21 crew member, Nicole Stott produced the first watercolor paintings in space.

Canadian astronaut Chris Hadfield played a guitar on the ISS to accompany his rendition of David Bowie's "Space Oddity." His music video filmed in space went viral back on Earth.

Astronaut Karen Nyberg's stuffed toy dinosaur floats on the ISS. Nyberg actually hand-crafted the doll for her son using only materials she found on the orbiting space station.

culture moved onto the ISS, a distinct new culture was forming on the orbiting outpost. "We've been living on this spaceship for 15 years. We're not in the newness phase, we are not in the plant-the-flag-and-say-we're-here-phase—we're actually in the we-left-Earth phase," said Canadian Space Agency astronaut Chris Hadfield in 2015, who performed his own take on David Bowie's "Space Oddity" on board the space station. "Whoever we were back there has now moved to here and the reality of this place, the weightlessness, the views, the tasking, the separation . . . that culture is becoming separate and distinct from culture on Earth."

OUTREACH FROM ORBIT

Tim "TJ" Creamer, who in 2015 became the first astronaut to join the ranks of NASA's flight directors in Mission Control, set a different type of record on January 22, 2010 with a message just 140 characters long. "Hello Twitterverse! We r now LIVE tweeting from the International Space Station -- the 1st live tweet from Space! :) More soon, send your ?s" Creamer wrote, becoming the first astronaut to post to the social media network Twitter directly from space. Since that tweet, astronauts and cosmonauts have used Twitter, Facebook, Instagram, and other similar services virtually to

◀ Astronauts living on the ISS take part in many live video "downlinks," connecting with the media, students, and the public to share their experience aboard the space station.

▶ Educator-astronaut Joe Acaba takes part in "Story Time From Space," an educational outreach project that films ISS crew members reading from science-themed kid's books.

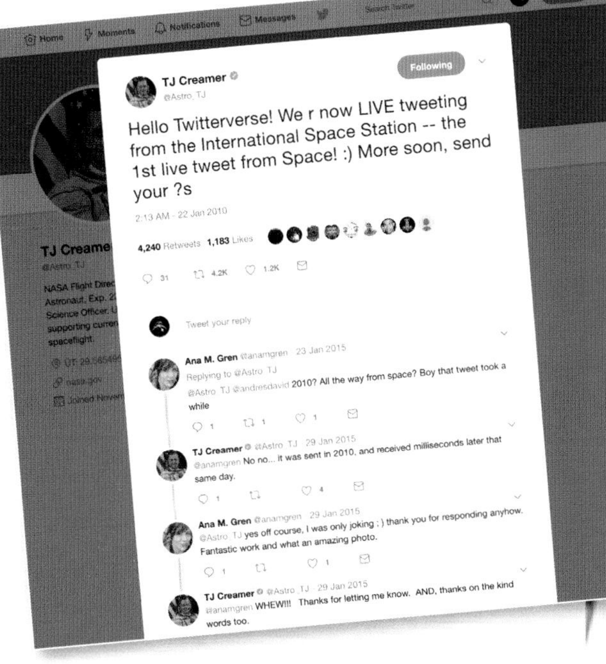

◀ Despite being hundreds of miles from Earth and home, ISS crew members still have access to the internet through their laptop and tablet computers. Astronauts can therefore send and receive email, use social media, and, crucially, video chat with family and friends on the ground.

▲ The first-ever live "tweet" posted in space was made on January 22, 2010, by T. J. Creamer aboard the ISS. The message was retweeted more than 4,000 times.

▼ Scott Kelly offers his "proof" to the Reddit community before an Ask Me Anything Q&A session on the ISS in 2016.

bring the public aboard the ISS. Social media enabled the crew members to interact directly with their followers, answering questions, sharing photos as they were taken, and chiming in on discussions between others on Earth. A natural extension of earlier missions' use of ham radio and video downlinks to reach into classrooms, offices, and homes with a glimpse at life in space, social media added a two-way exchange previously reserved to only select invited groups. The result has been a more engaged public, with astronauts' social media accounts often reaching into the hundreds of thousands, if not millions of followers. "Good luck, captain — and make sure to Instagram it," said US President Barack Obama to Scott Kelly, the first NASA astronaut to spend a year in space aboard the ISS, during the President's 2015 State of the Union address.

Chapter Six

THE SPACE STATION IN POP CULTURE

In spite of a few early appearances, the space station was a latecomer to science fiction. Spaceships got far more attention in both text and art. This is probably because spaceships travel to places while space stations do not. They may orbit an exotic planet or star, but going around in circles isn't as romantic as traveling to distant places. Possibly for this reason, when space stations *did* start making major appearances in science fiction, they were viewed, in effect, as characters themselves.

(Left) The cover art for Murray Leinster's 1953 novel *Space Platform*, which suggested that a space station would be constructed on Earth and launched into space.

(Above) *Captain "Space" Kingley* was the hero of a series of young adult books published in England in the early 1950s. Here we see the captain and his spaceship with space station X-3 hovering in the distance.

The Space Station in Pop Culture

MANY OF THE CLASSIC STORIES SET IN SPACE STATIONS USE THE BACKGROUND OF THE STATION AS A JUMPING-OFF POINT FOR THEIR PLOTS. EARLY IN THIS BOOK, WE ALREADY MENTIONED THREE IMPORTANT EARLY FICTIONAL SPACE STATIONS—"THE BRICK MOON," THE MARTIAN POLAR STATION FROM *ON TWO PLANETS*, AND ASTROPOL—IN WHICH THE EXISTENCE OF THE STATION ITSELF WAS AN IMPORTANT, AND EVEN PIVOTAL, PART OF THE PLOT.

In 1926, German author Karl Laffert published a novel, *World Fire*, that had been inspired by Hermann Oberth's suggestion that a giant orbiting mirror could be used to control the Earth's weather. In the story, which was set around 1958, an "ether station" has been established in Earth orbit by a world government. On board are engineers tasked with regulating the Earth's climate by means of the station's giant mirrors. Although not a work of fiction, J.D Bernal's *The World, the Flesh and the Devil* (1929) speculated on the possibility of an Earth surrounded by a ring of inhabited satellites. The setting of much of Frank K. Kelly's story "Famine on Mars" (1934) was also a space station in Earth orbit. Interestingly enough, Kelly had both men and women crew members.

IMAGINING THE SPACE STATION

Although Hermann Oberth had coined the term *Weltraumstation* in 1923, Manly Wade Wellman was probably the first to use the English term "space station" in fiction, in his short story "Space Station No. 1" (1936). Wellman's space station was located in the same orbit as Mars and acted as a refueling depot for spacecraft traveling between the Earth and Jupiter. Space stations also made impressive appearances in *Vandals of the Void* (1930) by J.M. Walsh and *The Power Planet* (1931) by Murray Leinster. Both were inhabited by hundreds of crew members, with Leinster's station taking a cue from Oberth and acting as a power-generating station:

This 1955 edition in a popular series described a space station assembled from the nose cones of ferry rockets. The station orbited at an altitude of 37,496 km (23,300 mi) above the Earth.

The Power Planet, of course, is that vast man-made disk of metal set spinning about the Sun to supply the Earth with power. Everybody learns in his grammar-school textbooks of its construction just beyond the Moon and of its maneuvering to its present orbit by a vast expenditure of rocket fuel.

"Only forty million miles from the Sun's surface, its sunward side is raised nearly to red heat by the blazing radiation. And the shadow side, naturally, is down to the utter cold of space. There is a temperature drop of nearly seven hundred degrees between the two sides, and Williamson cells turn that heat-difference into electric current, with an efficiency of 99 percent. Then the big Dugald tubes—they are twenty feet long on the Power Planet—transform it into the beam which is focused on the Earth and delivers something over a billion horsepower to the various receivers that have been erected. The space station itself is ten miles across, and it rotates at a carefully calculated speed so that the centrifugal force at its outer edge is very nearly equal to the normal gravity of Earth. So that the nearer its center one goes, of course, the less is that force, and also the less impression of weight one has.

In his 1931 story "The Prince of Space," Jack Williamson anticipated the space colonies envisioned in the 1970s by Gerard O'Neill. His orbiting habitat was described as a vast cylinder a mile wide, its interior lined with lawns, gardens, homes, and parks. Rotation on the colony's long axis provided gravity, while elaborate machinery provided heat, light, and oxygen:

"The City of Space is in a cylinder," Captain Smith said. "Roughly five thousand feet in diameter, and about that high. it is built largely of meteoric iron which we captured from a meteoric swarm making navigation safe and getting useful metal at the same time. The cylinder whirls constantly, with such speed that the centrifugal force against the sides equals

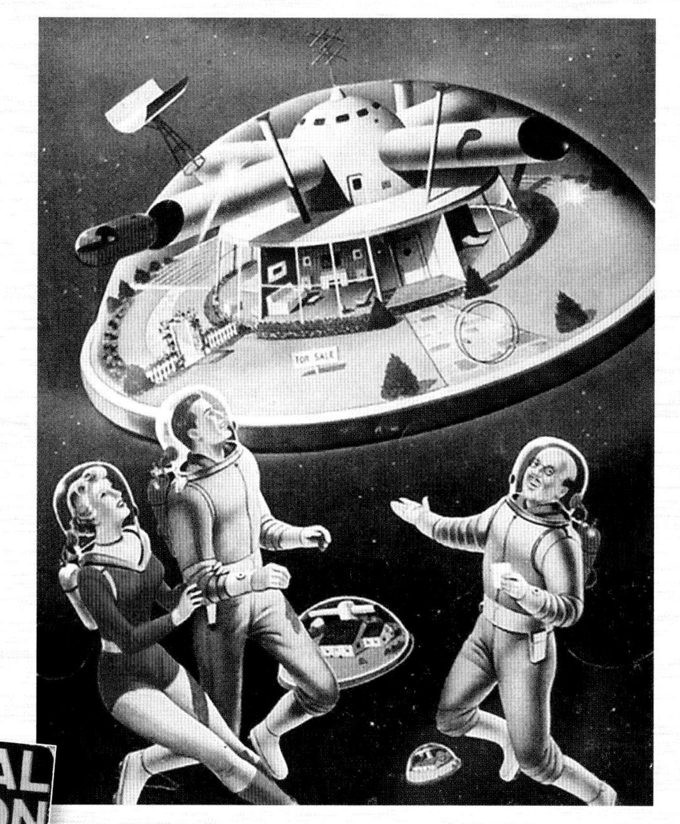

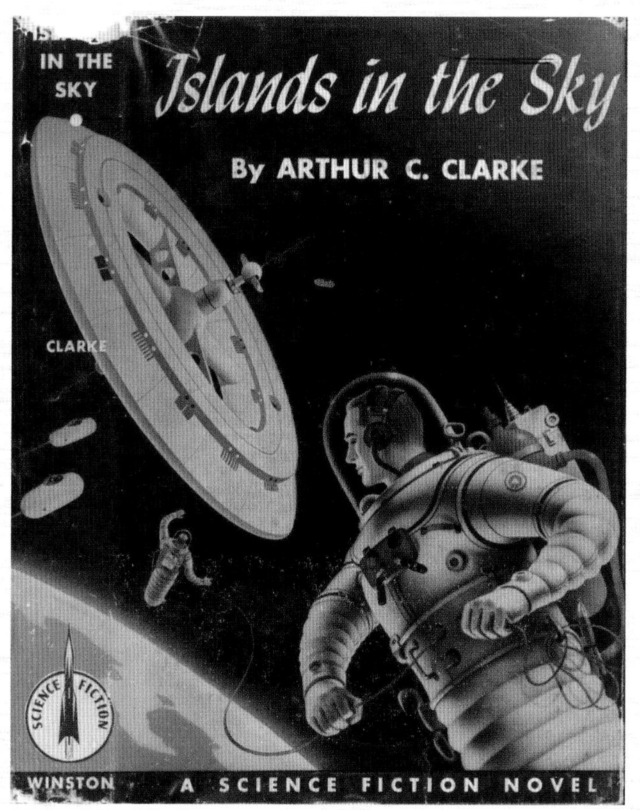

Arthur C. Clarke's 1952 juvenile novel *Islands in the Sky* realistically described the construction of a space station and what daily life aboard would be like.

Artist Alex Schomburg parodied the post-war housing boom in this illustration of the ultimate suburban home, set within a space station.

James White's Sector General series—a series of 12 science-fiction novels—centered around a giant, orbiting space hospital.

the force of gravity on the Earth. The city is built around the inside of the cylinder—so that one can look up and see his neighbor's house, apparently upside down, a mile above his head. We enter through a lock in one end of the cylinder." R.F. Starzl and Everett C. Smith's "The Metal Moon" (1932) was a vast, hollow sphere, similarly containing an Earth-like environment. Another spherical habitat, home to more than a million inhabitants, figured prominently in "Factory in the Sky" (1941) by Basil Wells.

TAKING CENTER STAGE

One of the first authors to make use of the space station as an integral plot element was George O. Smith, an American science-fiction writer who was particularly prolific during the 1950s and the 1960s. His Venus Equilateral series, beginning with *Astounding* (1942), was set on a space station located at one of the Trojan points in the orbit of Venus. These are the points in the orbit of any planet that lie 60° ahead of the planet and 60° behind. Any object located there will remain in a stable position. The purpose of Smith's station is to act as a communications link between Earth, Venus, and Mars.

"Terra Station" in Robert Heinlein's *Space Cadet* (1948) is a wheel-shaped space station not dissimilar to the BIS Ross-Smith design, although this version is the size of a small city. Heinlein also set much action in a city-sized space station in "Space Jockey" (1947). In 1952, Arthur C. Clarke published *Islands in the Sky*, a young adult novel dedicated largely to a description of daily life aboard a space station orbiting the Earth. The station itself clearly based on the 1949 BIS design. One of the major functions of Clarke's station is to provide support for a network of communications satellites.

Murray Leinster's *Space Platform* (1953) concerned itself with a large, toroidal space station similar to von Braun's designs. Strangely enough, where most authors assumed that such a station would be constructed in orbit from materials sent from the Earth, Leinster's station is built here on Earth and launched into space in one piece. Numerous other pre–space flight novels and stories featured space stations, including *Space Station No. 1* (1957) by Frank Belknap Long, *Station in Space* (1958) by James Gunn, and *The Wheel in the Sky* (1954) by Rafe Bernard.

MILITARY AND MEDICAL

The military potential of an orbiting space station had occasionally been touched upon by early writers, though this reached its heyday during the postwar "Red Scare." It was made clear that even the *Collier's* space station performed the important function of acting as an orbiting observatory of terrestrial military activities. The real possibility of using space stations as launching platforms for nuclear weapons was not lost on writers, either. For example, in Cyril Kornbluth's 1955 novel *Not This August*, an armed space station is used to threaten Russia with annihilation unless it abandons its occupation of the United States.

Beginning in 1962, James White's immensely popular Sector General series of novels was set around Sector 12 General Hospital, an enormous space station located in deep space. It is designed to treat a wide variety of alien life forms suffering from unearthly diseases and requiring unusual life support. Even the hospital staff itself represents dozens of species from many different worlds, all requiring—like the patients—special accommodation for their environments and behaviors. The hospital station itself is an enormous cylinder consisting of 384 floors.

Donald Kingsbury's *The Moon Goddess and the Son* (1986) concerns itself with the construction of a space station (as well as describing the technologies necessary for achieving cheap access to low Earth orbit), while Allen Steele's 1989 novel *Orbital Decay* describes the construction of a solar power satellite. Most the action takes place on board nearby Olympus Station, a rotating wheel that houses the construction crews and is known to its inhabitants as Skycan. It consists of 42 modules in addition to the central hub, each 8 m (24 ft) long by 2 m (6 ft) wide. These modules contain living quarters, computers, research laboratories, hydroponic gardens, life support, and so on. An inner torus provides a passageway connecting the modules. Two spokes connect the wheel with the central hub.

SPACE STATIONS IN THE MOVIES AND ON TV

In recent decades, most Hollywood space stations have been influenced by the real and utterly functional space stations that have been orbiting the Earth for the past several decades, such as those seen in films ranging from *Space Camp* (1986) to *Gravity* (2013). The appealing elegance of the space wheel is hard to abandon, with another recent appearance in the 2013 film *Elysium*, which featured an enormous rotating space habitat based on the Stanford Torus concept.

②

①

③

1. This 1959 Japanese film featured a handsomely decorative space station that was doomed by invading aliens. Here we see a Columbia Pictures theatrical poster for the 1960 US release of the film, translated as *Battle in Outer Space*.

2. Although a cheaply made B-picture, *Project Moonbase* (1953) included one of the first Hollywood depictions of a space station, a model created by futurist Jacques Fresco.

3. This classic von Braun–inspired space station appeared in an unknown film, probably from the early 1950s.

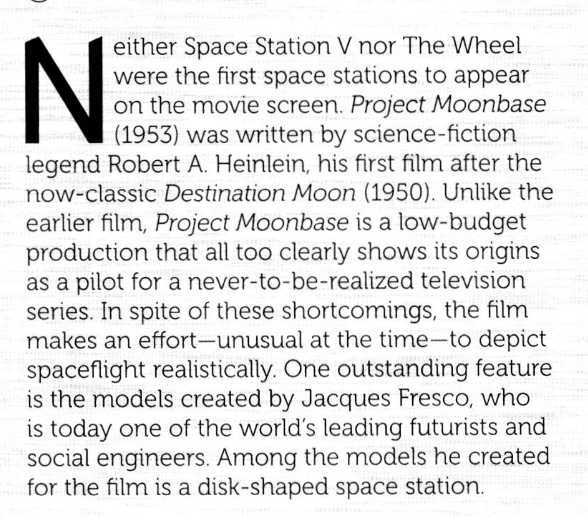

Neither Space Station V nor The Wheel were the first space stations to appear on the movie screen. *Project Moonbase* (1953) was written by science-fiction legend Robert A. Heinlein, his first film after the now-classic *Destination Moon* (1950). Unlike the earlier film, *Project Moonbase* is a low-budget production that all too clearly shows its origins as a pilot for a never-to-be-realized television series. In spite of these shortcomings, the film makes an effort—unusual at the time—to depict spaceflight realistically. One outstanding feature is the models created by Jacques Fresco, who is today one of the world's leading futurists and social engineers. Among the models he created for the film is a disk-shaped space station. Unusual for the time, it did not rotate and a not entirely successful attempt was made to depict life aboard a zero-gravity space station.

CELLULOID DESIGNS

To take advantage of the public interest in spaceflight generated by the launch of Sputnik in 1957, producer Roger Corman rushed *War of the Satellites* into theaters. Although the film's $70,000 budget precluded any elaborate sets or special effects, much of the story is set aboard a small, spherical, crewed artificial Earth satellite.

In *Mutiny in Outer Space* (1965), by contrast, a toroidal space station is overrun by an alien fungus that had been found on the Moon. In addition to the station on which most of the action is set, the plastic space station model kit designed by Elwyn Angle also makes an appearance. It was not the first time this kit had been featured in a movie. It also showed up in *Assignment Outer Space*, an Italian science fiction epic from 1960. The large number of plastic model kits of spacecraft that existed at the time were a goldmine for budget-conscious film producers.

Similar in plot was Japan's *Green Slime* (1968), in which space station Gamma 3—a von Braunian wheel—is overrun by the titular monster (or monsters, as it turned out). Japanese filmmakers seemed to be particularly fond of the wheel-shaped space station; variations on the concept appeared in

4. *Rocky Jones, Space Ranger* was one of the most popular children's science fiction television series during the 1950s. The space station that featured in one episode may be the first ever to appear on TV.
5. Artist and model-maker Morris Scott Dollens created this evocative scene for his unrealized 1950s motion picture, *Dream of the Stars*.
6. This impressive space station had a prominent appearance in the Soviet science fiction movie *Road to the Stars* (1957), directed by Pavel Klushantsev.
7. The popularity of *Star Trek* inspired an endless procession of merchandise, including this model kit of the K-7 Space Station.

"I KNOW I'VE MADE SOME VERY POOR DECISIONS RECENTLY, BUT I CAN GIVE YOU MY COMPLETE ASSURANCE THAT MY WORK WILL BE BACK TO NORMAL."

HAL 9000 IN *2001: A SPACE ODYSSEY*

numerous science-fiction movies produced between the late 1950s and throughout the 1970s and 1980s, such as *Battle in Outer Space* (1959).

Wheel-shaped stations figured prominently in several early Soviet space films, especially in the semi-documentary *Road to the Stars* (1957). A slightly less traditional space station appeared two years later in *Nebo Zovyot* (1958; The Sky Calls), in which a rotating wheel is combined with a large non-rotating structure used as a docking platform for spacecraft.

The Japanese-American science-fiction film *Solar Crisis* (1990) included a space station known as "Sky Town" that was an almost perfect duplicate of the classic *Collier's* space station, blown up, however, to the size of a small country.

MOVIE ICONS

Of the non-classical space stations to appear on the big screen perhaps the best known is the iconic Death Star from *Star Wars* (1977). The enormous artificial planet was meant to be 120 km (75 mi) in diameter with a crew of approximately 1.7 million (to say nothing of 400,000 droids). By contrast, the 1979 James Bond film *Moonraker* featured a large space station of a strikingly different design. Although depicted as rotating to provide artificial gravity, it is composed of a complex combination of tubes and spheres that appears awkward and ungainly,

1. The 2002 film *Solaris* was based on the classic 1961 science fiction novel by Stanislaw Lem. Almost the entire film takes place on board a space station orbiting the possibly sentient planet Solaris.

but at least was an attempt to appear original. NASA space station designs have influenced a number of Hollywood space stations in the past few decades . . . especially so since the development of the ISS. The ill-fated ISS-inspired space station in *Gravity* (2013) is probably the most familiar of these.

TV SPACE STATIONS

Space stations have featured less prominently on TV, despite the fact that television in general has fully embraced space as a commercial and popular topic. The 1962 British television puppet series *Space Patrol* featured a von Braun–inspired wheel in many of its episodes and was one of the earliest regular appearances of a space station on TV. The *Solarnauts* (1967), the pilot for a proposed but never produced British TV series, opened with a huge, rotating space habitat that combined von Braun's wheels with an enormous domed city at the end of one axis.

The American television series *Men Into Space* (1959–60) included the *Ares* space station (in reality, a design proposed by Lockheed) in an episode entitled, naturally enough, "Building a Space Station." While the original *Star Trek* series (1966–69) included space stations in several episodes, they largely served only as background settings. Space station K-7, for example, was best known for its infestation of "tribbles," an alien

"THINK OF IT. FIVE YEARS AGO NO ONE HAD EVER HEARD OF BAJOR OR DEEP SPACE 9 AND NOW ALL OUR HOPES REST HERE."

A CHARACTER IN *STAR TREK: DEEP SPACE NINE*

2. The title space station in *Elysium* (2013), produced, written, and directed by Neill Blomkamp, is an enormous rotating colony accessible only to those rich and powerful enough to afford to live there.

3. The eponymous Deep Space Nine was a habitat established by the Federation of Planets to guard the entrance to a wormhole. The *Star Trek: Deep Space Nine* series ran in the 1990s.

species. Subsequent *Star Trek* spinoffs, such as *Star Trek: Enterprise* (2001–05) and *Star Trek: The Next Generation* (1987–94), also occasionally featured space stations, one of which appeared in both the latter TV series and the later movies.

One of the only television series to take place almost entirely aboard a space station was *Star Trek: Deep Space Nine*, which ran from 1993 to 1999. The station was a full-scale habitat with a permanent population of 300. It was an outpost situated near the mouth of a stable wormhole, a device allowing for endless plot developments. An even larger space settlement was the eponymous *Babylon 5*, home to 250,000 humans and aliens in the *Babylon 5* television series, which ran from 1994 to 1999.

③

THE WHEEL

Conquest of Space (1955) was producer George Pal's follow-up to his immensely successful *Destination Moon* (1950). Although not as well received by the public or science fiction fans, it was still one of the earliest attempts to depict space flight realistically.

Much of the action for the first act of the film *Conquest of Space* takes place aboard "the Wheel," a space station modeled after the design by Wernher von Braun.

One of the wheel-shaped station's first appearances on motion picture screens—indeed, one of the first depictions of any sort of space station—was in George Pal's *Conquest of Space* (1955). The station, affectionately known as "the Wheel" to its crew, was consciously based on Wernher von Braun's *Collier's* magazine series. In fact, the entire film was essentially a condensation of the space program outline by von Braun. In one impressive special effects scene, the station is shown being serviced by a winged shuttle arriving from the Earth, while at the same time a spacecraft constructed for a mission to Mars orbits near the station. In fact, the film illustrates the two main functions expected of an orbital station up to that time: its primary job of acting as a platform for Earth observation—such as tracking weather—and as a base for the assembly of spacecraft destined for the Moon and planets. A sign of the times in which the movie was produced is the tacit assumption that such a space station would be entirely under military control.

SPACE STATION V

The appeal of the wheel-shaped space station continued as late as 1968 in the form of Space Station V in Stanley Kubrick's *2001: A Space Odyssey*. This was an enormous 560 m (1,836 ft) double ring—like two von Braun stations attached on the same axis—that was depicted in the film as still being under construction. Rotating to provide artificial gravity, it was fully equipped to handle travelers and tourists making the journey from the Earth to the Moon, and its facilities include lounges, restaurants, and a hotel.

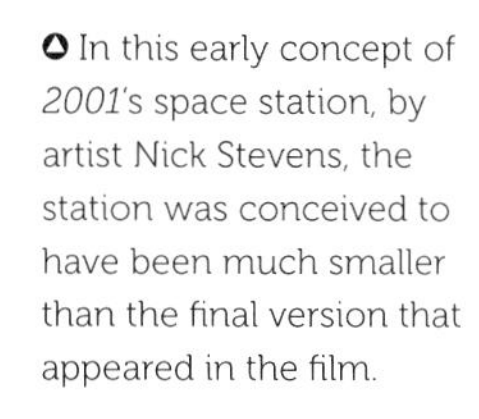

In this early concept of *2001*'s space station, by artist Nick Stevens, the station was conceived to have been much smaller than the final version that appeared in the film.

A publicity poster for *2001* shows the Orion spacecraft exiting. In the half century since its release, *2001*'s Space Station V has become one of the most iconic of all movie spacecraft.

SPACE STATION TOYS AND BOOKS

Space station toys are largely absent from the shelves of toy shops, probably for the simple reasons that spaceships, by their very nature, are more exciting. Even the prolific postwar Japanese makers of tin wind-up and battery-powered toys largely ignored space stations in favor of robots and spaceships. There were a few exceptions, however.

Lindberg's MARS PROBE SPACE PROGRAM
4 NEW AND EXCITING plastic construction kits
USS TRANSPORT
MARS PROBE SPACE TRANSPORT KIT NO.1149
$1.50 each
MARS PROBE LANDING MODULE KIT NO.1147
MARS PROBE COMMUNICATION SATELLITE KIT NO.1150
USSS
MARS PROBE SPACE STATION KIT NO.1148
ASSORTMENT NO.MP1147-50
RETAIL PRICE $1.50 EACH • PACK 2 DOZEN KITS • WEIGHT 20lbs. PER CARTON
AVAILABLE INDIVIDUALLY OR AS AN ASSORTMENT
New York Show Room: S. M. Coplon & Son, 200 Fifth Ave., NYC
LINDBERG PRODUCTS INC., 8050 N. MONTICELLO AVE. SKOKIE, ILLINOIS 60076
the LINDBERG line®
ESTABLISHED SINCE 1933

①

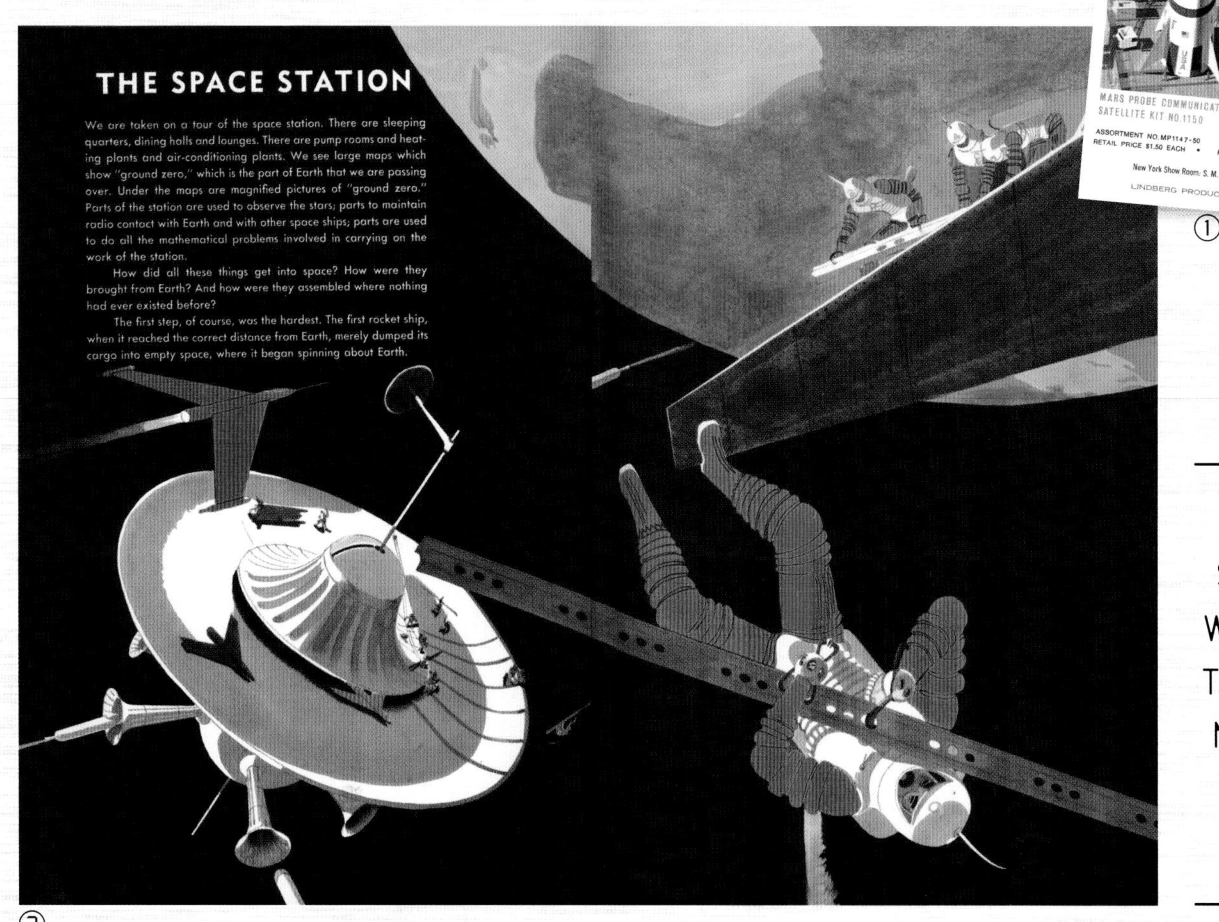

THE SPACE STATION

We are taken on a tour of the space station. There are sleeping quarters, dining halls and lounges. There are pump rooms and heating plants and air-conditioning plants. We see large maps which show "ground zero," which is the part of Earth that we are passing over. Under the maps are magnified pictures of "ground zero." Parts of the station are used to observe the stars; parts to maintain radio contact with Earth and with other space ships; parts are used to do all the mathematical problems involved in carrying on the work of the station.

How did all these things get into space? How were they brought from Earth? And how were they assembled where nothing had ever existed before?

The first step, of course, was the hardest. The first rocket ship, when it reached the correct distance from Earth, merely dumped its cargo into empty space, where it began spinning about Earth.

②

"ONCE THE SPACE STATION IS FINISHED, WE MAY START OUT ON THE JOB OF SENDING A MANNED SHIP TO THE MOON."

FROM *TOMORROW THE MOON*, A 1959 CHILDREN'S BOOK FROM WALT DISNEY

Toymaker Horikawa produced two wheel-shaped tin space station toys in the 1950s and 60s, while Nomura created one in the 1960s. In the early 1950s, British toymaker Johillco created several different "satellite space stations," including the Interplanetary Space Station X/47, which was made of masonite. In spite of their names, all of the Johillco "space stations" appeared to be what we would call today a "space base"—that is, a station based solidly on the surface of a planet or moon. For instance, its beautifully Art Moderne Station Saturnia and Martian Satellite Station sets came equipped with dinosaurs and space men to battle with the station's staunch defenders.

PLASTIC KITS

When spaceflight finally became a reality, or at least a near-reality, the plastic model kit industry—which until that time had been largely limited to models of ships, cars, and aircraft—discovered that it had an entirely new and exciting market to exploit. Models of rockets and spacecraft, both real and speculative, appeared on the shelves of hobby shops by the scores. Among all of these were a number of space stations. The first of these was a model space station designed by von Braun for the "Man in Space" segments of the *Walt Disney's Wonderful World of Color* TV show. It was produced in 1955 by Strombecker, which went on to create several other spacecraft inspired by the Disney series.

Strombecker, perhaps encouraged by the success of the Disney/von Braun kits, looked to other space travel experts for inspiration. One of these was Krafft Ehricke, a German expatriate engineer then employed by Convair. Among the spacecraft Ehricke had designed that Strombecker recreated in miniature was the "Convair Manned Observational Satellite." Ehricke's idea was to construct a space station from pre-existing components—largely the fuel tanks and other structural members of supply rockets—which lent all of his designs not only a visual consistency, but an elegance not shared by many other spacecraft designs.

Being well aware of the popularity of not only spaceflight in general, but the *Collier's*

1. Originally offered in the late 50s, this von Braun-inspired collection of model kits included, of course, a take on his famous space station.
2. This illustration from *The Big Book of Space* (1953) included a space station inspired by the Smith-Ross design.
3. A Buck Rogers jigsaw puzzle, featuring an impressively hostile and well-armed space station, was produced in 1952 by Milton Bradley.
4. This illustration appeared in *The Dawning Space Age*, published by the Civil Air Patrol in the mid-1950s.
5. Engineer and rocket expert G. Harry Stine designed the Pilgrim Observer Space Station specifically for its release as a model kit.

③

④ ⑤

> "SCIENCE FICTION IS THE MOST IMPORTANT LITERATURE IN THE HISTORY OF THE WORLD, BECAUSE IT'S THE HISTORY OF IDEAS, THE HISTORY OF OUR CIVILIZATION BIRTHING ITSELF . . . SCIENCE FICTION IS CENTRAL TO EVERYTHING WE'VE EVER DONE, AND PEOPLE WHO MAKE FUN OF SCIENCE FICTION WRITERS DON'T KNOW WHAT THEY'RE TALKING ABOUT."
>
> RAY BRADBURY

1. The completed Lindberg space station was a large, handsomely detailed model that is still a prize item among collectors of sci-fi memorabilia.
2. The Lindberg space station model—inspired by the popularity of the von Braun *Collier's* series—first appeared in 1958 and is still available today.
3. This battery-powered Japanese-made plastic and tinplate toy came equipped with not only a revolving antenna and blinking lights, but "space noise" as well. It was designed by S. H. (Horikawa).
4 and 5. The first space station to be offered in kit form was a spin-off from the popular Walt Disney *Tomorrowland* television series devoted to space travel. Unlike many others, this toy was actually designed by Wernher von Braun himself.

series in particular, the Lindbergh company produced three model kits that bore a startling and wholly unauthorized resemblance to the spacecraft von Braun had designed for the magazine. These consisted of a three-stage launch vehicle, a lunar lander and, of course, a space station modeled closely on the classic wheel created by von Braun and Chesley Bonestell.

Probably the most important of the space station kits sold in the 1950s was the space station produced by Revell. Designed by aerospace engineer Elwyn E. Angle, it was certainly the most elaborate of all the kits, boasting a fully detailed interior and working parts. This model, along with several of the Lindbergh kits, even found themselves featured by budget-conscious producers in two or three science fiction films of the era.

AUTHENTIC DESIGNS

The first "real" space station kit—in that it was based on a design seriously proposed by Convair for the US Air Force—was the Atlas Orbiting Laboratory, which was issued by the Hawk company in 1958. It illustrated Convair's idea that an existing Atlas rocket could be converted into a space station, in much the same way that the SIVB stage of the Saturn rocket was used as the basis for Skylab. Space stations largely lay fallow for nearly two decades after that. In 1970, rocket expert G. Harry Stine designed the "Pilgrim Observer Space Station" for MPC, though, strictly speaking, it was really an interplanetary spacecraft. It was not until 1981 that model makers turned again to the aerospace industry, this time for the "Space Operations Center" from Revell. Based on a design produced by NASA's Johnson Space Center, it closely resembled today's ISS. Since then, there have been numerous models available for hobbyists, such as the Mir space station kits offered by RealModels and Heller, the ISS from Revell and Heller, and even a kit of the Russian space station featured in the science fiction film *Armageddon* (1998).

SPACE STATION
MAN MADE EARTH OF SPACE AGE
NON STOP MYSTERY ACTION
BATTERY DRIVEN MOTOR
MADE IN JAPAN
New Action Toy
BLINKING LIGHTES
SPACE NOISE
SEE THROUGH ACTIONS
REVOLVING ANTENNA
SPACE STATION
MAN MADE EARTH OF SPACE AGE
NON STOP MYSTERY ACTION
BATTERY DRIVEN MOTOR
③
WALT DISNEY'S space station STROMBECKER
© WALT DISNEY PRODUCTIONS
World Rights Reserved
S-1
EXPLODED VIEW
YOUR MODEL WILL LOOK LIKE THIS
WALT DISNEY'S "SPACE STATION"
THE SPACE STATION S-1 is wheel shaped and measures 200 feet across. It will be set up as an advanced base in the orbit 1,000 miles above the earth and will act as headquarters for the final ascent to the moon. The outside structure of the Station is divided into three main parts. The outside rim is used as living and working quarters for a 50 man crew. The inner section or the hub is used as an approach to the station through an air lock. The three large spokes are elevator shafts and the small pipes are used as condensers for the turbo-generators and air conditioning plant. The turbo-generator supplies the station with electricity through heat given off by an atomic reactor. The wheel will rotate three times a minute and as a result produces an artificial gravity for the crew.
The inside of the wheel is divided into nine sections:
1. Headquarters and Communications.
2. Weather Observation.
3. Military — here with the use of optical and radar telescopes every spot on the globe can be observed.
4. Emergency Hospital.
5. Astronomy Division — here planning will be done for the trip around the moon.
6. Calculating machines.
7. Maintenance facilities and air conditioning equipment.
8. Botanical and zoological laboratory.
9. Living quarters.
The whole structure will be prefabricated and tested on the ground — then dismantled and transported in sections by unmanned cargo rockets and re-assembled again when the cargo rockets reach their destination. Since everything is weightless in space it is just a matter of fitting the various sections of the Space Station together and immediately covering it with thin metal plates to protect it against meteors.
MADE IN U.S.A. BY
Strombeck-Becker Mfg. Co., MOLINE, ILLINOIS
DETAILED ASSEMBLY
④
⑤

THE REVELL SPACE STATION

Probably the most important of the imaginary space station kits sold in the 1950s was designed by aerospace engineer Elwyn E. Angle and produced by Revell. It was not only extraordinarily detailed, including the interior of the station, but came with an "Operation Manual" that outlined the details of its construction, functions, and management. It describes the station's role as an astronomical observatory, weather and Earth observation post, communications relay, way station for interplanetary exploration, and a laboratory for scientific studies. The Angle-Revell station was to be about 24 m (80 ft) long and 11 m (36 ft) in diameter. By comparison, Skylab was 21 m (82 ft) by 6.6 m (22 ft). With a crew of 20, Revell station would orbit the Earth at an ideal altitude of 10,380 km (6,450 mi), making a full orbit every six hours. The space station would be assembled in Earth orbit from prefabricated parts carried into orbit by a modified version of the XSL-01 launch vehicle—another unique plastic model kit designed by Angle for Revell. Once the station was complete, it would be regularly serviced by shuttles ferrying supplies and personnel from Earth. Angle made the interesting suggestion that his space station could be converted to an interplanetary spacecraft by attaching an "auxiliary power package...an ionic or some similar low thrust device" to the station.

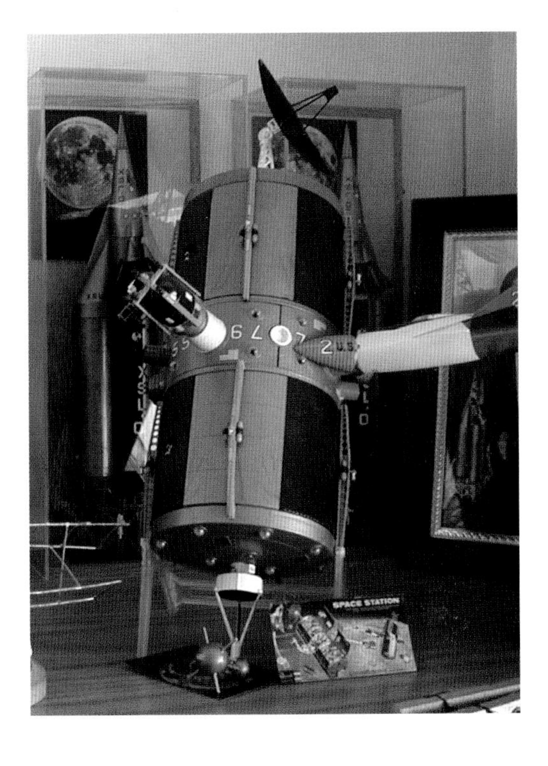

On account of its level of scientific and engineering detail, and now its rarity, Elwyn Angle's space station (seen here is one from his personal collection) is today considered the holy grail of space model collectibles.

Such was the deep consideration behind the Elwyn Angle space station kit, that it came with an eight-page "Operations Manual" that outlined the station's construction, functions, and maintenance.

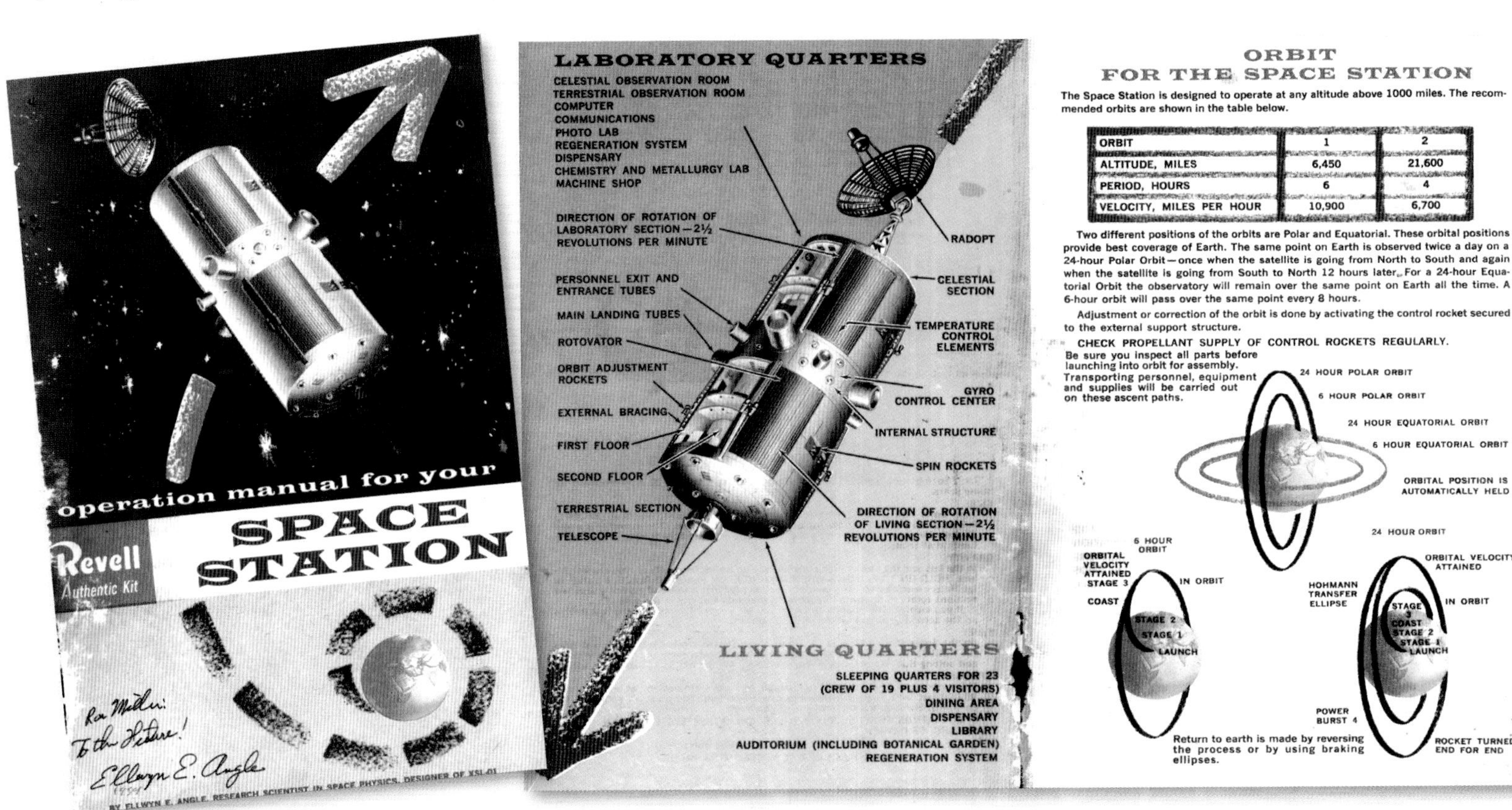

operation manual for your
SPACE STATION
Revell
Authentic Kit

LABORATORY QUARTERS
CELESTIAL OBSERVATION ROOM
TERRESTRIAL OBSERVATION ROOM
COMPUTER
COMMUNICATIONS
PHOTO LAB
REGENERATION SYSTEM
DISPENSARY
CHEMISTRY AND METALLURGY LAB
MACHINE SHOP

DIRECTION OF ROTATION OF LABORATORY SECTION—2½ REVOLUTIONS PER MINUTE
PERSONNEL EXIT AND ENTRANCE TUBES
MAIN LANDING TUBES
ROTOVATOR
ORBIT ADJUSTMENT ROCKETS
EXTERNAL BRACING
FIRST FLOOR
SECOND FLOOR
TERRESTRIAL SECTION
TELESCOPE
RADOPT
CELESTIAL SECTION
TEMPERATURE CONTROL ELEMENTS
GYRO CONTROL CENTER
INTERNAL STRUCTURE
SPIN ROCKETS
DIRECTION OF ROTATION OF LIVING SECTION—2½ REVOLUTIONS PER MINUTE

LIVING QUARTERS
SLEEPING QUARTERS FOR 23 (CREW OF 19 PLUS 4 VISITORS)
DINING AREA
DISPENSARY
LIBRARY
AUDITORIUM (INCLUDING BOTANICAL GARDEN)
REGENERATION SYSTEM

ORBIT FOR THE SPACE STATION

The Space Station is designed to operate at any altitude above 1000 miles. The recommended orbits are shown in the table below.

ORBIT	1	2
ALTITUDE, MILES	6,450	21,600
PERIOD, HOURS	6	4
VELOCITY, MILES PER HOUR	10,900	6,700

Two different positions of the orbits are Polar and Equatorial. These orbital positions provide best coverage of Earth. The same point on Earth is observed twice a day on a 24-hour Polar Orbit—once when the satellite is going from North to South and again when the satellite is going from South to North 12 hours later. For a 24-hour Equatorial Orbit the observatory will remain over the same point on Earth all the time. A 6-hour orbit will pass over the same point every 8 hours.

Adjustment or correction of the orbit is done by activating the control rocket secured to the external support structure.

CHECK PROPELLANT SUPPLY OF CONTROL ROCKETS REGULARLY.

Be sure you inspect all parts before launching into orbit for assembly. Transporting personnel, equipment and supplies will be carried out on these ascent paths.

24 HOUR POLAR ORBIT
6 HOUR POLAR ORBIT
24 HOUR EQUATORIAL ORBIT
6 HOUR EQUATORIAL ORBIT
ORBITAL POSITION IS AUTOMATICALLY HELD
6 HOUR ORBIT
ORBITAL VELOCITY ATTAINED STAGE 3
COAST
IN ORBIT
STAGE 2
STAGE 1
LAUNCH
24 HOUR ORBIT
ORBITAL VELOCITY ATTAINED
HOHMANN TRANSFER ELLIPSE
IN ORBIT
STAGE 3
COAST
STAGE 2
STAGE 1
LAUNCH
POWER BURST 4
ROCKET TURNED END FOR END

Return to earth is made by reversing the process or by using braking ellipses.

THUNDERBIRD 5

The immensely popular British TV series *Thunderbirds Are Go* (2015) featured the impressive space station Thunderbird 5 among its array of futuristic aircraft and spacecraft. Hovering 36,000 km (22,400 mi) above the Earth in a geostationary orbit, its primary functions were to monitor Earth for potential disasters and listen for distress calls. Its crew then monitored and oversaw any rescue operations.

Thunderbird 5 as it appeared in the original TV series. It was a geo-stationary satellite whose crew monitored radio broadcasts from Earth for distress signals.

A detail of the Thunderbird 5 miniature that was specially created for the 2004 live-action feature film.

THE SPACE STATION IN COMIC BOOKS

The presence of a space station in comic books and newspaper comic strips harked back to the earliest appearances of the space station in science fiction: as background, a convenient setting for the action, but rarely as an active participant. There have been, of course, movie and TV spinoffs that featured space stations, such as the Dell adaptations of the Disney "Man in Space" TV series and a *2001: A Space Odyssey* promotional comic book issued by the Howard Johnson restaurant chain.

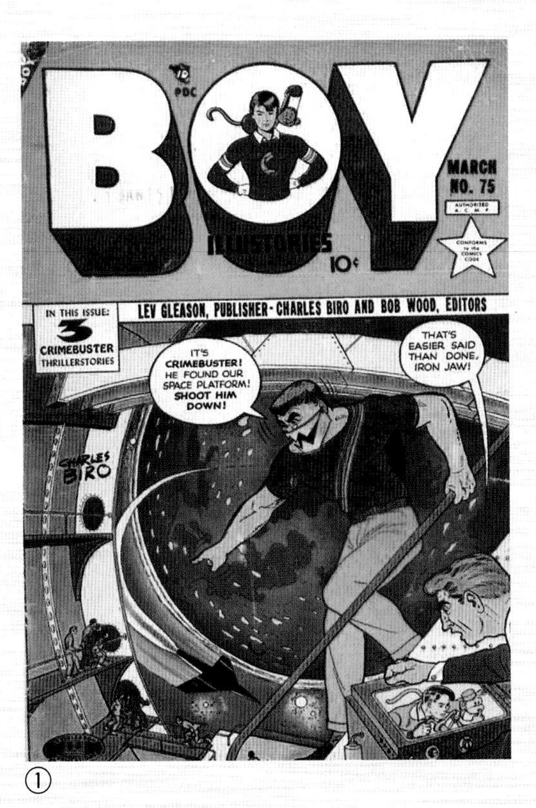

①

②

1. Even before Wernher von Braun popularized the concept of the space station, the idea was exploited by comic book writers as a convenient setting for their stories, such as we see in this *Boy Illustories* comic from March 1952.

2. Scientific accuracy was the least of the concerns for comic book artists and writers, as made clear in this 1952 offering from *Strange Worlds* with its outlandish alien creatures and ray guns.

Two notable exceptions were *Dan Dare, Pilot of the Future*, an immensely popular British comic book series, and *Space Family Robinson*, based on the popular 1960s television series, *Lost in Space*. The latter ran for 59 issues, from 1962 to 1982. It might be something of stretch, however, to include *Space Family Robinson*. The titular family did live aboard "Space Station One," which had been launched in 2002 and positioned in the asteroid belt. But it was in fact a spacecraft capable of interplanetary flight. It was not powered by rockets, however, using instead a form of magnetic drive, and was equipped with two "Spacemobiles" that allowed the Robinsons to explore the different worlds they visited.

The *Dan Dare* series first appeared in 1950 and has continued in one form or another until today. Although space stations did not play a commanding role in the series, they did have a presence and were thought out in some detail. These included the McHoo Asteroid Base, which first appeared in April 1950. Although perhaps not strictly a space station, the use of a small asteroid as a way station and deep space observatory certainly fits most of the functions of a space station. More in the traditional mode was the M.E.K.I Space Station, a hockey puck–shaped structure that was designed to function largely as a facility for spacecraft repair and refueling and for cargo transfer. In spite of its shape, the station did not rotate to provide gravity—this was created by artificial gravity generators. The J-Series space stations acted primarily as passenger transit facilities, and were equipped with restaurants and lounges.

> "IT'S CRIMEBUSTER! HE FOUND OUR SPACE PLATFORM! SHOOT HIM DOWN!"
>
> IRON JAW IN *BOY ILLUSTORIES*, 1952

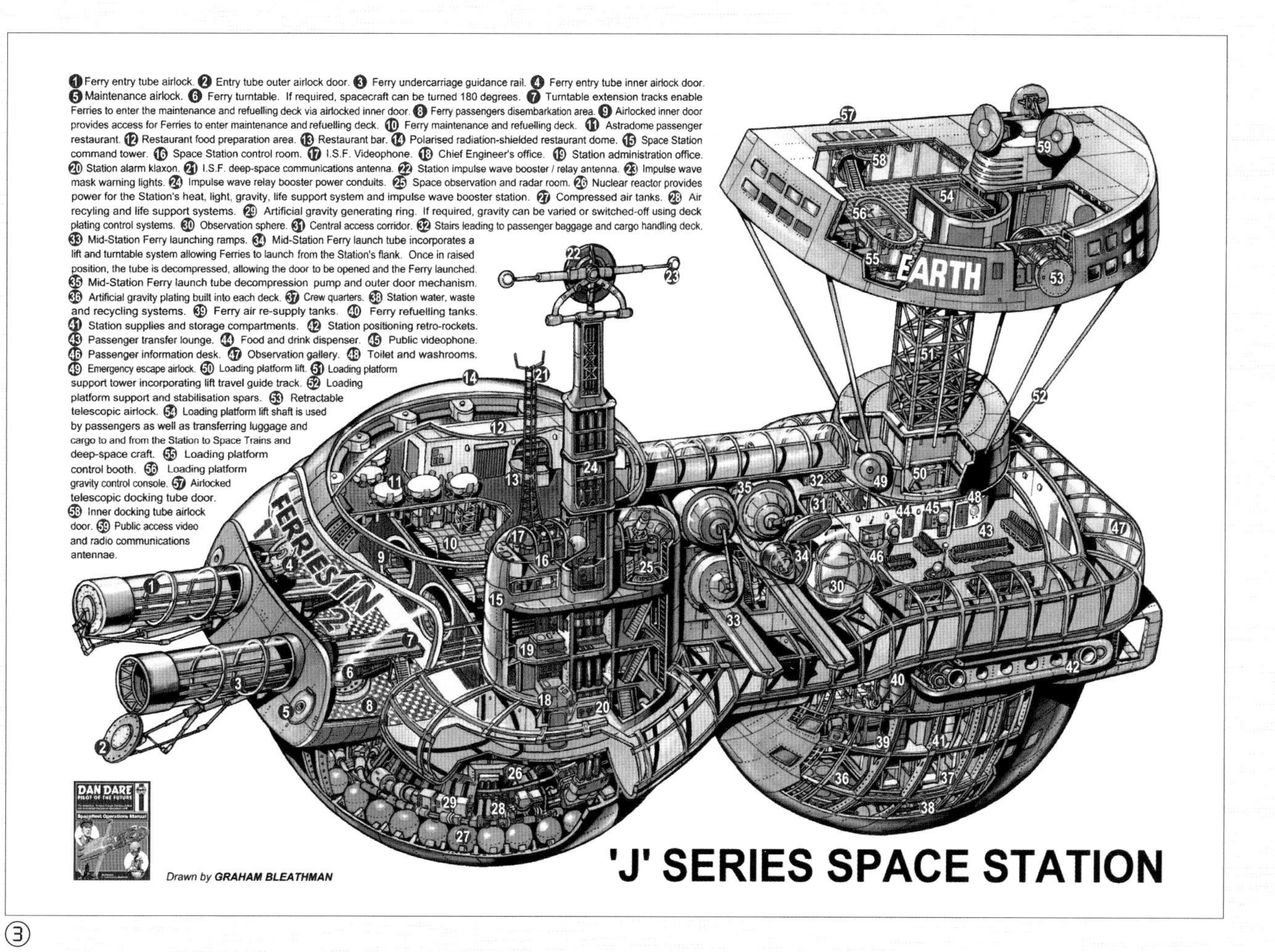

③

3. The popular and now iconic British comic book series *Dan Dare* has featured a plethora of spacecraft designs over its more than 60-year history. Included among them is the "J" series, a wayside stop for planet-bound space tourists.

4. As well as being an impressive cinematic experience, *2001: A Space Odyssey* was filled with product placements. Among them was the restaurant chain Howard Johnson's, which produced this combination children's menu and comic book tie-in to the film.

5. An illustrated panel from the Howard Johnson's *2001* children's menu was based on Robert McCall's famous poster painting for the film. The poster itself is now in the collection of the Smithsonian National Air and Space Museum located in Washington, DC.

④

⑤

①

②

③

④

1. Space Station One (the large object being attacked) was a mobile spacecraft equipped with hydroponic gardens to provide food and oxygen as well as a pair of shuttle craft to allow visits to the surfaces of different planets.
2. The space station illustrated in *Space Adventures* (1953) was clearly inspired by the recently published and immensely popular *Collier's* space series, which continued to influence artists over the following decades.
3. In "Marvel Two-In-One" (1977), it's up to Spider-Man and the Thing to save the entire universe from a villain headquartered in his enormous space base. . .
4. . . .while an equally puzzling object baffles Super Pup and the occupants of a classic wheel-shaped space station in this installment of *Space Comics*, 1954.

"WHETHER OR NOT HIS DESIGN EVER BECOMES ANYTHING BUT A PAPER PLANE, HE HAS STIMULATED THE POPULAR IMAGINATION."

HAROLD LELAND GOODWIN, WRITING ABOUT WERNHER VON BRAUN IN *THE SCIENCE BOOK OF SPACE TRAVEL* (1954)

5. In 1952, *Lost Worlds* made a valiant if not entirely successful attempt to explain the functions of a future space station—including reviving Hermann Oberth's space mirror idea.

6. A 1962 issue of the *Drift Marlo* comic book devoted a page to explaining the details of an actual space station proposed by the USAF and designed by Krafft Ehricke.

7. The influence of the *Collier's* series was in full force when *The Illustrated Story of Space* appeared in 1959. It includes not only von Braun's Wheel, but his winged ferry rocket as well.

⑤

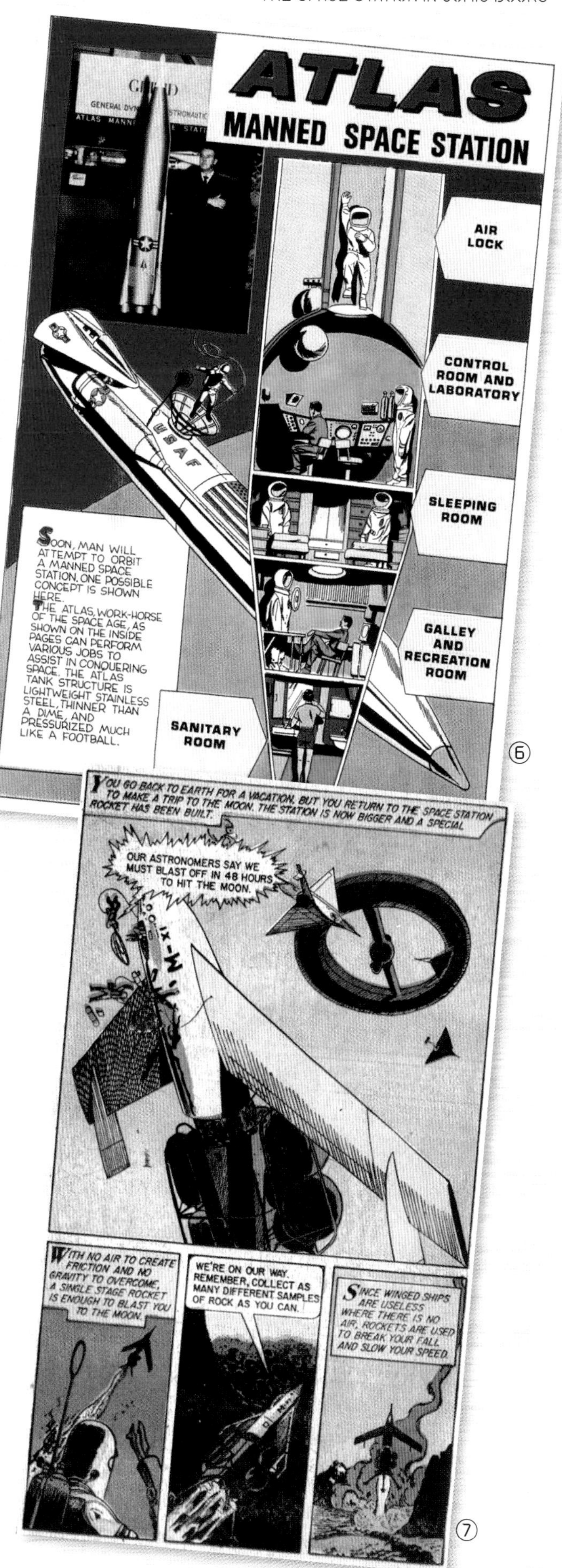

⑥

⑦

Chapter Seven

THE FUTURE OF THE SPACE STATION

During the twentieth century, the space station was thought of as a station in orbit around Earth. But in the coming centuries, space stations will be designed to take humans to other planets of the solar system and into deep space. Some scientists and engineers have even envisaged generational ships that could carry humans to the nearby stars of the galaxy. Space stations could consequently be at the forefront of humankind's habitation of worlds beyond our own.

(Left) NASA's Orion capsule is being developed and should be in operation in the mid-2020s. One proposed mission would be to take a modified Orion to visit the satellites of Mars. The mission would require additional habitation, propulsion, and airlock modules.

(Above) Commercial developers like Space Exploration Technologies (Space-X) have proposed developing "Jumbo Passenger Rockets," roughly analogous with the 1970s "Jumbo Jet" airliners. Here a Jumbo approaches the ISS, with tourists on board.

The Future of the Space Station

THE FOCUS OF MUCH FUTURE THOUGHT ABOUT SPACE STATIONS IS USING THEM TO BETTER UNDERSTAND AND DEVELOP SCIENCE IN READINESS FOR DEEPER EXPLORATION. THE SPACE STATION IS A VIABLE, AND IN MANY INSTANCES THE ONLY, TEST BED AVAILABLE IN THE NEAR TERM FOR VALIDATING SPACE ENVIRONMENT MODELS OR FOR ADVANCING TECHNOLOGIES.

The design of systems that cope with space and make a more habitable and effective working environment are critical success factors for future exploration missions. For space engineers and scientists, areas for further test, assessment, and improvement via space station missions include:

(1) The natural environment of space—more accurate models of the microgravity, radiation, meteoroid, and electrical charge environments. Existing models can be refined that better describe the environment in specific locations, such as at different altitudes or in relation to orbital latitude or day/night transitions.
(2) The operating environment—the fundamental design standards for human-operated space facilities should be revisited based on space station experience.
(3) Specification of requirements for human physiology and space biology based on ISS experience. Human psychology, cognition, and behavioral performance are all critical areas of study for space exploration. Performance stressors can effect factors such as response time, alertness, precision of movement, sensory perception, isolation. A station represents a unique facility to develop training strategies that will allow the astronauts to identify and counteract problem areas.
(4) Robotics requirements—the usefulness and value of robotics systems working with or in lieu of humans.
(5) Sensors and manipulators—an identification of systems required for inspection or operation, and the revision of human interface systems to reflect contemporary progress in rich multimodal human interface systems for visual, haptic, speech, etc. These can be applied to problems of in-space assembly, anomaly response, virtual, synthetic or augmented vision system, and the like.
(6) Assembly operations—improving the efficiency, speed, and precision of in-space assembly systems at large-scale (e.g. cranes, docking mechanisms), mid-scale (e.g., anthropomorphic robots), or small-scale (dexterous and/or micro manipulators) can help to produce more effective or more reliable deployment or assembly systems for space, lunar, or planetary missions.
(7) Mission training and operations—new, more efficient systems for Earth-proximity that can reduce ground support requirements, and new systems for more distant planetary missions that make the flight crew more self-reliant in cases where communications have lag or may have unscheduled interference.
(8) Systems functionality, durability and safety certification or test requirements, documentation and processes—well-defined, streamlined, and simplified processes should reduce the time, difficulty, or complexity of flight acceptance and safety certification.

In this NASA image by Pat Rawlings, a solar clipper converts solar power to propulsion. The method would permit a relatively lightweight and inexpensive spacecraft to carry humans to Mars at a high velocity.

> "SPACE IS FOR EVERYBODY. IT'S NOT JUST FOR A FEW PEOPLE IN SCIENCE OR MATH, OR FOR A SELECT GROUP OF ASTRONAUTS."
>
> CHRISTA MCAULIFFE

A station affords a unique opportunity to serve as an engineering test bed for hardware and operations critical to the exploration tasks. The spacecraft that will support future missions beyond low Earth orbit will likely be a large, complex system that must be maintained by its crew. Like a station, it will not return to the

A Space Exploration Technologies (Space-X) Dragon unmanned automated cargo spacecraft is berthed to the ISS.

A commercial passenger module is being prepared to leave Earth orbit for the Moon in this NASA view by Pat Rawlings, Chris Cordingley, and Bill Gleason.

Brian Duffy (back), Janice Voss (left), and Nancy Currie-Gregg inside the commercial Spacehab module on shuttle STS-57. Spacehab was commercially designed, built, and operated and proved much less expensive to operate than the similar government Spacelab.

ground for servicing, and provisioning of spares will be severely constrained by transportation limitations. Although significant technical support can be provided by ground personnel, all hands-on maintenance tasks will have to be performed by the ship's crew.

PLANETARY MISSION SUPPORT

The information gleaned from space station missions is not just about refining existing technologies, processes, and behaviors. It is also invaluable for the research and development of brand new technologies that could revolutionize our future in space. Such systems could include: propulsions systems, such as solar electric drives or nuclear power; advanced power systems, including solar arrays, ultra-efficient batteries, solar dynamic turbo power generation, or fuel cells, to improve longevity and efficiency; advanced life-support systems; additive manufacturing and machining; systems for the in-situ production of tools, components, and structures using more compact raw material stock; advanced habitation systems; artificial gravity systems; and systems for counteracting human physiological degradation in response to long duration zero-gravity; EVA concepts and suits.

As a result of international experience, engineers and designers have already realized fundamentally different approaches in the following design areas:

(1) Environmental control and life-support systems—US systems generally rely on common system element redundancy, while Russian systems generally rely on dissimilar design and redundancy.
(2) System control and monitoring systems—US systems have relied upon active ground monitoring and control. US systems must generally be operated through the spacecraft data management system. Russian systems generally are more independent, automated, or rely on crew support. Exploration systems may be less dependent on ground support due to communications challenges and data/artificial intelligence system improvements.
(3) Module architectures—The architecture of US and Russian modules of the ISS are also very different. The US architecture was designed with the intent that systems would be maintained and changed over several decades. As a result most of the major system components inside of the US modules are built into 'standard' equipment racks, each about the size of a large refrigerator. This enhances repair and maintenance and since racks can be moved from one module to another, major components can be moved from one location to another, through hatches, within the ISS. Russian systems are generally designed into a particular module, mounted directly to the module walls or structure. Smaller hatches in the Russian segment preclude the transfer of standard racks from the United States segment into the Russian segment.

LONG-DURATION MISSIONS

For long-duration planetary missions, environmental control and life-support systems have to be designed to sustain the environment for years. Systems have to be accessible for the on-board crew to maintain and repair, and modularity has to enable components to be replaced. Because the operation of the environmental control system is critical to supporting life, every failure must be considered possible, and even in the case of multiple simultaneous failures, crew survivability must be ensured.US systems are usually designed with backups that are similar to the primary system; for instance, there may be two identical units with one reserved for emergency use.

The Russian approach to ensuring that a critical life support capability is maintained is usually to have backup capabilities that are a different type of system. For instance, to provide breathing oxygen, both the primary US and Russian systems use chemical electrolysis of water. The US system can be taken apart and serviced in orbit by the crew; critical replacement units are kept in storage on-board. The primary Russian electrolysis system, called the Elektron, is backed up by stored oxygen tanks and by candles that can be burned to create oxygen. In the case an Elektron unit fails in orbit, one of the backup systems will be used until a new Elektron unit can be launched on a resupply ship.

As we shall see in this chapter, it is not just the Americans and the Russians who are forging ahead with investigations in space stations and related technology—China and the Europeans are also playing their part. For it is evident that those who master the science of the space station, could be the masters of deeper space and distant planets.

FROM THE PAST TO THE FUTURE

The concept of the space station as a way station and testbed for future exploration is older than the space program itself. The concept is also multinational—Germans, Russians, and Americans all envisioned that a space station was the first step to going much deeper into the universe. It would allow testing of life support systems before taking off on a missions to other planets, missions that would take many months, or even many years, to complete.

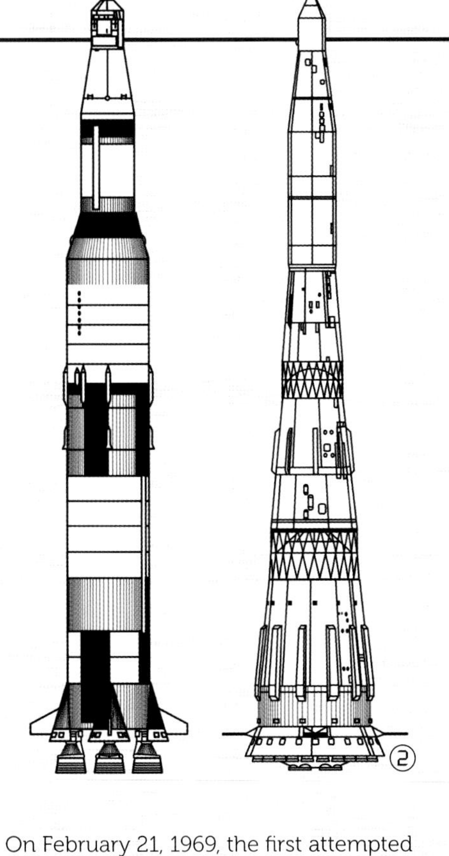

1. President George Bush announces a new space exploration initiative to return to the Moon and explore Mars on the occasion of the Apollo 11 20th anniversary.
2. The Moon rockets: the US Saturn V and Russia's N-1. The Saturn rocket was always successful, while N-1 never made a successful flight.

The long-term plans for space stations have always been ambitious. The Russians intended that their space station would be the prototype of a vehicle that could orbit the Moon and make flybys of Venus and Mars. The N-1 rocket, known by many as the Russian Moon rocket, was actually first envisioned in the 1950s for launching a large Earth-orbiting space station. The N-1 was approved for development in 1959 and was expected to be in operation in 1963. Its intended missions included the launch of heavy space stations with five-person crew, orbited at 400 km (248 mi). The stations would conduct military reconnaissance and basic research. The N-1 was also be used to send a crew of two or three on a circumlunar flight, including entry into orbit around the Moon. Using systems developed and tested on the space station, MK interplanetary spacecraft, which looked very much like the DOS (Salyut) long-duration space stations, would carry a two- or three-person crew on two- to three-year flyby missions of Venus and Mars.

SPACE STATION ROLES

In June 1960, Soviet Chief Designer Korolev reached agreement with the military on the future missions of the orbital station:

- Strategic reconnaissance
- Combat operations against enemy spacecraft
- Strikes against any point on Earth
- Military communications
- Military applications
- Air-to-air defense against enemy ballistic missiles
- Study of the space environment
- Radiation studies
- Study of the Sun
- Study of the Earth and planets
- Astronomical observations
- Weather observation
- Biological research

There would be two or three of the large space stations orbiting simultaneously, controlling a wide variety of military assets in space. Anti-ballistic missile interceptors and combat Sputniks would control the orbital space environment between 300 and 2,000 km (186–1,243 miles) altitude. It wasn't until five years after the inception of the N-1 program, in 1964, that the Soviets decided to use the N-1 rocket for Moon landing missions, competing with the US Apollo program.

Almost as soon as he began, Korolev ran into competition from other design bureaus for the large development budgets being supported by the Soviet military. Premier Khrushchev told Korolev to come to agreement with his competition on the design and missions. He never came to a complete agreement, but was able to put together enough support and was authorized to start development of the N-1 in September 1962. Work on the launch complex at Baikonur began one year later. At the time, it was anticipated that the first N-1 flight would take place in 1965, but due to delays the first vehicle did not reach the launch pad until 1968.

On February 21, 1969, the first attempted launch ended in failure after a minute of flight due to contaminants in the fuel. Three more attempts would be made. On July 3, 1969, again as a result of contamination in the fuel, the rocket exploded just after launch, destroying the launch pad. Two years later, on June 27, 1971, a third launch attempt ended in failure as the rocket spun out of control about 50 seconds into flight. The spin was caused by changes in the launch guidance introduced because of the previous failures. This failure, though, was not the result of engine failures or contamination, and the first-stage engines worked well, so it seemed the rocket's development was progressing.

In 1971 the first DOS/Salyut space station reached orbit and was crewed for a mission of 23 days by a three-man crew. Meanwhile, several more DOS and military Almaz stations were in development. In 1972, construction began of Korolev's Multiple Module Orbital Complex (MOK). This would include two large-diameter space station modules, several smaller specialized research modules, and free-flying satellites that would orbit nearby. The two large modules would be launched on two N-1 rockets.

On November 2, 1972, the fourth N-1 was launched. The first-stage engines ran well but halted seven seconds before planned. In preparation for staging, the engine thrust was being moderated and at that point one engine exploded, terminating the flight. Still, contentious designers from different Soviet design bureaus argued in favor of different

③ ④

3. Gravitational contours in cis-lunar space, the gravity is balanced at the five marked Lagrange points. Placed at these points minimal fuel would be required for the station to remain in position.
4. The Modular Pressurized Rover would extend the range of surface missions. Note how the robotic arms and an airlock aid in collecting samples. NASA design by G. Kitmacher.

engines and significant changes to the N-1. In May 1974, the Soviet Politburo decided to terminate the N-1 program.

As the first two N-1 rocket launches were attempted in February and July 1969, the US Apollo Program was in the final stages of preparation for the first crewed Moon landing which would occur on July 20, 1969. At the time of the Apollo 11 Moon landing, a US Space Task Group defined the space station's role in the post-Apollo US space program:

> we can consider manned bases in Earth orbit, lunar orbit or on the surface of the Moon or manned missions to Mars. . . . A space station module would be the basic element of future manned activities. . . . The space station will be permanent structure, operating continuously, supporting 6–12 occupants. . . . The same space station module could provide a permanent manned station in lunar orbit from which expeditions could be sent to the surface.

The space station would be used to develop the systems and operations required for years-long planetary missions. Using the space station, life scientists would research the ways that crews operated in isolation during such long-duration missions. Technologists would develop and test highly reliable life-support systems and power supply systems required to support a planetary expedition. But for the time being, in 1972, NASA was directed to focus its attention on establishing a cost effective Earth-to-orbit transportation system, the space shuttle. A space station would be deferred to later. Lunar bases or planetary expeditions would follow.

SIGSPACE

More than a decade later, in 1983, President Reagan chartered the Senior Interagency Group on Space (SIGSpace) to develop the space station program. The SIGSpace study said that:

> The space station can provide the necessary first step for future historical advances in space. It is a necessary first step before returning to the Moon or for the mining of materials, or for sending a manned expedition to Mars for on-site exploration, or for conducting a manned survey of the asteroids, or for constructing a manned scientific and communications facility in geosynchronous orbit, or for building a complex of advanced scientific and industrial facility in low Earth orbit.

Some urged President Reagan to approve the space station in the context of longer-range and more dramatic goals. Reagan said that sufficient information was not available to support such goals, but that the decision to proceed with the space station would preserve the pursuit of more ambitious goals in the future. The president felt a decision not to proceed with the space station would preclude the establishment of the future infrastructure for commercialization, industrialization, or exploration. He believed that a space station could stimulate the commercial development of space, much as the railroads opened the western frontier earlier in US history. Reagan also believed that economic growth, productivity, and new jobs could be traced to US investment in advanced technology.

The space station got the formal go-ahead in January 1984. "The Space Station would be a laboratory in space where discoveries could be made; and it is the next logical step." But without a longer-term goal, there was no consensus among the decision makers about a clear, consistent and universally accepted purpose.

During the first five years of the space station, active consideration was given to the construction of assembly facilities for future exploration. Later, under President George H. W. Bush, they would support the visionary Space Exploration Initiative (SEI). But SEI was curtailed in the early 1990s. This left the space station missing a key element of its rationale. This translated into garbled program objectives and created opportunities for political debates that threatened the ISS program's funding. Although the ISS was built as a cooperative international project, debates over the longer-term goals of the human space flight program have continued throughout the Administrations of George W. Bush, Barack Obama, and Donald Trump. Long-term goals have changed with each presidential administration. From lunar bases to missions to Mars to asteroid recovery missions to a deep space gateway: a space station situated at a Lagrangian point, a point of equal gravitation between the Earth, the Moon, and the Sun located 400,000 km (248,500 mi) away from planet Earth.

NEW INITIATIVES

NASA has begun work on a program to leave low-Earth orbit and to develop a miniature space station that will operate in deep space. A new rocket and spacecraft are planned for the late 2020s, and a Deep Space Gateway (DSG) could be established. The DSG has been envisioned as a miniature space station not nearly the size or complexity of the ISS, but based on several ISS-derived modules. It would be situated at a Lagrangian point or in lunar orbit.

The DSG's location in the equi-gravisphere between the Sun, the Moon, and Earth would isolate the astronauts in a location from which the Earth and Moon would be seen only as small disks at a great distance. Communications between there and Earth would lag several seconds, slowing conversations. For such a mission, the astronauts would also be completely reliant on fail-safe life support systems. Significant problem areas to be negotiated include the degradation of human physiology in zero-gravity, solar radiation, and galactic cosmic rays. Returning to the Earth would also require critical trajectories and several days' passage at tremendous velocity. As envisioned since Tsiolkovsky and Oberth more than a century ago, the DSG could serve as a testbed for propulsion system technologies like advanced electric or nuclear propulsion. DSG might also be developed into a Deep Space Transport (DST), traveling well beyond the Moon and perhaps taking the first humans on a flyby of the nearest planets, Mars, or Venus. The DST would then be relied upon to return the astronauts and the spacecraft to Earth, after which the DST would be refurbished and reused.

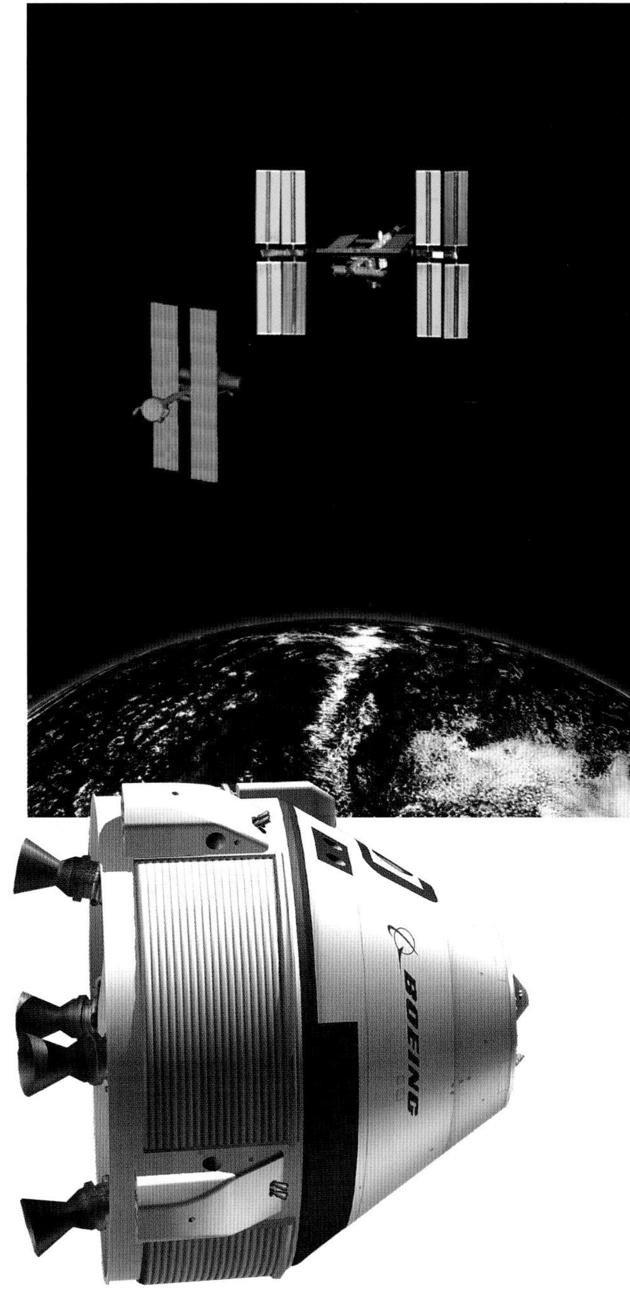

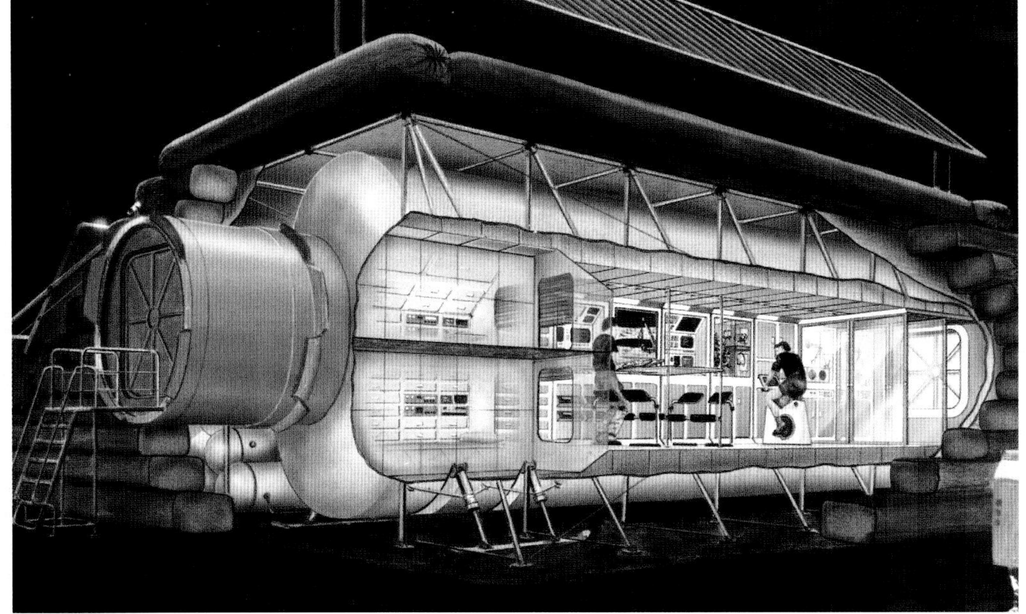

When the ISS was being designed, it was envisioned that its modules and interior racks would form the building blocks for future space bases. This is the First Lunar Outpost, based on a ISS-type module, as envisioned by Gary Kitmacher and John Ciccora for NASA.

Here a trans-Mars vehicle, with similar design and systems characteristics to the ISS, leaves the ISS and Earth orbit.

The Boeing CST-100 Starliner is being developed commercially as a crew carrier to be used to ferry astronauts to the ISS and low-Earth orbit.

▲ The Orion is a crew capsule being developed by NASA that will be able to fly to lunar distances. Here it approaches a Lunar Orbit Platform Gateway with a crew of up to four. Orion will operate by the mid-2020s, and NASA has proposed the Gateway as its destination.

▶ NASA has begun early design work for a deep-space planetary habitat that would be able to carry astronauts to the planet Mars in the 2030s. This view shows a prototype of one of the habitats.

MIR II

The Soviets planned several space stations. The pinnacle was Mir, launched in 1986, but there was a tentative plan to expand Mir with a second Base Block—this was called Mir 1.5. There was also a plan for Mir II, which would have consisted of a Base Block (DOS8) plus four new modules, Spektr, Priroda, FGB, and Nauka. Another version of Mir II would have combined some of these modules with a large truss and much larger solar power generators. The later versions were indefinitely deferred when the NASA–Mir Program began. For the NASA–Mir Program, NASA signed a contract to add two new modules to Mir: Spektr was redesigned for compatibility with the US Spacelab. Priroda was redesigned for compatibility with Spacehab and the shuttle. A Russian docking module was delivered by the shuttle. The other half of Mir II became part of the ISS: the Base Block (DOS 8) more commonly referred to as the Service Module, FGB, and Nauka.

The last module to berth to Mir was the Priroda. Like all Russians modules it was designed for automated autonomous operation.

The Soviet, and later Russian, Mir Orbital Station. The first module was launched in 1986 and the last in 1996. Mir operated for 15 years.

Beginning with Mir's core module, many Russian modules are designed with spherical nodes with up to six berthing ports to accommodate additional modules. The Uslovoy "Prichal" module was designed for use on the ISS and could be used to link ISS to future Russians space stations.

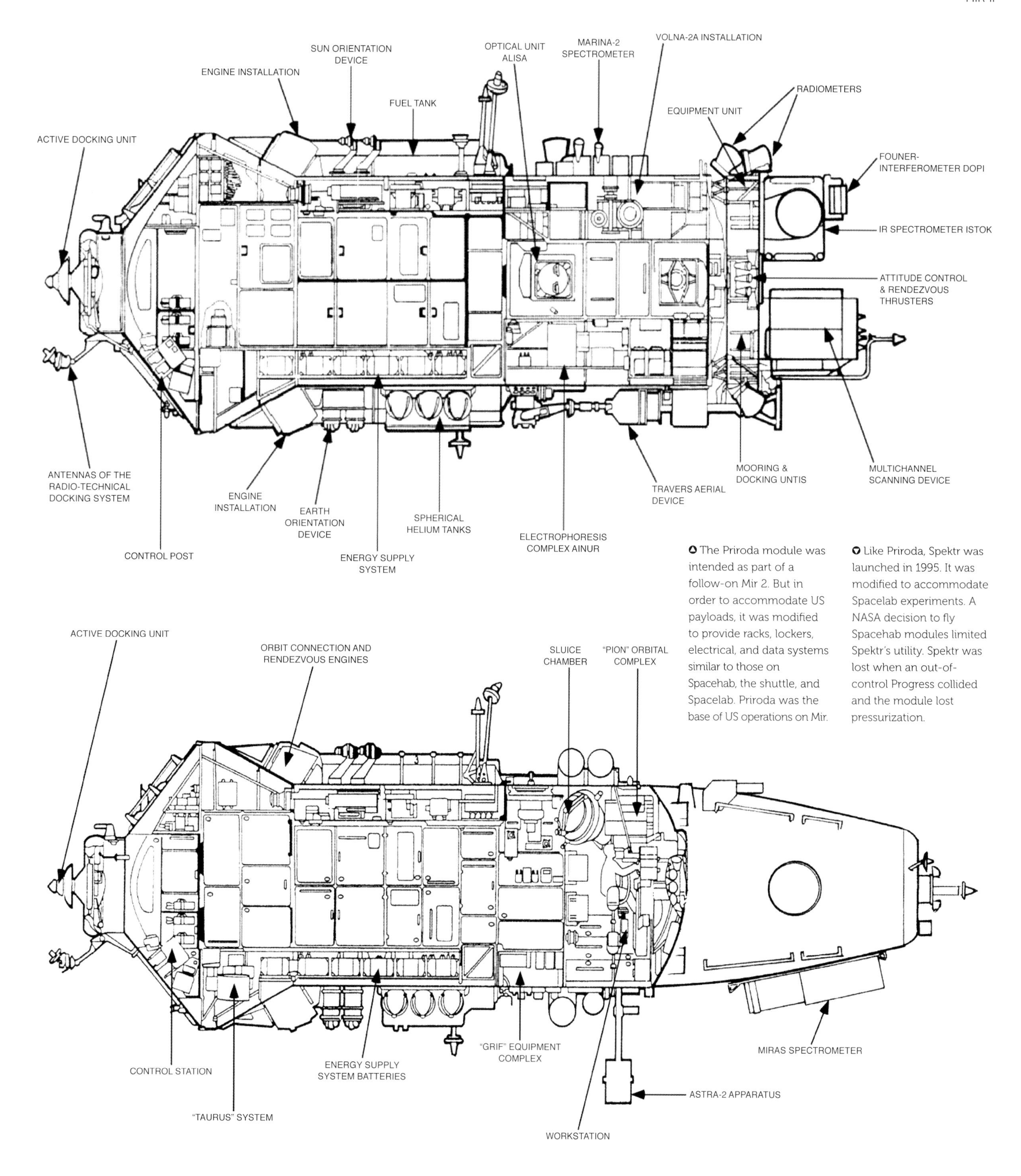

▲ The Priroda module was intended as part of a follow-on Mir 2. But in order to accommodate US payloads, it was modified to provide racks, lockers, electrical, and data systems similar to those on Spacehab, the shuttle, and Spacelab. Priroda was the base of US operations on Mir.

▼ Like Priroda, Spektr was launched in 1995. It was modified to accommodate Spacelab experiments. A NASA decision to fly Spacehab modules limited Spektr's utility. Spektr was lost when an out-of-control Progress collided and the module lost pressurization.

TIANGONG

China considers the development of industry in space to be an important part of the nation's future economic and technological strategy. After earlier abortive attempts to start a human space flight program, China succeeded when they collaborated with the Russians in the 1990s and adapted elements of the Soyuz and Mir designs to meet their needs.

1

2

1. A Shenzhou crewed spacecraft slowly approaches the first Tiangong station.
2. Similar Long March rockets have been used to launch Shenzhou crewed spacecraft and Tiangong space station modules. A larger and heavier Tiangong station awaits the availability of the more powerful booster.
3. The crew of Shenzhou 9 floats inside the Tiangong station. From left are Liu Wang, Liu Yang, and Jing Haipeng.
4. The first Chinese man in space, Yang Liwei, lies in the reentry capsule of Shenzhou-5 spacecraft. Yang Liwei orbited for more than 21 hours.
5. Shenzhou has some similarities with the Russian Soyuz but the reentry module is larger
6. On a screen in the Beijing Aerospace Mission Control Center, the Shenzhou 10 crewed spacecraft as it docks with the first Chinese space station, Tiangong 1, on June 13, 2013.

3

In 1998 China selected its first *taikonauts*. After training for five years, Yang Liwei, became the first Chinese man launched into orbit in the spacecraft Shenzhou-5 (Divine Vessel).

Shenzhou is based primarily on the design of the Russian Soyuz, but has some significant differences in construction and layout. Soyuz typically weighs 4,200 kg (9,240 lb) compared to Shenzhou's 7,800 kg (17,160 lb). Liwei's flight meant that China became the third nation, after Russia and the US, to place a human in orbit in their own spacecraft and on their own rocket. On the third Shenzhou flight, China conducted a successful spacewalk.

In 2011, China launched Tiangong-1 (Heavenly Palace), a module to test systems for a future space station. Both the Tiangong and Shenzhou are launched by the same type of rocket and so are constrained to similar launch masses, approximately 8,500kg (18,500lb). Although roughly similar in outline to the DOS modules used for Salyut, Mir, and the ISS Service Module, which weigh approximately 20,400kg (44,800 lb), Tiangong is smaller and lighter at 8,500 kg (18,500 lb). Tiangong-2 was launched five years after the first. A crew of two stayed on board for 30 days, performing experiments and testing systems. Months later, while unmanned, a resupply craft called Tianzhou (Heavenly Vessel) docked automatically multiple times and refueled Tiangong. Tianzhou weighed considerably more than Tiangong and was the heaviest Chinese satellite. According to plans published in 2008, Tiangong-2 was to be visited by a second crew. The Chinese have a new booster rocket in development that will be able to carry 23,000 kg (50,600 lb) to orbit, more than twice their current capability. Initial tests of the Long March 5 rocket were problematic and booster testing is continuing. Once the new Long March 5 is operational, it will likely be used to assemble a larger space station than Tiangong.

4

5

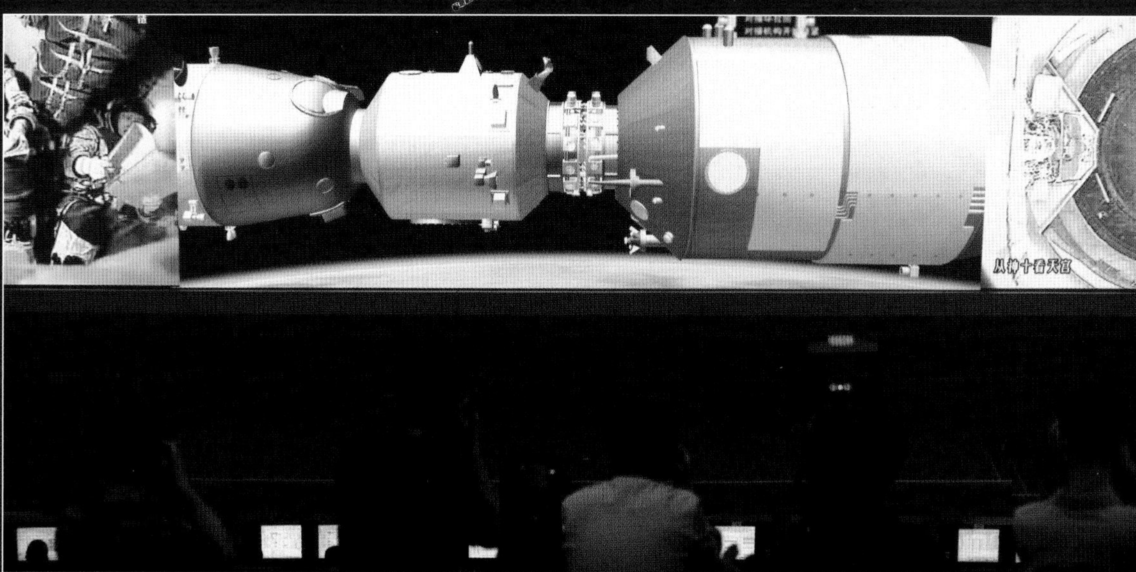

6

COMMERCIALIZATION

American interest in space commercialization goes back to the National Aeronautics and Space Act of 1958. It states that NASA should "seek and encourage, to the maximum extent possible, the fullest commercial use of space." In the 1980s, President Reagan stepped up interest in space commercialization. With the space shuttle, the promise of microgravity research was bright; companies were preparing to begin pharmaceutical production on the shuttle, and this provided a rationale for presidential approval of the space station. US commercial space policy was elevated to the same significance as civil and national security space. Reagan directed the government to purchase commercial capabilities, encourage private investment, and minimize commercial space regulation. Subsequent administrations continued the pressure.

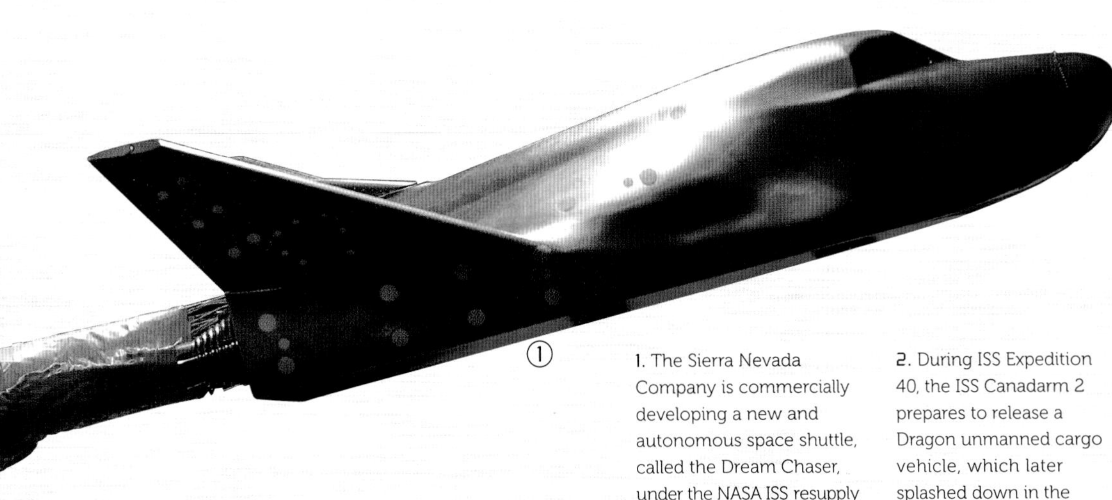

1. The Sierra Nevada Company is commercially developing a new and autonomous space shuttle, called the Dream Chaser, under the NASA ISS resupply services contract.

2. During ISS Expedition 40, the ISS Canadarm 2 prepares to release a Dragon unmanned cargo vehicle, which later splashed down in the Pacific.

3. The ISS Expedition 34 crew skillfully captures the commercially developed unmanned Dragon cargo capsule using the Canadarm 2 system on March 3, 2013.

4. The Dragon space capsule was designed from the outset to carry a crew. This interior view of the crewed Dragon shows the seats that accommodate seven crewmembers.

Early commercialization attempts were unsuccessful. The *Challenger* accident halted pharmaceutical production plans. Space Industries Inc. (SII), however, had planned the Industrial Space Facility (ISF) as a man-tended space platform—it would be a small man-tended station relying on the shuttle for servicing. It never progressed because NASA refused support agreements, and SII officers felt NASA viewed the ISF as a threat to the ISS. Later attempts were more successful.

NASA sponsored dozens of university researchers developing experiments for flight on the space shuttle. A backlog of experiments awaiting flights formed. Shuttle experiment capacity was very restrictive. NASA hired Spacehab, Inc. under the Commercial Middeck Augmentation Module contract. Spacehab built commercially owned modules and established a small workforce. Spacehab proved faster, more efficient, and less expensive than alternative programs, and was adopted for most Shuttle missions to Mir because of the short turnaround time and constrained budgets. Later, Spacehab became a primary logistics carrier for shuttle missions to the ISS.

When the shuttle program ended after the Columbia accident, there was pressure to develop commercial replacements. SpaceX was the first private company to design, develop, and operate rockets and spacecraft going to orbit and returning to Earth from the ISS. Their Dragon capsule offered services previously only provided by governments. When the company began returning their rockets and reusing them, costs came down even further. Other companies won NASA contracts to deliver ISS cargo, such as Orbital ATK's Cygnus spacecraft and Sierra Nevada's Dream Chaser miniature shuttle. Boeing and SpaceX won contracts to carry crew to and from the ISS. These companies design, develop, and operate their own vehicles. As with the other private technology companies, costs on these projects have been significantly below typical government costs for similar services.

COMMERCIAL USERS

NASA has used commercial methods to purchase new services on the ISS. Water is the most expensive commodity in space, but water could be scavenged from shuttles when they were flying; shuttle fuel cells generated electrical power and produced excess water in the process. When shuttle flights stopped, water had to be carried as cargo, replacing other needed supplies. On the ISS, as astronauts breathe, they produce unwanted carbon dioxide, and as the regenerative life-support system filters and purifies waste water, an unwanted byproduct is hydrogen. A sabatier reactor, however, combines

④

carbon dioxide and hydrogen to produce water. Rather than NASA paying to develop the Sabatier system, a contract was awarded for water production services, with the contractor responsible for hardware development and performance.

Operation of the ISS itself could be one of the new commercialization initiatives. NASA has said that it will turn low Earth orbit over to commercial companies in the future as NASA's exploration program moves beyond LEO. Several companies are developing new privately owned and commercially operated systems like ISS add-on modules, new space stations, and future vehicles.

NEW SPACE STATION

Another alternative is a new space station. The ISS currently operates to government set budget plans. Currently ISS is planned through 2024, but an extension to 2028 is in discussion. Each international partner must approve the continuation. The greater the uncertainty and the longer uncertainty continues, the more difficult it is to find users.

Bigelow Aerospace has built a series of inflatable modules. They can providing more interior volume for the same amount of mass. A test module—the Bigelow Expandable Activity Module (BEAM)—has been in place on ISS. It remains berthed, in use as a storage area. In the future, Bigelow hopes to provide an independently operating destination station for commercial vehicles bringing tourists or researchers to orbit. They hope to lease modules or habitats to biotechnology, pharmaceutical, entertainment, government, or university tenants or to tourists.

Axiom Space is also planning a new space station to replace the ISS. They intend to provide an orbital base for research, manufacturing, advertising, and tourism. Their business plan is first to launch a module to provide commercial operations on the ISS. As ISS reaches its end of life, the module will be used as a moving van and will become the cornerstone of a new commercial station that will replace ISS. Axiom will train astronauts, including both their own and guests, and integrate payloads, and manage the operation of the payloads and their modules.

Almost all of the new commercial providers are certain they can operate more efficiently than NASA's bureaucratic government processes by using standard business practices. At least six industry sectors have been identified that are believed will be interested in future commercialization of LEO: operations, space-based manufacturing, technology development, systems testing, tourism, advertising, and image branding.

INFLATABLES

The idea of using inflatable balloon-like structures for space stations goes back to von Braun's 1952 proposal for a 76 m (250 ft) diameter rotating inflatable wheel-shaped space station. The first inflatable structure used for a crewed spaceflight was the airlock used for Alexei Leonov's first spacewalk in March 1965, built by the same factory that made the cosmonaut's spacesuits. In the early 1960s, NASA worked with Goodyear Aircraft Corporation on an inflatable ring or torus that could fit a six-person crew. It would rotate slowly in space, so that its occupants could enjoy the benefits of artificial gravity. Virtually all space station designers at the time believed artificial gravity was absolutely necessary for any long-term stay in space. Several models of the design were built.

Scientists studied inflatable habitats for use as bases on the Moon in the late 1980s and this led to a NASA patent for the Transhab; a collapsible, inflatable space station module. Robert Bigelow, a hotelier, sought to place tourist accommodations in orbit. He bought rights to the NASA patent and developed a series of module prototypes. Two were placed in orbit for testing and one, called the Bigelow Expandable Activity Module, was berthed to the ISS Analyses have shown the inflatable modules have are better for

A Bigelow Aerospace Expandable Activity Module (BEAM). Another BEAM launched inside an unmanned Dragon and reached ISS in April 2016. After a year, NASA contracted with Bigelow to use BEAM as a storage module until 2020.

In 1962 NASA's Virginia Langley Research Center worked with Goodyear to build a mock-up inflatable space station. Here NASA Administrator James Webb observes the mock-up.

Engineers at NASA Johnson Space Center developed the Transhab module concept. Bigelow purchased the rights to the patent and developed the BEAM modules.

Inflatable satellites were used early in the space program, including the airlock use during the first Soviet spacewalk. In 1989 NASA envisioned an inflatable Lunar Base—the dome seen here. It would be inflated after site preparation, using the First Lunar Outpost on the right to start.

At Bigelow Aerospace in Las Vegas, Nevada, several large inflatable B330 modules are tested. They provide ample space: two of these modules have roughly the same capacity as the entire ISS pressurized volume. The modules could also be used as an Earth-orbiting hotel.

micro-meteoroid protection; rigidity, flammability, and off-gassing; durability; lifespan; crack propagation prevention; and the effects of atomic oxygen. Bigelow modules are far more radiation proof than aluminum structures, which create secondary radiation effects that can be very harmful to humans when outside of Earth's radiation protection in low Earth orbit. An alternative to collapsible or inflatable fabric structures are metallic structures that maintain their shape through pressurization. The Atlas ICBM, in order to reduce mass, had no internal framework. It was an inflatable thin skinned stainless steel shell that would collapse if it lost pressurization. Many designers have proposed using such spent rocket shells as the basis of inhabitable space modules.

Space modules could serve as medical laboratories for the study of humans and their ability to function on long space missions. Researchers could study the effects of space environment on materials, equipment, and powerplants. The station could also be used to develop new stabilization, orientation, and navigational techniques. With telescopes and cameras on board, a further purpose would be as an orbiting astronomical observatory and for Earth-oriented meteorological, geographical, or military reconnaissance surveys.

COLONIES IN SPACE

In the mid-1970s, the space shuttle would fly within a few years, and the promise of inexpensive routine space travel meant that lots of people would have access to space. The shuttle was going to be the first leg of a transportation system that could take people, hardware, and raw materials into high Earth orbits and to. NASA was ready to begin major space construction projects to make use of the Space Transportation System's capabilities.

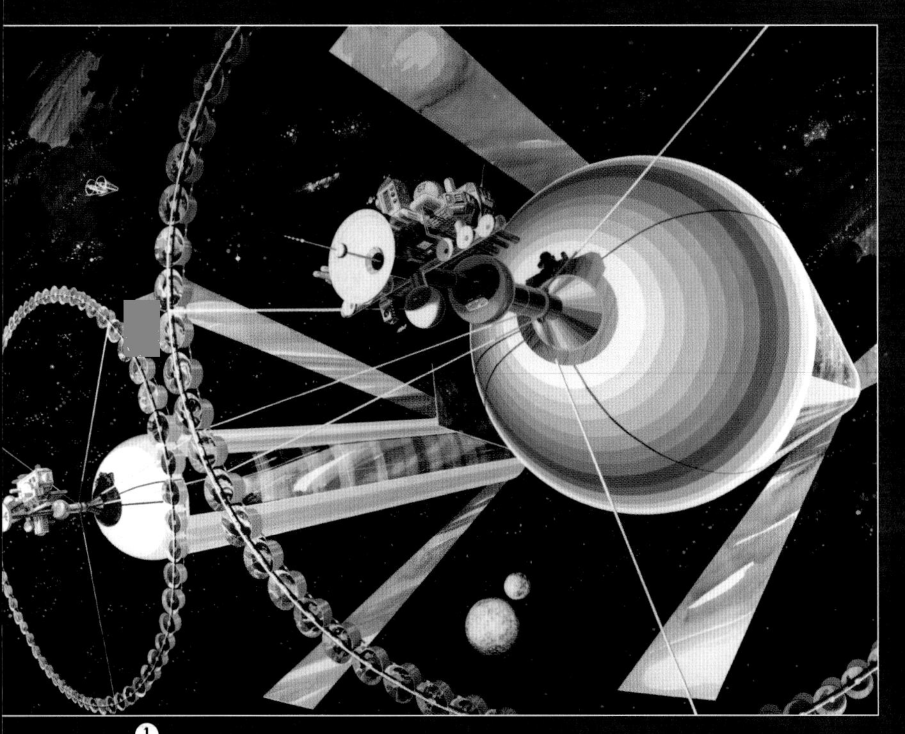

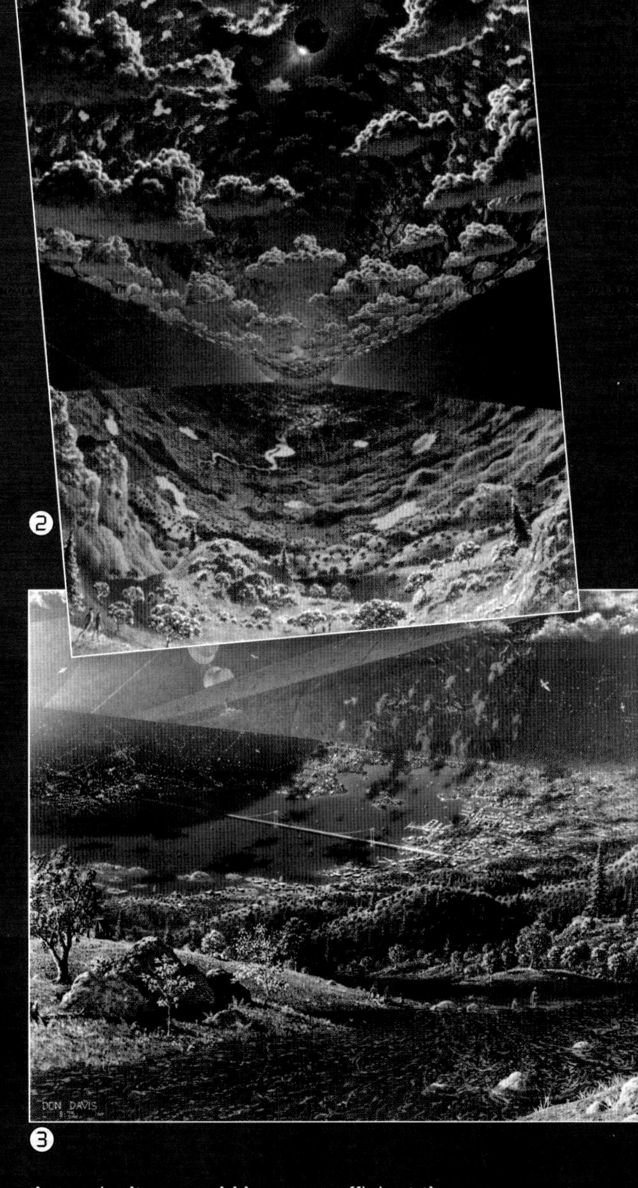

1. The 1970s Gerard O'Neill space colonies. Each is a cylinder. The two would be attached and rotate in opposite directions, providing artificial G for the inhabitants and countering gyroscopic forces. Large mirrors would reflect the Sun into the cylinder interior to provide daylight conditions.
2. The interior of the O'Neill space colony cylinder, with clouds and weather.
3. The end of one of O'Neill's colonies is the interior of a hemisphere 8 km (5 miles) in diameter. This view, as envisaged by Don Davis for NASA, shows a bridge similar to San Francisco's Golden Gate as a size reference.

At a time when the nation was in the midst of an oil crisis, with long lines of vehicles and owners waiting for gas, attention turned to solar power satellites. They would need to be huge structures with solar arrays, capturing the energy of the Sun and beaming it down to Earth via microwave. An unlimited and inexpensive supply of energy from the Sun could supply all the electrical needs of Earth.

The sunlight present in space is enormous and could provide tremendous energy. The power could be used at the colony or transmitted to Earth. Microwave beams would be benign with zero emissions and less ground surface area would be required than for solar panels. The beams would pass through the atmosphere with little or no hindrance.

LIVING AND WORKING

The space colony would provide a facility in space where the inhabitants would be responsible for building the solar power satellites. The raw material from which the colonies and power stations would be manufactured would come from the Moon. It is much less expensive to bring materials from the lunar surface than from the Earth because of the reduced lunar gravity. Materials from the lunar rocks would supply the chemicals to build aluminum and glass structures.

Tens of thousands of people would live and work on space colonies. The colonies would be large enough and rotate fast enough that gravity would be simulated and residents would not experience disorientation. Food and water would initially be brought from Earth, but eventually each colony would become self-sustaining. An internal surface area of 100 acres could sustain about 10,000 people. By using the unlimited power of the Sun and increasing carbon dioxide levels, because there is unlimited power, light without nighttime, and with the lack of any storms that would devastate crops, the agriculture could be more efficient than on Earth itself.

DESIGNS

The Bernal Sphere Island 1 was based on a design that originated in 1929. A hollow sphere would provide the population's main habitat with agricultural crop growth cylinders each equal in length to the sphere's diameter, on two sides. At either end, systems would convert sunlight into power and provide cooling, communications, and mirrors for lighting, and docking ports for visiting vehicles. The entire system would rotate for artificial gravity. Different size spheres could house different sized populations. A sphere 500 m (1,640 ft) in diameter could support 10,000 people, a 1.8 km (1.1 mi) sphere could accommodate 30,000, or a 16-km (10-mi) diameter could house 140,000.

The Stanford Torus was essentially an enlarged doughnut-shaped design similar to Noordung's 1928 proposal but much larger.

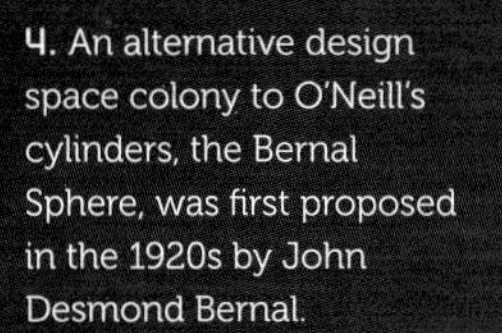

4. An alternative design space colony to O'Neill's cylinders, the Bernal Sphere, was first proposed in the 1920s by John Desmond Bernal.

5. The interior view of the Bernal sphere colony, home to thousands of people. Island 2 was a larger Bernal sphere with a diameter of 1,800 m (5,905 ft).

6. The Stanford Torus was a third design for a large space colony. This was doughnut or torus shaped, but much larger at 1.8 km (1.2 miles) in diameter.

7. Agriculture would maintain the colony's environment and produce food. In the Stanford Torus, crops would be grown on multiple levels.

8. A Stanford Torus colony in the final stages of construction. In this view, we see the interior, and radiation shields have been installed.

9. A cutaway view of the Stanford Torus. A rotation rate of once per minute would provide 1 g of artificial gravity for the people inside.

The torus was more than 1.8 km (1 mile) in diameter and consisted of a 130 m (427 ft) thick tube. A solar power station 10 km (6 miles) away would beam-in electrical power. A large reflector would ensure proper lighting. Supply vehicles and spaceliners carrying people would dock to the center hub and elevators would take people to the rim. The torus would rotate once a minute to generate artificial gravity at the perimeter of the torus.

The O'Neill Island 3 was a 32-km (20-mile) long, 13-km (8-mile) diameter cylinder design, in which 1,300 km² (500 square miles) of interior surface area would provide living space for tens of thousands of people. Windows in the cylinder would allow the entry of sunlight, while long reflectors would reflect sunlight in through the windows.

SURVIVAL

Beyond the ability to support a workforce building solar power satellites, proponents argued that space colonies could provide large-scale human life support systems that could ensure the survival of humanity in case of massive war or pestilence on Earth. Proponents said the colonies would be easier to get to than the Moon or a planet and would be easier to build in space where there is no gravity. Space colonies would also be easier to maintain, since there is no night and thus working hours are optimized. A lunar base would require a solar-powered system during the day, but an alternative system, either storage batteries or nuclear power, for lunar night. On the lunar surface, massive machines would be required for building site preparation. A space colony could also be a step towards a galaxy-class starship that could take human settlers to planets of other stars far across the Milky Way galaxy.

Threats to life on a space colony would include solar radiation and galactic cosmic rays. Protection might be provided in the form of a magnetic radiation shield that would provide protection in a similar manner to the Van Allen radiation belts that surround Earth. An alternative is mass shielding, in which soil launched from the Moon would insulate the colony from radiation and cosmic rays.

Material for building the space colonies would come from a mining operation on the Moon. Lunar soil would be accelerated by an electromagnetic mass driver to lunar escape velocity. At the construction location, processing units would separate the lunar minerals. Those required for construction would be processed through a 3D printing unit. Leftover materials could then be used for radiation shielding.

It has been said that colonization of space is technically possible, although the challenges remain formidable. In order to make it feasible, launch costs from Earth need to be significantly reduced first, and then large-scale manufacturing in space needs to be demonstrated.

CONCLUSION: WHAT IS A SPACE STATION?

That might seem to be an odd question to ask at the end of a book about space stations, but it bears exploring given the visions and realities illustrated on the preceding pages. The Oxford English Dictionary defines "space station" as a "large artificial satellite used as a long-term base for crewed operations in space." While certainly not inaccurate, that definition seems to tell only part of the story.

Dividing the phrase into its component words, "space," of course, describes the location . . . "station," therefore, must denote the purpose. One definition of "station" is "a place or building where a specified activity or service is based." Another is a "place on a railway line where trains regularly stop so that passengers can get on or off." Yet a third cites "station" as being the "site at which a particular species, especially an interesting or rare one, grows or is found." As humanity has discovered during its first 60 years in pursuit of establishing space stations, all three of those definitions could apply. And yet our definition is still not complete.

On October 22, 2012, NASA astronaut Sunita "Suni" Williams marked the 100th day of her second stay on board the ISS. Inside the orbiting outpost—which is found along a line, of sorts, where visiting spacecraft regularly stop—Williams floated within a room filled with the equipment and supplies needed to support a specific activity: scientific research. But when asked that day to describe where she was, Williams defined her surroundings as something more than a site or place or base of operations. "It's definitely our home at this point in time," she remarked. Williams and her crewmates were the 33rd contingent to reside on the ISS, extending a streak of continuous human occupancy in space. Where rockets and spaceships make spaceflight possible, the space station makes orbit—and more broadly, outer space—our home.

Space stations have and will continue to serve as humanity's home away from home, whether they are circling Earth or some other celestial body. They are a place not just to work or serve an activity, but to live and reside. "Once I got on board the space station, it really felt like I was visiting an old home," said NASA astronaut Scott Kelly, the first American to spend a year in space aboard the ISS.

The first space stations were modest in their offerings. A far cry from the early visions of rotating cities and vast latticework structures in the sky, the Salyut and Skylab space stations first launched by Russia and the United States were one-room laboratories, capable of supporting crews on multi-

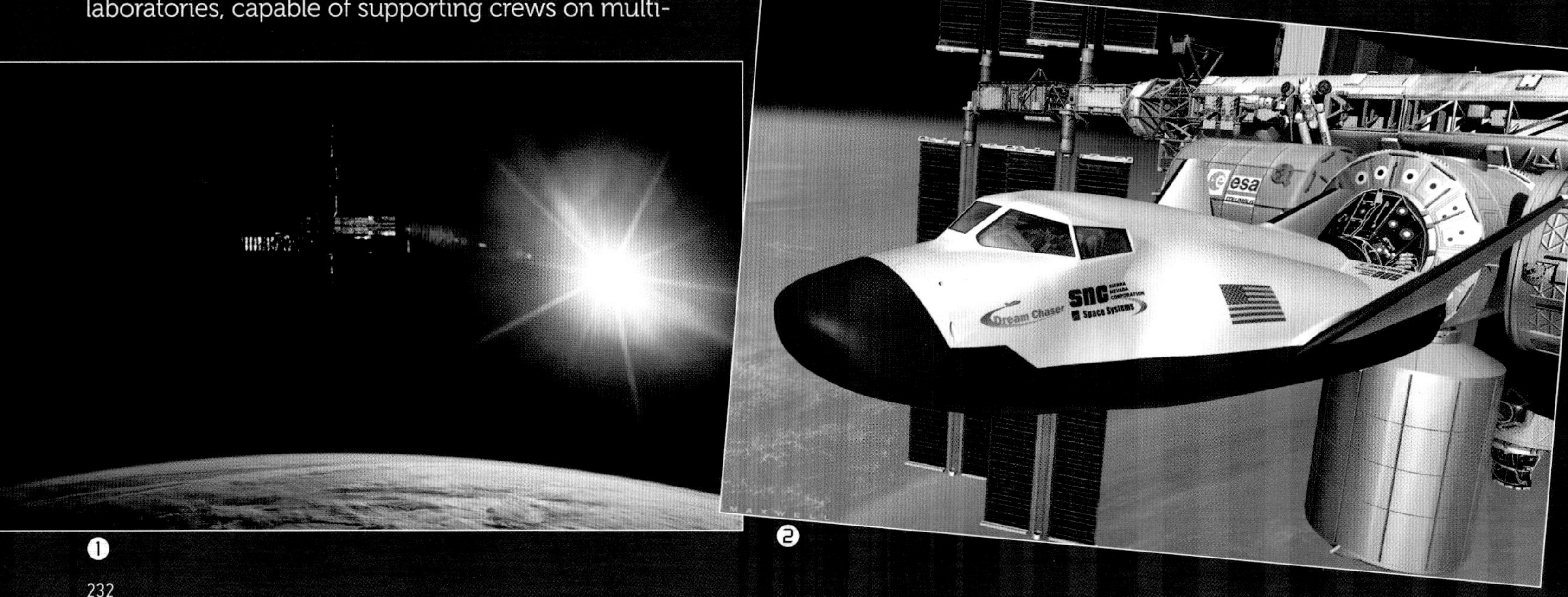

1. Only the reflective parts of Russia's Mir space station are clearly visible in this dramatic photograph, taken during rendezvous operations by the space shuttle *Discovery* and the Mir space station.
2. An artist's concept of the Dream Chaser spacecraft being developed by Sierra Nevada Corp. (SNC) of Centennial, Colorado, for NASA. NASA selected Sierra Nevada to develop the small shuttle for carrying cargo to and from the ISS under the Commercial Cargo Program (CCP). The goal of CCP is to drive down the cost of space travel. SNC plans to eventually man-rate the vehicle for carrying more people routinely into space at lower cost than ever before.
3. NASA astronaut Tracy Caldwell Dyson, Expedition 24 flight engineer, looks through a window in the Cupola of the ISS, observing the movement of clouds across the oceans of Earth.

❸

month stays but with few of the creature comforts enjoyed by those on the ground. They were homes, but precarious ones. Russia's Mir space station, with its interlocking modules, demonstrated that humanity could assemble an outpost from multiple parts on orbit and staff it with crews from more than one country. The ISS built upon that, literally and figuratively, growing to the size of a five-bedroom house. It became the ultimate "home office," combining the facilities found in a world-class laboratory with the comforts of a modern home. .

Now the ISS is serving as a model for where humanity will next call home. "It's a precursor to what it's going to be like when we start living on the Moon. The 'Lunians,' or whatever we're going to call them, or the Martians, when we first start living on Mars," noted Canadian astronaut Chris Hadfield. As humanity looks to expand not just its reach but its encampments beyond Earth, the concept of a space station is becoming the staging platform from where we descend to a planetary surface or the gateway from where we will depart for points further out in our solar system and beyond.

So what is a space station? It is humanity's home away from our current home, Earth, and preceding our next home, far out in the galaxy. The station we pioneered in Earth orbit will become the means by which we leave our planet behind and find new places to live.

INDEX

RESOURCES / ACKNOWLEDGMENTS

MUSEUMS

Cosmosphere International SciEd Center and Space Museum: http://cosmo.org
NASA Kennedy Space Center: www.kennedyspacecenter.com
National Museum of the US Air Force: http://www.nationalmuseum.af.mil
Smithsonian National Air and Space Museum: https://airandspace.si.edu
Space Center Houston, Houston, Texas: www.spacecenter.org

WEBSITES

collectSPACE: www.collectspace.com
Center for the Advancement of Science in Space (CASIS): www.iss-casis.org
Dreams of Space: http://dreamsofspace.blogspot.co.uk
Encyclopedia Astronautica: http://www.astronautix.com
Gagarin Research and Test Cosmonaut Training Center: http://www.gctc.su
NASA History: https://history.nasa.gov/
NASA Images: https://www.nasa.gov/multimedia/imagegallery/index.html
NASA Image Archive: https://www.nasa.gov/content/nasa-images-archive
NASA Image and Video Library: https://images.nasa.gov
NASA Johnson Space Center: https://www.nasa.gov/centers/johnson/home/index.html
NASA Johnson Space Center Oral History Program: https://www.jsc.nasa.gov/history/oral_histories/oral_histories.htm
NASA Marshall Space Flight Center Oral History Collections: http://libguides.uah.edu/c.php?g=263653&p=1760003
RSC Energia: https://www.energia.ru/english
Roscosmos: http://en.roscosmos.ru
Space.com: www.space.com

SELECT BOOKS

Anderson, Clayton. *The Ordinary Spaceman: From Boyhood Dreams to Astronaut*. Lincoln, NE: University of Nebraska Press, 2015.

Ansari, Anousheh and Homer Hickam. *My Dream of Stars: From Daughter of Iran to Space Pioneer*. New York, St. Martin's Press, 2010

Chladek, Jay. *Outposts on the Frontier: A Fifty-Year History of Space Stations*. Lincoln, NE: University of Nebraska Press, 2017.

Foale, Colin. *Waystation to the Stars: The Story of Mir, Michael and Me*. London: Headline, 1999.

Garan, Ron. *The Orbital Perspective: Lessons in Seeing the Big Picture from a Journey of 71 Million Miles*. Oakland, CA: Berrett-Koehler, 2015.

Garriott, Owen, Joseph Kerwin, and David Hitt. *Homesteading Space: The Skylab Story*. Lincoln, NE: University of Nebraska Press, 2008.

Garriott, Richard and David Fisher. *Explore/Create: My Life in Pursuit of New Frontiers, Hidden Worlds, and the Creative Spark*. New York: William Morrow, 2017.

Hadfield, Chris, *You Are Here: Around the World in 92 Minutes: Photographs from the International Space Station*. Boston, MA: Little, Brown and Company, 2014.

Hale, Edward Everett. *The Brick Moon*. Barre, MA: Imprint Society, 1971.

Kelly, Scott. Endurance: *A Year in Space, A Lifetime of Discovery*. New York: Knopf, 2017.

Kitmacher, Gary H., *Design of the Space Station Habitable Modules*. Reston, VA: American Institute of Aeronautics and Astronautics, Inc., 2002. Available at http://www.spacearchitect.org/pubs/IAC-02-IAA.8.2.04.pdf

Kitmacher, Gary H., *NASA-MIR: Development, Integration and Operation of Systems of the Priroda Module of the Mir Orbital Station* (Reston, VA: American Institute of Aeronautics and Astronautics, Inc., 2002. Available at http://www.spacearchitect.org/pubs/IAC-02-T.P.01.pdf

Kitmacher, Gary H. *Reference Guide to the International Space Station*. Washington, DC: NASA, 2007, 2010, 2015. Available at https://www.nasa.gov/pdf/508318main_ISS_ref_guide_nov2010.pdf

Laliberté, Guy. *Gaia*. New York: Assouline Publishing, 2011.

Launius, Roger. *Space Stations: Base Camps to the Stars*. Washington, DC: Smithsonian Books, 2003.

Lebedev, Valentin. *Diary of a Cosmonaut: 211 Days in Space*. Moscow: Nauka/Zhizn, 1983.

Lebedev, Vladimir and Yuri Gagarin. *Survival in Space*. London: Bantam, 1969.

Linenger, Jerry. *Off the Planet: Surviving Five Perilous Months Aboard the Space Station Mir*. New York: McGraw Hill, 1999.

Mark, Hans. *Space Station: A Personal Journey*. Durham, NC: Duke University Press, 1987.

McCurdy, Howard. *Space Station Decision: Incremental Politics and Technological Choice*. New York: Johns Hopkins University Press. 1990.

Moore, Patrick. *Earth Satellites*. New York: WW Norton, 1956.

Nixon, David. *International Space Station: Architecture Beyond Earth*. London: Circa Press, 2016.

Noordung, Hermann, *The Problem of Space Travel*. Washington, DC: NASA, 1995.

Oberth, Hermann. *Man Into Space*. New York: Harper & Bros., 1957.

Olsen, Gregory. *By Any Means Necessary: An Entrepreneur's Journey into Space*. Princeton, NJ: GHO Ventures, 2010.

Parazynski , Scott and Susy Flory. *The Sky Below: A True Story of Summits, Space, and Speed*. Seattle, WA: Little A, 2017

Parkinson, Bob and R.A. Smith. *High Road to the Moon*. London:British Interplanetary Society, 1979.

Peake, Timothy. *Hello, is This Planet Earth? My View from the International Space Station*. London: Century, 2016.

Pettit, Don. *Spaceborne*. Chicago, IL: Press Syndication Group, 2016.

Pratt, Fletcher and Jack Coggins. *By Spaceship to the Moon*. New York: Random House, 1952.

Romick, Darrell. *Concept for a Manned Earth-Satellite Terminal Evolving from Earth-to-Orbit Ferry Rockets*. Akron, OH: Goodyear Aircraft Corp., 1956.

Shayler, David. *Around the World in 84 Days: The Authorized Biography of Skylab Astronaut Jerry Carr*. Burlington, Canada: Apogee Books, 2008.

Siddiqi, Asif *A. Challenge to Apollo: The Soviet Union and the Space Race, 1945–1974*. Washington, DC: NASA, 2000. Available at https://history.nasa.gov/SP-4408pt1.pdf

US Congress. *Civilian Space Stations and the U.S. Future in Space*. Washington, DC: US Congress, Office of Technology Assessment, OTA-STI-241, November 1984.

Virts, Terry. *View From Above*. Washington, DC: National Geographic, 2017.

Von Braun, Wernher et al.. *Across the Space Frontier*. New York: Viking Press, 1952.

Williams, Jeffrey. *The Work of His Hands*. St. Louis, MO: Concordia, 2010.

MAGAZINES / ARTICLES

Air & Space
Journal of the British Interplanetary Society
Aviation Week & Space Technology
Quest

ACKNOWLEDGMENTS

The authors would like to thank the following individuals for their assistance with this publication:
Rob Godwin, Griffin Media/Apogee Books: www.cgpublishing.com
Dr. Charles Lundquist, NASA Marshall Space Flight Center, University of Alabama, Huntsville
Robert F. Thompson, NASA
Mary N. Wilkerson, Mori Associates, NASA Johnson Space Center

PICTURE CREDITS

The publisher would like to thank the following people and organizations. Every effort has been made to acknowledge the pictures. However, the publisher welcomes any further information so that future editions may be updated.

Page number and position are indicated as follows: T = top, L = left, R = right, C = center, CL = center left, CR = center right, B = bottom, BL = bottom left, BR = bottom right, etc.

Front cover: Image courtesy of Lockheed Martin Corporation.

Back Cover: All NASA, except the following: British Interplanetary Society: TL.
Image courtesy of Lockheed Martin Corporation: BR.
From the authors' collections: TC; BL.

Endpapers: NASA

All images public domain or from the authors' collections, except the following:

Alamy/AF Fotographie: 25: BL
Alamy/Chronicle: 25: TL
Alamy/Entertainment Pictures: 200: CL; 209: TR
Alamy/Granger Historical Archive: 61: CR; 118: B, TR; 119: BL; 120: CL
Alamy/ITAR-TASS News Agency: 58: TR; 73: BL; 102: CL
Alamy/Keystone Pictures USA: 100–101: T
Alamy/Lake Erie Maps and Prints: 15: T
Alamy/Moviestore Collection: 25: BR
Alamy/Pictorial Press Ltd: 60: CR
Alamy/Science History Images: 72: BR
Alamy/SOTK2011: 50: TR
Alamy/Sputnik: 44

amanderson2: 134: CL

David Baker: 91: BL

Boeing: 174: T

Bonestell LLC: 30: B

Brian Brondel (CCBY-SA 3.0): 14: BC

British Interplanetary Society: 22: CR

Canadian Space Agency/Chris Hadfield: 191: BR
Canadian Space Agency (CSA): Logo: 166

Reproduced by kind permission of The Dan Dare Corporation Ltd/Illustration by Graham Bleathman: 211: B

Dille Family Trust: 205: T. Buck Rogers® is a registered trademark owned by the Dille Family Trust and used under license from the "Trust". Copyright © 2018 Dille Family Trust. All Rights Reserved

Elephant Book Company Ltd/Nigel Partridge: 15: BR

European Space Agency: 178: T
European Space Agency (ESA): Logo: 166
European Space Agency/NASA: 175: BR
European Space Agency/S. Corvaja: 179: BR

Getty Images/Bettmann: 52–53; 54: CL; 62: CR; 113: T; 130: BL
Getty Images/Ron Case: 84
Getty Images/Sun Hao: 224: CL
Getty Images/Hulton Archive: 13: BR
Getty Images/Keystone-France/Gamma-Rapho via Getty Images: 53: CL
Getty Images/Keystone/Hulton Archive: 101: TR
Getty Images/Joe McNally: 130: TR
Getty Images/MCT: 61: TR
Getty Images/Popular Science via Getty Images: 36: CL
Getty Images/Walter Sanders: 55: CL
Getty Images/Sovfoto/UIG via Getty Images: 60: TR; 72: TR; 73: TR; 101: BL; 103; 104; 105: BR
Getty Images/SSPL: 121: T; 131: BL
Getty Images/Stocktrek: 127: T
Getty Images/STR/AFP: 224: TR, BR
Getty Images/Ullstein Bild: 53: BR
Getty Images/VCG via Getty Images: 225: TR, BR

David A. Hardy: 37: TR

Don Davis for NASA: 230: TR; 231: C

Frank Henriquez: 93: B

Japan Aerospace Exploration Agency (JAXA): Logo: 166

© Estate of Jeffrey D. Jones: 16: C

Laika ac: 47: TL

Library of Congress: 38: CR

Image courtesy of Lockheed Martin Corporation: 214

Marvel Comics: 212: BL

Matthew J. Cotter: 203: TL

McDonnell-Douglas (December 1967). Courtesy John B. Charles: 91: T

ML Watts: 49: BL

NASA: 2; 6–7 (all images NASA/Nicole Stott); 8; 41: BR; 43: T; 46; 47: TR, TC, CR; 51: TL; 52: BL; 54: TR; 55: T; 58: BR; 59: T, BL, BR; 62: TR; 63: BL; 65–71 (all images); 74–75 (all images); 78–83 (all images); 85: TL, TR; 86: TR, BR; 87–89 (all images); 92: TR, BL; 93: TL; 96: TR; 97: T; 98: TR, B; 99: CL, BR; 100: BR; 106–112 (all images); 113: BR; 114–117 (all images); 120: TR; 121: CL; 122–126 (all images); 127: BR; 128–129 (all images); 130: BR; 131: T; 132: CL; 133; 134: TR; 135–164 (all images); 164–165; 165: R (NASA/Carla Cioffi), TL (NASA/Bill Ingalls); 166: BL, Logo; 167–172 (all images); 173: B; 174: BL, BR; 176–177 (all images); 178: B; 179: T; 180–190 (all images); 191: T, BL; 192: TR, BL; 193: TR, BL; 215; 216: (NASA/Artwork by Pat Rawlings); 217–223 (all images); 226–233 (all images)

National Air & Space Museum, Smithsonian Institution Washington, D.C./Photos by Eric F. Long: 51: R (SI 98-15012); 56: B (NASM-2006-2092.03); 63: TR (NASM-2006-2092.03)

National Archives, Washington, D.C.: 64: TR

National Reconnaissance Office: 90: BR

Naval Research Laboratory: 62: BR

REX Shutterstock/AP: 64: CL
REX Shutterstock/ITV: 209: B
REX Shutterstock/MGM/Stanley Kubrick Productions/Kobal 203: BR
REX Shutterstock/Moviestore Collection: 201: BR
REX Shutterstock/Snap Stills: 200–201

Rick Guidice for NASA: 230: CL; 231: TL, TC, CR.

Roscosmos: Logo: 166

Courtesy of The San Diego Air & Space Museum: 51: CL; 94: TR, BL, BR; 95: TR, CL, BR.

Science Photo Library/Detlev van Ravenswaay: 105: T; 225: BL.
Science Photo Library/Sputnik: 60: CL; 100: BL; 102: TR

Shutterstock/Lukasz Janyst: 12: B
Shutterstock/jaroslava V: 48: BL

Sierra Nevada Corporation: 175: T

SpaceX: 173: T

Thinkstock/Goodshot: 119: T

Twitter/collectSPACE: 193: CL

US Post Office/Bureau of Engraving and Printing: 45

USAF: 56: TR; 57: BR

SKYLAB

A. COMMAND AND SERVICE MODULE
1 SPS ENGINE
2 RUNNING LIGHTS (8 PLACES)
3 SCIMITAR ANTENNA
4 DOCKING LIGHT
5 PITCH CONTROL ENGINES
6 CREW HATCH
7 PITCH CONTROL ENGINES
8 RENDEZVOUS WINDOW
9 EVA HANDHOLDS
10 EVA LIGHT
11 SIDE WINDOW
12 ROLL ENGINES (2 PLACES)
13 EPS RADIATOR PANELS
14 SM RCS MODULE (4 PLACES)
15 ECS RADIATOR

B. MULTIPLE DOCKING ADAPTER
1 AXIAL DOCKING PORT ACCESS HATCH
2 DOCKING TARGET
3 EXOTHERMIC EXPERIMENT
4 INFRARED SPECTROMETER VIEWFINDER
5 ATMOSPHERE INTERCHANGE DUCT
6 AREA FAN
7 WINDOW COVER
8 CABLE TRAYS
9 INVERTER LIGHTING CONTROL ASSEMBLY
10 L-BAND ANTENNA
11 PROTON SPECTROMETER
12 RUNNING LIGHTS (4 PLACES)
13 INFRARED SPECTROMETER
14 FILM VAULT 4
15 FILM VAULT 1
16 SO82 (A&B) CANISTERS
17 M512/M479 EXPERIMENT
18 AREA FAN
19 COMPOSITE CASTING
20 FILM VAULT 2
21 TV CAMERA INPUT STATION
22 UTILITY OUTLET
23 M168 STS MISCELLANEOUS STOWAGE CONTAINER
24 REDUNDANT TAPE RECORDER
25 RADIAL DOCKING PORT
26 10-BAND MULTISPECTRAL SCANNER
27 TV CAMERA INPUT STATION
28 TEMPERATURE THERMOSTAT
29 RADIO NOISE BURST MONITOR
30 ATM C&D CONSOLE

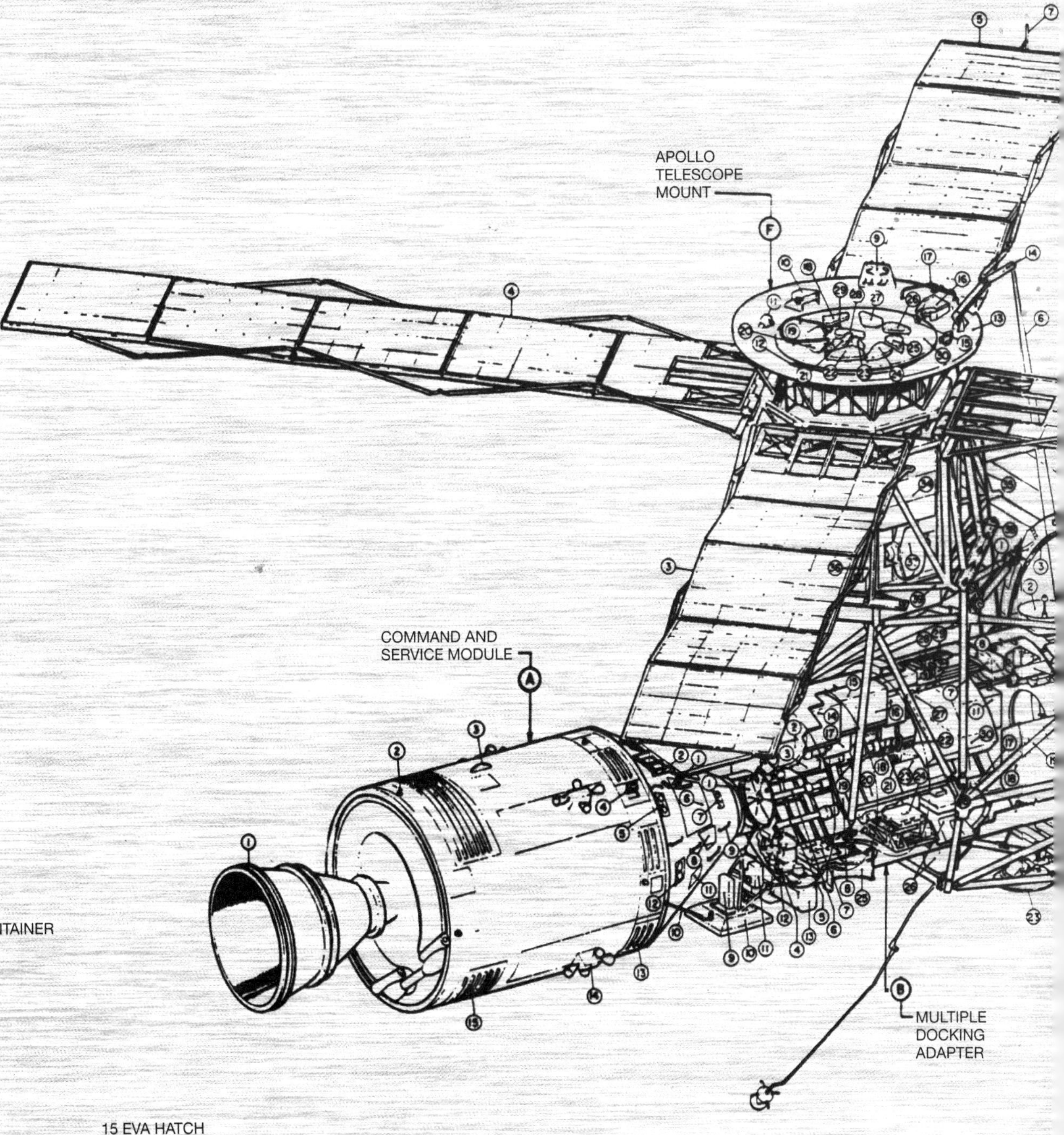

C. AIRLOCK MODULE
1 DEPLOYMENT ASSEMBLY REELS AND CABLES
2 SOLAR RADIO NOISE BURST MONITOR ANTENNA
3 HANDRAILS
4 DO21/DO24 SAMPLE PANELS
5 (REMOVED)
6 CLOTHESLINE (EVA USE)
7 PERMANENT STOWAGE CONTAINER
8 STA IVA STATION
9 NITROGEN TANKS (6 PLACES)
10 OXYGEN TANKS (6 PLACES)
11 MOLECULAR SIEVE
12 CONDENSATE MODULE
13 ELECTRICAL FEEDTHRU COVER
14 ELECTRONICS MODULE 1
15 EVA HATCH
16 AIRLOCK INSTRUMENTATION PANEL
17 MOLECULAR SIEVE
18 STS C&D CONSOLE
19 ATM DEPLOYMENT ASSEMBLY
20 BATTERY MODULE (2 PLACES)
21 EVA PANEL
22 AIRLOCK INTERNAL HATCHES (2 PLACES)
23 S193 MICROWAVE SCATTEROMETER ANTENNA
24 RUNNING LIGHTS (4 PLACES)
25 HANDRAILS
26 STUB ANTENNAS (2 PLACES)
27 THERMAL BLANKET
28 DISCONE ANTENNA (2 PLACES)